INDUSTRIAL MATHEMATICS
with Charts, Formulas
and Tables

INDUSTRIAL MATHEMATICS

with

Charts, Formulas and Tables

by

HARRY W. MUNI
Professional Engineer

PRENTICE-HALL, INC.   Englewood Cliffs, N. J.

Prentice-Hall International, Inc., *London*
Prentice-Hall of Australia, Pty. Ltd., *Sydney*
Prentice-Hall of Canada, Ltd., *Toronto*
Prentice-Hall of India Private Ltd., *New Delhi*
Prentice-Hall of Japan, Inc., *Tokyo*

Library of Congress Cataloging in Publication Data

Muni, Harry W   (date)
    Industrial mathematics with charts, formulas, and
tables.

    1. Industrial engineering--Mathematics. I. Title.
T57.M85   658.5'001'51   72-5740
ISBN 0-13-461558-1

Printed in the United States of America

A Word About the Author

HARRY W. MUNI is a Registered Professional Engineer in the State of New Jersey. His industrial experience includes twenty years in various industries such as petroleum, light and heavy metal fabrication and assembly, wire and cable, electronics, and foundry. This experience includes line and staff positions varying from Plant Manager to Assistant to Vice President and Management Consultant. He has had many articles published in various professional management publications. Mr. Muni taught for several years at Lafayette College and Rutgers University on the subjects of Organization, Production and Inventory Control, Plant Layout, Materials Handling, and Work Measurement. His industrial plant training experience includes such subjects as Production Control and Work Measurement for first line supervision through top line management levels.

Mr. Muni holds a Master's degree in Mechanical Engineering, with a Management Major, from Polytechnic Institute of Brooklyn. The author is a member of the Industrial Management Society, International Materials Management Society, and the New Jersey Professional Engineers Society.

A WORD FROM
THE AUTHOR

The primary purpose of this book is to bring together, in one volume, the most frequently used mathematical formulas, charts and tables in order to help management personnel at all levels in all different functions improve the profit picture of their industrial and business enterprises. Mathematical quantitative measurements have been and are still being developed successfully to gain this objective and to avoid, as much as possible, pure subjective thinking in major management profit decision areas. The formulas and procedures included in this book have demonstrated their practical applications in managing a successful profit enterprise. This book can be used in any type of enterprise regardless of its size or activity.

Every formula presented herein is well defined by symbols and their definitions. Further, each formula is followed by a practical example of its use with a statement of a problem, followed by a solution or a procedure to be followed to arrive at a solution. The reader will be able to easily recognize the approach for the use of the formulas contained in this book to help solve his particular problems in his line of business.

The use of mathematical formulas involves references to various tables of mathematical relationships. This book provides all of the tables needed with the formulas. Graphs are included throughout the book where practical to complement the tables and formulas.

Where mathematical formulas or procedures are not available, examples and solutions are offered. Examples of these situations include chapter 6, "Critical Path Scheduling," and chapter 14, "Flow Process Charts." These subjects are described in sufficient detail to permit the reader to apply the same principles to problems in his business.

Some of the outstanding features of the book include:

- *Span of Control.* All necessary formulas and several examples are presented to determine the number of levels of supervision needed under different organizational conditions. Thus, the reader is offered a quantitative evaluation of organizational structure.

- *Production and Inventory Control.* The economics of quantity to manufacture or buy, inventory costs, purchase quantity discounts, etc., are carefully examined through mathematical formulas to evaluate the best course of management decision. Based upon past practices, one may evaluate more precisely the quantities of production and inventory to be used in terms of cost, lead time for ordering, and other factors.

- *Inspection and Quality Control* are discussed with descriptions of control charts and sampling tables that insure desired levels of quality of incoming and outgoing products coupled with the economic costs. Formulas are described with examples as well as associated sampling tables.

- *Cost Control Procedures* are described in great detail, including business volume break-even of sales and production costs, economic replacement study of equipment, budgeting, etc. The reader will have at his convenience the techniques that are used in this important area for effective management planning and control to gain maximum profits.

- *Plant Facilities* are evaluated for location and layout and materials handling. As to location, the book describes factors for intangible benefits that should be considered over and above tangible benefits in selecting a plant site. As to layout and materials handling, mathematical equations are presented that include estimating floor space requirements, space utilization, etc., to aid in evaluating alternates in considering cost factors.

- *Statistical Mathematics* is described by graphs, nomographs, basic probability considerations, and the like. This discussion is designed to cover the subjects most usually encountered by line and staff management personnel. The nomographs are especially useful for quick and easily used references in solving basic problems.

- *Work Measurement* is discussed in depth, including examination of techniques for methods improvement, for establishing labor standards, manpower planning, work sampling, wage incentives, etc. The discussion of manpower planning, as an example, is especially useful not only for current needs but for future years' planning of manpower requirements. Overtime requirements are considered in these calculations. The techniques of work sampling describe a detailed procedure to determine productive *vs.* non-productive manpower utilization, including rationale — for both factory and office operators. On the subject of wage incentives, only the most current plans are described. The reader is in position to evaluate these plans, make comparisons, and to determine which plan, if any, is desirable for his operations. The total subject of work measurement is presented in a compact form to encompass the most frequently encountered functions.

- In the field of *Personnel,* job evaluation and merit rating are evaluated, using — in so far as possible—quantitative measures. Specific plans and examples are offered in an attempt to reduce these subjects to mathematical relationships including hourly rated, clerical, and management personnel positions.

- *Building Services* includes a basic review of the fundamentals of electricity, heat, water and air. The most commonly encountered problems in these activities are described with the necessary basic formulas and examples.

- The *Appendix* offers a review of the basic mathematics used in industry and business. This includes algebra, logarithms, trigonometry, and interest factors.

I have attempted to cover these subjects in concise terms with clear definitions of formulas and procedures and examples of their use. This book, then, is a complete reference manual coupled with a comprehensive set of tables for management decisions—based upon the use of mathematics.

Harry W. Muni, P.E.

–

TABLE OF CONTENTS

List of Exhibits **13**

1. **How to Use This Book** **19**

   Use of Charts, Formulas and Tables
   Examples

Part I—Management Controls

2. **Span of Control** **25**

   Calculation of Number of Levels of Line Supervision
   Calculation of Number of Line Supervision Personnel
   Payroll Savings

3. **Production Control** **29**

   Economical Production Lot Quantity—Production and Inventory Costs
   Economical Production Lot Quantity—Production, Inventory Costs
     and Purchase Quantity Discounts

4. **Inventory Control** **31**

   Reorder Point
   Average Inventory
   Average Holding Costs
   Average Out-of-Stock Costs
   Total Weekly Holding and Out-of-Stock Costs

7

5. **Inspection and Quality Control** 35

Control Charts by Attributes
Single Sampling Lot Tolerance Tables
Double Sampling Lot Tolerance Tables
Sampling Tables for Average Outgoing Quality

6. **Critical Path Scheduling** 41

Formula
Example
Solution

7. **Manufacturing Economics** 43

Business Volume Break-Even Point
Contribution Ratio
Economic Replacement Study — Annual Cost Method

8. **Budgetary Cost Control** 46

Conventional Budgeting
Flexible Budgeting—Low-High Method

Part II—Plant Facilities

9. **Plant Location** 51

Choice of Plant Site—Intangible Cost Factors
Plant Location Survey

10. **Plant Layout and Materials Handling** 56

Estimating Department Floor Area Requirements
Building Construction Outline
Cross Department Relationships
Space Utilization

Part III—Statistical Mathematics

11. Graphs and Nomographs **63**

Definitions—Graphs
Definitions—Nomographs
Nomograph—Addition and Subtraction
Nomograph—Multiplication and Division
Nomograph—Excess Air in a Furnace

12. Probability—Poisson Distribution **71**

Formula Definitions
Poisson Distribution
Symbols and Definitions

13. Mathematical Averages **77**

Arithmetic Average
Median Average
Mode Average

Part IV—Work Measurement

14. Flow Process Charts **81**

Definition and Purpose
Examples

15. Work Standard **85**

Definition
Formula
Examples

16. Manpower Planning **88**

Formula
Examples

17. Work Sampling **93**

    Definition
    Formula
    Examples
    Use of Random Numbers

18. Learning Curves **99**

    Definition
    Formulas
    Examples

19. Wage Incentives **103**

    Piece Work Plan
    100 Per Cent Premium Plan
    Halsey Plan

Part V—Personnel

20. Job Evaluation **109**

    Hourly Rated Jobs
    Formulas
    Examples

21. Merit Rating **112**

    Clerical Jobs
    Management Jobs

Part VI—Building Services

22. Electricity **117**

    Ohm's Law
    Series Connections

Parallel Connections
Electrical Power
Electrical Horsepower

23. Heat, Water and Air **122**

Heating and Cooling
Fuel Consumption

Appendix

I. Algebra **127**

Basic Laws and Definitions
Polynomials (Addition, Subtraction, Multiplication and
    Division Formulas)
Factoring Polynomials
Axioms of Equality
Simultaneous Equations
Quadratic Equation Formula

II. Fundamentals of Logarithms **135**

Definition
Natural and Common Logarithms
Exponential and Logarithm Relations
Components of Logarithms
Interpolation of Tables
Uses of Logarithms

III. Geometry **139**

Definition and Scope
Triangle
Quadrilateral
Parallelogram
Trapezoid
Circle
Cylinder
Sphere

IV. Trigonometry **146**

    Definition and Scope
    Pythagorean Theorem
    Trigonometric Ratios—Right Triangle
    Trigonometric Relationships—All Triangles

V. Slide Rule **152**

    Operations and Scope
    Reading the Scales
    Multiplication and Division—C and D Scales
    Trigonometry
    Powers of a Number
    Logarithms

VI. Use of Interest Factors **155**

    Symbols and Definitions
    Single-Payment—Compound-Amount Factor
    Single-Payment—Present-Worth Factor
    Equal-Payment—Compound-Amount Factor
    Equal-Payment—Sinking-Fund Factor
    Equal-Payment—Present-Worth Factor
    Equal-Payment—Capital-Recovery Factor

Tables

Table I: Important Constants 161

Table II: Powers and Roots 162

Table III: Reciprocals, Circumference and Areas of Circles 202

Table IV: Factorials 212

Table V: Common Logarithms 213

Table VI: Natural Logarithms 218

Table VII: Interest Factors for Annual Compounding Interest 232

Table VIII: Cumulative Poisson Distribution 242

Table IX: Exponential and Hyperbolic Functions 249

Table X: Table of Random Numbers 256

Table XI: Natural Sines, Cosines, and Tangents of Angles from 0° to 90° 298

Index **305**

LIST OF EXHIBITS

Figure 6.1   Arrow Diagram for Project P 41

Figure 9.1   Primary Factors of Site Selection (in Rank Order) 54

Figure 9.2   Secondary Factors of Site Selection (in Rank Order) 55

Figure 9.3   Shifting Site Selection Factors 55

Figure 9.4   Reason for Constructing New Plant 55

Figure 9.5   Setting of New Plants 55

Figure 9.6   Rating of Company's Planning Methods 55

Figure 11.1   Temperature Degrees—Fahrenheit vs. Centigrade 64

Figure 11.2   Circumference of Circle 65

Figure 11.3   Area of Circle 67

Figure 11.4   Addition and Subtraction 68

Figure 11.5   Multiplication and Division 69

Figure 11.6   Nomograph—Excess Air in a Furnace 70

Figure 12.1   Cumulative Poisson Distribution 73

Figure 12.2   The Exponential Distribution 73

Figure 12.3   Average Delay in a Waiting Line System 76

Figure 14.1   Flow Process Chart of the Material Type (Office Form)
with Preprinted Headings and Captions 82

Figure 14.2   Flow Process Chart with Preprinted Symbols Showing Present
and Proposed Methods with a Summary of Cost Savings 83

Figure 16.1   The WJR Company—Total Required Productive Man Hours
Per Month 90

Figure 16.2   The WJR Company—Standard Manpower Requirements 92

Figure 18.1   90% Learning Curve 100

Figure 18.2   90% Learning Curve—Log-Log Graph 101

Figure 18.3   85% Learning Curve 102

Figure 18.4   85% Learning Curve—Log-Log Graph 102

Figure IIIa   Triangle 140

Figure IIIb   Quadrilateral 141

Figure IIIc   Parallelogram 141

Figure IIId   Rhombus 142

Figure IIIe   Rectangle 142

Figure IIIf   Square 142

Figure IIIg   Trapezoid 142

Figure IIIh   Circle 143

Figure IVa   Right Triangle 146

Figure IVb   All Triangles 149

Figure Va   Slide Rule 153

Figure Vb   Reading C and D Scales 153

Figure Vc   Multiplication and Division—C and D Scales 153

Figure Vd   Trigonometry—SRT, S and D Scales 154

Figure Ve   Powers of a Number 154

Figure Vf   Logarithm Scale 154

Table 5.1   Single Sampling Table for Lot Tolerance Per Cent Defective
             (LTPD) = 7.0% 36

Table 5.2   Double Sampling Table for Lot Tolerance Per Cent Defective
             (LTPD) = 7.0% 38

Table 5.3   Single Sampling Table for Average Outgoing Quality Limit
             (AOQL) = 3.0% 38

Table 5.4   Double Sampling Table for Average Outgoing Quality Limit
             (AOQL) = 3.0% 40

Table 6.1   Cost Table 42

Table 6.2   Indirect Costs 42

Table 9.1   The JRM Company—Intangible Plant Site Cost Factors 52

Table 9.2   Intangible Plant Cost Factors for SLR, CCL and AMM Locations 52

Table 12.1   Symbols and Definitions 72

Table 12.2 Probability of Delay 75

Table 16.1 The WJR Company—Total Required Standard Productive Man
Hours Per Month 89

Table 16.2 The WJR Company—Available Straight Time Productive Hours
Per Month 90

Table 16.3 Total Available Straight Time Productive Hours 91

Table 16.4 Overtime Hour Requirements 91

Table 16.5 Standard Manpower Requirements 92

Table 20.1 The UDP Company—Job Rating Plan for Hourly Rated Jobs ... 109

Table 21.1 The UMB Company—Merit Rating Plan for Clerical Jobs 112

Table 21.2 The ASM Company—Merit Rating Plan for Management Jobs ... 113

Table I Important Constants 161

Table II Powers and Roots 162

Table III Reciprocals, Circumference and Areas of Circles 202

Table IV Factorials 212

Table V Common Logarithms 213

Table VI Natural Logarithms 218

Table VII Interest Factors for Annual Compounding Interest 232

Table VIII Cumulative Poisson Distribution 242

Table IX Exponential and Hyperbolic Functions 249

Table X Table of Random Numbers 256

Table XI Natural Sines, Cosines, and Tangents of Angles from 0° to 90° ... 298

INDUSTRIAL MATHEMATICS
with Charts, Formulas
and Tables

1. HOW TO USE
THIS BOOK

USE OF CHARTS, FORMULAS AND TABLES

As this handbook is designed for use by management personnel on a professional level, it is assumed that they have the basic background in mathematics to use a given formula with accompanying charts and tables. In all cases, for problems to be solved within the subjects discussed, a formula is given and its terms defined, followed by an example and solution with specific references to charts and tables used to find the solution.

For the most part, the charts, formulas and tables are self-explanatory. Where necessary, explanations are given to avoid any possible erroneous interpretation.

Most of the tables are located following the appendix. They are referred to throughout the book in problems having formulas requiring these data. These tables are complete and carry figures to a sufficient number of digits to insure accurate solutions. Table II., Powers and Roots, page 162, is an example of a table referred to several times in the book.

In addition, there are other tables shown within individual chapters since their use is limited to that chapter. An example of this type is table 5.1, which is the first table in chapter 5. Charts, formulas and figures also follow this numbering system where depicted in the individual chapters.

EXAMPLE 1:

A typical problem encountered by industrial management will serve as an illustration of the value of this handbook in finding a solution that is accurate yet simple when using mathematics.

The JWR Company is analyzing its production run schedules. The company recognizes that there has to be some kind of a balance between production and inventory costs in order to realize minimum costs. For illustrative purposes, an analysis will be made of the production of Product L. The basic purpose of the analysis, then, is to determine the most economic production run.

Referring to the index under Economical production lot quantity, the following is found in chapter 3, "Production Control," page 29;

$$Q = \sqrt{\frac{2RS}{i}}$$

where

Q = Economical production lot quantity units
R = Annual production requirement units
S = Setup costs per order
i = Annual inventory cost per unit

For Product L., the following data has been determined:

Q = Unknown
R = 800
S = 14
i = 0.14

The solution is as follows:

$$Q = \sqrt{\frac{2 \times 800 \times 14}{0.14}}$$
$$= \sqrt{160,000}$$
$$Q = 400 \qquad \text{(Table II.)}$$

This subject is further enlarged upon to include other variables such as purchase quantity discounts. The appropriate formulas and solutions are described. The same approach is used throughout this handbook where different conditions and/or variables are involved.

EXAMPLE 2:

The temperature equivalents between the English system expressed in degrees Fahrenheit and the Metric system is given by the following formula as depicted in chapter 11, "Graphs and Nomographs," page 63:

$$F = (1.8 \times C) + 32$$

where

F = Degrees Fahrenheit
C = Degrees Centigrade

For a Centigrade value (C) of 10, determine the equivalent value in the English system in degrees Fahrenheit (F).

The solution is as follows:

$$F = (1.8 \times 10) + 32 = 50$$

EXAMPLE 3:

Business volume break-even point.

Formula:

(depicted in chapter 7, "Manufacturing Economics," page 43)

$$BEP = \frac{F_c}{1 - \frac{V_c}{S}}$$

where

BEP = Break-even point (no profit-no loss point)
F_c = Fixed costs
V_c = Variable costs
S = Sales

Example:

For a sales volume (S) of \$1,250,000, fixed costs of \$210,000 (F_c) and variable costs of \$850,000 (V_c), calculate the break-even point (BEP).

Solution:

$$BEP = \frac{\$210,000}{1 - \frac{850,000}{1,250,000}} = \frac{210,000}{1 - 0.68} = \frac{210,000}{0.32} = \$656,250$$

EXAMPLE 4:

Wage Incentives

Formula:

100 PER CENT PREMIUM PLAN, Chapter 19, "Wage Incentives," page 103.

$$E_p = H_r \times A_t + H_r (SH_n \times n - A_t)$$

where

E_p = Total earnings
H_r = Hourly base rate
A_t = Actual hours worked
SH_n = Standard hour rate per unit
n = Number of units

Example:

The SLJ Company guarantees an hourly base rate (H_r) of $2.60/hr. In a given day an operator made 36 units (n). The standard hour rate per unit (SH_n) is 0.25 hour per unit. Compute the earnings for this day.

Solution:

$$E_p = \$2.60/\text{hour} \times 8 \text{ hours} + \$2.60/\text{hour} \times (0.25) \text{ hours/unit} \times 36 \text{ units} - 8$$
$$(\text{hours})$$
$$= \$20.80 + \$2.60 \ (9 \text{ hours} - 8 \text{ hours})$$
$$= \$20.80 + \$2.60 \ (1 \ \text{ hour})$$
$$E_p = \$23.40$$

Part I
MANAGEMENT CONTROLS

2. SPAN
OF CONTROL

CALCULATION OF NUMBER OF LEVELS OF LINE SUPERVISION

Formula:[1]

$$L = \frac{\text{Log P} - \text{Log F}}{\text{Log R}} + 1$$

where

L = Number of levels of line supervision including the top executive in immediate charge of the unit

P = Number of primary operatives

F = Span of control for first line supervision

R = Span of control above first line supervision

Example 1:

For an organization of 7000 primary operatives (P), a span of control of 10 for first line supervision (F) and a span of control of 5 for above first line supervision (R), how many levels of line supervision (L) are required?

Solution:

$$L = \frac{\text{Log 7000} - \text{Log 10}}{\text{Log 5}} + 1$$

---

1. Ralph C. Davis, "The Influence of the Unit of Supervision and the Span of Executive Control on the Economy of Line Organization Structure," (Columbus, Ohio; The Bureau of Business Research, Ohio State University, Copyright 1941 by The Ohio State University), p. 6.

$$= \frac{3.845098 - 1.000000}{0.698970} + 1 \quad \text{(Table V.)}$$

$$= \frac{2.845098}{0.698970} + 1$$

$$= 4.07 + 1$$

$$L = 5.07 \quad \text{or } L = 5$$

Example 2:

For an organization of 12,000 primary operatives (P), a span of control of 15 for first line supervision (F), and a span of control of 4 for above first line supervision (R), how many levels of line supervision (L) are required?

Solution:

$$L = \frac{\text{Log } 12,000 - \text{Log } 15}{\text{Log } 4} + 1$$

$$= \frac{4.079181 - 1.176091}{0.602060} + 1 \quad \text{(Table V.)}$$

$$= \frac{2.903090}{0.602060} + 1$$

$$= 4.82 + 1$$

$$L = 5.82 \quad \text{or } L = 6$$

CALCULATION OF NUMBER OF LINE SUPERVISION PERSONNEL

Formula:[2]

$$E = \frac{P \ (1 + 1/R + 1/R^2 + 1/R^3 + \ldots 1/R^{L-1})}{F}$$

where

E = Number of line supervision personnel
P = Number of primary operatives
R = Span of control above first line supervision
F = Span of control for first line supervision
L = Number of levels of line supervisors including the top executive in immediate charge of the unit

Example 1:

For an organization of 7,000 primary operatives (P), 5 levels of line supervision (L), a span of control of 5 for above first line supervision (R), and a span of control of 10 for first line supervision (F), how many line supervisors are required (E)?

---

2. *Ibid.*, p. 14.

Solution:

$$E = \frac{7000 \ (1 + 1/5 + 1/5^2 + 1/5^3 + 1/5^{\ 5-1})}{10}$$

$$= \frac{7000 \ (1 + 1/5 + 1/5^2 + 1/5^3 + 1/5^4)}{10}$$

$$= \frac{7000(1 + 1/5 + 1/25 + 1/125 + 1/625)}{10} \quad \text{(Table II.)}$$

$$= \frac{7000(1 + 0.2000 + 0.0400 + 0.0080 + 0.0016)}{10} \quad \text{(Table III.)}$$

$$= 700(1.25)$$

$$E = 875$$

Example 2:

For an organization of 12,000 primary operatives (P), 6 levels of line supervision (L), a span of control of 4 for above first line supervision (R), 15 for first line supervision (F), how many line supervisors are required (E)?

Solution:

$$E = \frac{12,000(1 + 1/4 + 1/4^2 + 1/4^3 + 1/4^4 + 1/4^{\ 6-1})}{15}$$

$$= \frac{12,000(1 + 1/4 + 1/4^2 + 1/4^3 + 1/4^4 + 1/4^5)}{15}$$

$$= \frac{12,000(1 + 1/4 + 1/16 + 1/64 + 1/256 + 1/1024)}{15} \quad \text{(Table II.)}$$

$$= \frac{12,000(1 + 0.2500 + 0.0625 + 0.0039 + 0.0001)}{15} \quad \text{(Table III.)}$$

$$= 800(1.3165)$$

$$E = 1053$$

PAYROLL SAVINGS

Formula:

$$S = N_1 - N_2$$

where

S = Difference in number of supervisory personnel
N_1 = Number of supervisory personnel required under P_1 plan
N_2 = Number of supervisory personnel required under P_2 plan

Example 1:

For an organization of 7000 primary operatives, 5 levels of line supervision, a span of control of 10 for first line supervision and a span of control of 5 for above first line supervision, 875 line supervisors (N_1) would be required under P_1 plan. For the same organization but with a span of control of 15 for first line supervision, 583 line supervisors (N_2) would be required under P_2 plan. Determine the difference in number of supervisory personnel (S).

Solution:

$$S = 875 - 583$$
$$S = 292$$

Example 2:

For an organization of 12,000 primary operatives, 6 levels of line supervision, a span of control of 15 for first line supervision and a span of control of 4 for above first line supervision, 1053 line supervisors (N_1) would be required under P_1 plan. For the same organization but with span of control of 20 for first line supervision, 790 line supervisors (N_2) would be required under P_2 plan. Determine the difference in number of supervisory personnel (S).

Solution:

$$S = 1053 - 790$$
$$S = 263$$

3. PRODUCTION
CONTROL

ECONOMICAL PRODUCTION LOT QUANTITY – PRODUCTION AND INVENTORY COSTS

Formula:

$$Q = \sqrt{\frac{2RS}{i}}$$

where

Q = Economical production lot quantity units
R = Annual production requirements in units
S = Setup costs per order
i = Annual inventory cost per unit

Example:

The JWR Company has an annual production requirement (R) of 600 units for product L. The setup cost per order (S) is $15 and the annual inventory cost per unit (i) is $0.20. Calculate the economical production lot quantity (Q).

Solution:

$$Q = \sqrt{\frac{2 \times 600 \times 15}{0.20}}$$
$$= \sqrt{90,000}$$
$$Q = 300 \qquad \text{(Table II.)}$$

ECONOMICAL PRODUCTION LOT QUANTITY — PRODUCTION, INVENTORY COSTS AND PURCHASE QUANTITY DISCOUNTS

Formula:

$$C_A = C_N - C_P$$

where

C_A = Difference in annual cost between no purchase discount price and purchase discount price

C_N = Annual cost with no purchase discount price

C_P = Annual cost with purchase discount price

and

P_N = Purchase unit cost with no discount price

P_D = Purchase unit cost with discount price

N_N = Annual number of orders with no purchase discount price

N_D = Annual number of orders with purchase discount price

n_n = Number of units per order with no purchase discount price

n_d = Number of units per order with purchase discount price

Therefore:

$$C_N = (R \times P_N) + (S \times N_N) + (P_N \times n_n \times i)$$

and,

$$C_P = (R \times P_D) + (S \times N_D) + (P_D \times n_d \times i)$$

Example:

Continuing with the previous example, the unit cost of Product L from its vendor is $5.00 ($P_N$). The vendor is now offering a unit price of $4.50 ($P_D$) for order quantities of 500 units or more. The JWR Company is making an analysis of the difference in annual costs for this product considering all the various costs of production, inventory and purchase quantity discount (C_A). Additional data have been determined as follows, namely: N_N = 4, N_D = 1, n_n = 300, and n_d = 60. Determine C_A.

Solution:

$$
\begin{aligned}
C_N &= (600 \times 5.00) + (15 \times 4) + (5.00 \times 300 \times 0.20) \\
&= 3000 + 60 + 300 \\
C_N &= \$3360 \\
C_P &= (600 \times 4.50) + (15 \times 1) + (4.50 \times 600 \times 0.20) \\
&= 2700 + 15 + 540 \\
C_P &= \$3255 \\
C_A &= \$3360 - 3255 \\
C_A &= \$105
\end{aligned}
$$

4. INVENTORY CONTROL

REORDER POINT

Formula:

$$ROP = L \times U_w$$

Where

ROP = Reorder point in units
L = Lead time in weeks
U_w = Average usage in weekly units

Example:

The LMM Company's average lead time for product M between placing an order and receiving the order (L) is three weeks. The average usage is 80 units per week (U_w). Compute the reorder point (ROP).

Solution:

$$ROP = 3 \times 80$$
$$ROP = 240$$

AVERAGE INVENTORY

Formula:

$$A_h = ROP/2$$

where

A_h = Average weekly inventory during lead time (L) in units

ROP = Reorder point in units

Example:

The LMM Company's reorder point (ROP) for product M is 240 units. Compute the average weekly inventory (A_h) during the reorder point period.

Solution:

$$A_h = 240/2$$
$$A_h = 120$$

AVERAGE HOLDING COSTS

Formula:

$$W_h = A_w \times C_h$$

where

W_h = Average weekly holding costs

A_h = Average weekly inventory during lead time (L) in units

C_h = Holding cost per unit per week

Example:

The LMM Company's average weekly inventory (A_w) is 120 units. The holding cost per unit per week (C_h) is $0.60. Compute the average weekly holding costs $($W_h)$.

Solution:

$$W_h = 120 \times \$0.60$$
$$W_h = \$72$$

AVERAGE OUT OF STOCK COSTS

Formula:

$$W_o = P_o \times C_o$$

where

W_o = Average weekly cost of out-of-stock inventory

P_o = Per cent of time out of stock

C_o = Average cost per order of out-of-stock inventory

Example:

The LMM Company is making an analysis of a past sample period of 50 reorders (N). In this period, with a lead time of three weeks (L), it was found that the time out of stock (P_o) was 14%. The average cost per order of out-of-stock inventory (C_o) is $150. Compute the average weekly cost of out-of-stock inventory.

Solution:

$$W_o = 14\% \times \$150$$
$$W_o = \$21$$

TOTAL WEEKLY HOLDING AND OUT-OF-STOCK COSTS

Formula:

$$W_T = W_h + W_o$$

where

W_T = Total costs per week
W_h = Average weekly holding costs
W_o = Average weekly costs of out-of-stock inventory

Example:

The LMM Company is making an inventory cost analysis of its product M. The following data have been determined:

Average use per week (U_w) = 80 units
Holding cost per unit (C_h) = $0.60
Average cost per reorder of out-of-stock inventory (C_o)
= $150

In addition, the LMM Company has experienced the following reorder data, by lead time in weeks (L), for a past sample period of 50 reorders (N) for its product M:

n_n = Number of instances for a past sample period of 50 reorders (N)
P_o = % of time out of stock *based upon usage data above*

L	n_n	P_o
2	2	48/50 = 96%
3	41	7/50 = 14
4	6	1/50 = 2
5	1	0 = 0

From the above data and *formulas previously described in this chapter,* compute the total cost per week (W_T) for various lead times (L).

Solution:

L	W_h	W_o =	W_T
2	$48	$144	$192
3	72	21	93
4	96	3	99
5	120	0	120

5. INSPECTION AND QUALITY CONTROL

CONTROL CHARTS BY ATTRIBUTES

Formula:

$$\text{UCL} = \bar{p} + \sqrt[3]{\frac{\bar{p}(1-\bar{p})}{n_l}}$$

$$\text{LCL} = \bar{p} - \sqrt[3]{\frac{\bar{p}(1-\bar{p})}{n_l}}$$

where

UCL = Upper control limit

LCL = Lower control limit

and

$$\bar{p} = n_r/(n_l \times n_t)$$

where

n_r = Total number of rejects found in the total lots inspected

n_l = Average number of units in each lot inspected

n_t = Total number of lots inspected

Example:

The ESG Company has decided to try the use of a control chart for one of its high volume products, U. The company recorded data for 40 lots (n_t) with an average number of 250 units (n_l) inspected in each lot. The total number of rejects 500 (n_r) was found in the total number of lots inspected. Compute the Upper control limit (UCL) and the Lower control limit (LCL).

Solution:

$$\bar{p} = 500/(250 \times 40) = 0.0500$$

$$\sqrt{\frac{\bar{p}(1-\bar{p})}{n_1}} = \sqrt{\frac{0.0500\ (1-0.0500)}{250}} = \sqrt{\frac{0.0475}{250}} = \sqrt{0.00019}$$

$$= 0.0138 \qquad \text{(Table II.)}$$

$$3 \times 0.0138 = 0.0414$$

$$\text{UCL} = 0.0500 + 0.0414 = 0.0914$$

$$\text{LCL} = 0.0500 - 0.0414 = 0.0086$$

SINGLE SAMPLING LOT TOLERANCE TABLES

Formula:

(See table 5.1.)[1]

SINGLE
SAMPLING
7.0%
LTPD

Single Sampling Table for
Lot Tolerance Per Cent Defective (LTPD) = 7.0%

Lot Size	Process Average 0 to 0.07%			Process Average 0.08 to 0.70%			Process Average 0.71 to 1.40%			Process Average 1.41 to 2.10%			Process Average 2.11 to 2.80%			Process Average 2.81 to 3.50%		
	n	c	AOQL %	n	c	AOQL %	n	c	AOQL %	n	c	AOQL %	n	c	AOQL %	n	c	AOQL %
1–25	All	0	0	All	0	0	All	0	0	All	0	0	All	0	0	All	0	0
26–50	24	0	0.80	24	0	0.80	24	0	0.80	24	0	0.80	24	0	0.80	24	0	0.80
51–100	28	0	0.95	28	0	0.95	28	0	0.95	28	0	0.95	28	0	0.95	28	0	0.95
101–200	30	0	1.0	30	0	1.0	49	1	1.3	49	1	1.3	49	1	1.3	65	2	1.4
201–300	31	0	1.1	31	0	1.1	50	1	1.4	70	2	1.5	85	3	1.6	85	3	1.6
301–400	32	0	1.1	55	1	1.4	70	2	1.6	90	3	1.7	105	4	1.8	125	5	1.8
401–500	32	0	1.1	55	1	1.4	75	2	1.6	90	3	1.8	110	4	1.9	140	6	2.0
501–600	32	0	1.1	55	1	1.4	75	2	1.7	95	3	1.8	125	5	2.0	145	6	2.1
601–800	32	0	1.1	55	1	1.4	75	2	1.7	110	4	2.0	130	5	2.1	160	7	2.2
801–1000	33	0	1.1	55	1	1.4	95	3	1.9	110	4	2.1	145	6	2.2	180	8	2.4
1001–2000	55	1	1.5	75	2	1.8	95	3	2.0	130	5	2.3	185	8	2.5	230	11	2.8
2001–3000	55	1	1.5	75	2	1.8	115	4	2.1	150	6	2.4	215	10	2.8	300	15	3.0
3001–4000	55	1	1.5	75	2	1.8	115	4	2.2	165	7	2.6	235	11	2.9	330	17	3.2
4001–5000	55	1	1.5	75	2	1.8	130	5	2.4	185	8	2.7	250	12	3.0	350	18	3.3
5001–7000	55	1	1.5	75	2	1.8	130	5	2.4	185	8	2.7	270	13	3.1	385	20	3.4
7001–10,000	55	1	1.5	95	3	2.0	150	6	2.5	200	9	2.9	285	14	3.2	415	22	3.6
10,001–20,000	55	1	1.5	95	3	2.0	150	6	2.5	220	10	2.9	320	16	3.3	470	25	3.7
20,001–50,000	55	1	1.5	115	4	2.2	170	7	2.6	235	11	3.1	355	18	3.5	530	29	3.9
50,001–100,000	55	1	1.5	115	4	2.2	185	8	2.7	270	13	3.1	370	19	3.5	530	29	3.9

TABLE 5.1

1. H. F. Dodge and H. G. Romig, *Sampling Inspection Tables* (2nd ed. New York, N.Y.: John Wiley & Sons, Inc., © 1959 by Bell Telephone Laboratories, Inc.), p. 185.

Example:

Determine the minimum total inspection for a lot size range from a) 401–500 with a lot tolerance fraction defective of 0 to 0.07% process average, and b) for a process average of 1.41 to 2.10%.

Solution:

From table 5.1, the following data are determined, namely:

a) $n = 32$, $c = 0$
b) $n = 90$, $c = 3$

DOUBLE SAMPLING LOT TOLERANCE TABLES

Formula:

(See table 5.2.)[2]

Example:

Determine the minimum total inspection for a lot size range from a) 401—500 with a lot tolerance fraction defective of 0 to 0.07% process average, and b) for a process average of 1.41 to 2.10%.

Solution:

From table 5.2, the following data are determined, namely:
a) $n_1 = 39$, $c_1 = 0$
   $n_2 = 21$, $n_1 + n_2 = 60$, $c_2 = 1$
b) $n_1 = 39$, $c_1 = 0$
   $n_2 = 76$, $n_1 + n_2 = 115$, $c_2 = 4$

SAMPLING TABLES FOR AVERAGE OUTGOING QUALITY

Tables 5.3[3] and 5.4[4] depict examples of the Dodge-Romig Single Sampling Tables and Double Sampling Tables, respectively, for an Average Outgoing Quality Limit (AO QL) of 3.0%.

---

2. *Ibid.,* p. 194.
3. *Ibid.,* p. 202.
4. *Ibid.,* p. 214.

TABLE 5.2

Double Sampling Table for Lot Tolerance Per Cent Defective (LTPD) = 7.0%

Lot Size	Process Average 0 to 0.07% Trial 1 n_1	c_1	Trial 2 n_2	n_1+n_2	c_2	AOQL in %	Process Average 0.08 to 0.70% Trial 1 n_1	c_1	Trial 2 n_2	n_1+n_2	c_2	AOQL in %	Process Average 0.71 to 1.40% Trial 1 n_1	c_1	Trial 2 n_2	n_1+n_2	c_2	AOQL in %
1–25	All	0	–	–	–	0	All	0	–	–	–	0	All	0	–	–	–	0
26–50	24	0	–	–	–	0.80	24	0	–	–	–	0.80	24	0	–	–	–	0.80
51–75	31	0	15	46	1	0.90	31	0	15	46	1	0.90	31	0	15	46	1	0.90
76–110	34	0	16	50	1	1.1	34	0	16	50	1	1.1	34	0	16	50	1	1.1
111–200	36	0	19	55	1	1.2	36	0	19	55	1	1.2	36	0	39	75	2	1.4
201–300	37	0	23	60	1	1.3	37	0	23	60	1	1.3	37	0	38	75	2	1.5
301–400	38	0	22	60	1	1.3	38	0	42	80	2	1.5	38	0	57	95	3	1.7
401–500	39	0	21	60	1	1.3	39	0	41	80	2	1.5	39	0	61	100	3	1.7
501–600	39	0	26	65	1	1.3	39	0	46	85	2	1.6	39	0	61	100	3	1.7
601–800	39	0	26	65	1	1.4	39	0	46	85	2	1.6	39	0	81	120	4	1.9
801–1000	39	0	26	65	1	1.4	39	0	46	85	2	1.6	39	0	86	125	4	2.0
1001–2000	40	0	45	85	2	1.7	40	0	65	105	3	1.9	40	0	105	145	5	2.2
2001–3000	40	0	45	85	2	1.7	40	0	65	105	3	1.9	65	1	100	165	6	2.3
3001–4000	40	0	45	85	2	1.7	40	0	65	105	3	1.9	65	1	115	180	7	2.4
4001–5000	40	0	45	85	2	1.7	40	0	85	125	4	2.1	65	1	115	180	7	2.5
5001–7000	40	0	45	85	2	1.7	40	0	85	125	4	2.1	65	1	135	200	8	2.5
7001–10,000	40	0	45	85	2	1.7	40	0	85	125	4	2.1	65	1	135	200	8	2.6
10,001–20,000	40	0	45	85	2	1.7	40	0	85	125	4	2.1	65	1	155	220	9	2.7
20,001–50,000	40	0	45	85	2	1.7	40	0	105	145	5	2.3	65	1	170	235	10	2.8
50,001–100,000	40	0	45	85	2	1.7	40	0	105	145	5	2.3	65	1	170	235	10	2.9

Lot Size	Process Average 1.41 to 2.10% Trial 1 n_1	c_1	Trial 2 n_2	n_1+n_2	c_2	AOQL in %	Process Average 2.11 to 2.80% Trial 1 n_1	c_1	Trial 2 n_2	n_1+n_2	c_2	AOQL in %	Process Average 2.81 to 3.50% Trial 1 n_1	c_1	Trial 2 n_2	n_1+n_2	c_2	AOQL in %
1–25	All	0	–	–	–	0	All	0	–	–	–	0	All	0	–	–	–	0
26–50	24	0	–	–	–	0.80	24	0	–	–	–	0.80	24	0	–	–	–	0.80
51–75	31	0	15	46	1	0.90	31	0	15	46	1	0.90	31	0	15	46	1	0.90
76–110	34	0	31	65	2	1.2	34	0	31	65	2	1.2	34	0	31	65	2	1.2
111–200	36	0	54	90	3	1.5	36	0	54	90	3	1.5	36	0	69	105	4	1.5
201–300	37	0	58	95	3	1.6	37	0	73	110	4	1.7	60	1	80	140	6	1.9
301–400	38	0	77	115	4	1.8	60	1	85	145	6	1.9	60	1	100	160	7	2.0
401–500	39	0	76	115	4	1.8	60	1	105	165	7	2.1	60	1	135	195	9	2.2
501–600	65	1	90	155	6	2.1	65	1	120	185	8	2.2	85	2	130	215	10	2.3
601–800	65	1	105	170	7	2.2	65	1	140	205	9	2.3	85	2	165	250	12	2.5
801–1000	65	1	110	175	7	2.2	85	2	140	225	10	2.5	105	3	180	285	14	2.7
1001–2000	65	1	150	215	9	2.5	105	3	175	280	13	2.8	145	5	230	375	19	3.1
2001–3000	85	2	165	250	11	2.7	105	3	210	315	15	3.0	165	6	300	465	24	3.3
3001–4000	85	2	185	270	12	2.8	105	3	250	355	17	3.1	180	7	335	515	27	3.4
4001–5000	85	2	185	270	12	2.9	125	4	245	370	18	3.2	180	7	370	550	29	3.6
5001–7000	85	2	205	290	13	3.0	125	4	265	390	19	3.3	200	8	385	585	31	3.8
7001–10,000	85	2	205	290	13	3.1	125	4	300	425	21	3.4	200	8	450	650	35	3.9
10,001–20,000	85	2	220	305	14	3.2	125	4	335	460	23	3.5	220	9	485	705	38	4.0
20,001–50,000	85	2	240	325	15	3.3	145	5	360	505	26	3.5	220	9	565	785	43	4.1
50,001–100,000	105	3	270	375	18	3.4	165	6	390	555	29	3.7	235	10	610	845	47	4.2

Trial 1: n_1 = first sample size; c_1 = acceptance number for first sample
"All" indicates that each piece in the lot is to be inspected
Trial 2: n_2 = second sample size; c_2 = acceptance number for first and second samples combined
AOQL = Average Outgoing Quality Limit

TABLE 5.3

SINGLE
SAMPLING

3.0%

A O Q L

Single Sampling Table for
Average Outgoing Quality Limit (AOQL) = 3.0%

Lot Size	Process Average 0 to 0.06%			Process Average 0.07 to 0.60%			Process Average 0.61 to 1.20%			Process Average 1.21 to 1.80%			Process Average 1.81 to 2.40%			Process Average 2.41 to 3.00%		
	n	c	p_t %	n	c	p_t %	n	c	p_t %	n	c	p_t %	n	c	p_t %	n	c	p_t %
1-10	All	0	–	All	0	–	All	0	–	All	0	–	All	0	–	All	0	–
11–50	10	0	19.0	10	0	19.0	10	0	19.0	10	0	19.0	10	0	19.0	10	0	19.0
51–100	11	0	18.0	11	0	18.0	11	0	18.0	11	0	18.0	11	0	18.0	22	1	16.4
101–200	12	0	17.0	12	0	17.0	12	0	17.0	25	1	15.1	25	1	15.1	25	1	15.1
201–300	12	0	17.0	12	0	17.0	26	1	14.6	26	1	14.6	26	1	14.6	40	2	12.8
301–400	12	0	17.1	12	0	17.1	26	1	14.7	26	1	14.7	41	2	12.7	41	2	12.7
401–500	12	0	17.2	27	1	14.1	27	1	14.1	42	2	12.4	42	2	12.4	42	2	12.4
501–600	12	0	17.3	27	1	14.2	27	1	14.2	42	2	12.4	42	2	12.4	60	3	10.8
601–800	12	0	17.3	27	1	14.2	27	1	14.2	43	2	12.1	60	3	10.9	60	3	10.9
801–1000	12	0	17.4	27	1	14.2	44	2	11.8	44	2	11.8	60	3	11.0	80	4	9.8
1001–2000	12	0	17.5	28	1	13.8	45	2	11.7	65	3	10.2	80	4	9.8	100	5	9.1
2001–3000	12	0	17.5	28	1	13.8	45	2	11.7	65	3	10.2	100	5	9.1	140	7	8.2
3001–4000	12	0	17.5	28	1	13.8	65	3	10.3	85	4	9.5	125	6	8.4	165	8	7.8
4001–5000	28	1	13.8	28	1	13.8	65	3	10.3	85	4	9.5	125	6	8.4	210	10	7.4
5001–7000	28	1	13.8	45	2	11.8	65	3	10.3	105	5	8.8	145	7	8.1	235	11	7.1
7001–10,000	28	1	13.9	46	2	11.6	65	3	10.3	105	5	8.8	170	8	7.6	280	13	6.8
10,001–20,000	28	1	13.9	46	2	11.7	85	4	9.5	125	6	8.4	215	10	7.2	380	17	6.2
20,001–50,000	28	1	13.9	65	3	10.3	105	5	8.8	170	8	7.6	310	14	6.5	560	24	5.7
50,001–100,000	28	1	13.9	65	3	10.3	125	6	8.4	215	10	7.2	385	17	6.2	690	29	5.4

TABLE 5.4

Double Sampling Table for
Average Outgoing Quality Limit (AOQL) = 3.0%

Lot Size	Process Average 0 to 0.06%					Process Average 0.07 to 0.60%					Process Average 0.61 to 1.20%				
	Trial 1		Trial 2			p_t %	Trial 1		Trial 2			p_t %	Trial 1		Trial 2			p_t %
	n_1	c_1	n_2	n_1+n_2	c_2		n_1	c_1	n_2	n_1+n_2	c_2		n_1	c_1	n_2	n_1+n_2	c_2	
1–10	All	0	–	–	–	–	All	0	–	–	–	–	All	0	–	–	–	–
11–50	10	0	–	–	–	19.0	10	0	–	–	–	19.0	10	0	–	–	–	19.0
51–100	16	0	9	25	1	16.4	16	0	9	25	1	16.4	16	0	9	25	1	16.4
101–200	17	0	9	26	1	16.0	17	0	9	26	1	16.0	17	0	9	26	1	16.0
201–300	18	0	10	28	1	15.5	18	0	10	28	1	15.5	21	0	23	44	2	13.3
301–400	18	0	11	29	1	15.2	21	0	24	45	2	13.2	23	0	37	60	3	12.0
401–500	18	0	11	29	1	15.2	21	0	25	46	2	13.0	24	0	36	60	3	11.7
501–600	18	0	12	30	1	15.0	21	0	25	46	2	13.0	24	0	41	65	3	11.5
601–800	21	0	25	46	2	13.0	21	0	25	46	2	13.0	24	0	41	65	3	11.5
801–1000	21	0	26	47	2	12.8	21	0	26	47	2	12.8	25	0	40	65	3	11.4
1001–2000	22	0	26	48	2	12.6	22	0	26	48	2	12.6	27	0	58	85	4	10.3
2001–3000	22	0	26	48	2	12.6	25	0	40	65	3	11.4	28	0	62	90	4	10.0
3001–4000	23	0	26	49	2	12.4	25	0	45	70	3	11.0	29	0	76	105	5	9.6
4001–5000	23	0	26	49	2	12.4	26	0	44	70	3	11.0	30	0	75	105	5	9.5
5001–7000	23	0	27	50	2	12.2	26	0	44	70	3	11.0	30	0	80	110	5	9.4
7001–10,000	23	0	27	50	2	12.2	27	0	43	70	3	11.0	30	0	80	110	5	9.4
10,001–20,000	23	0	27	50	2	12.2	27	0	43	70	3	11.0	31	0	94	125	6	9.2
20,001–50,000	23	0	27	50	2	12.2	28	0	67	95	4	9.7	55	1	120	175	8	8.0
50,001–100,000	23	0	27	50	2	12.2	31	0	84	115	5	9.0	60	1	140	200	9	7.6

Lot Size	Process Average 1.21 to 1.80%					Process Average 1.81 to 2.40%					Process Average 2.41 to 3.00%				
	Trial 1		Trial 2			p_t %	Trial 1		Trial 2			p_t %	Trial 1		Trial 2			p_t %
	n_1	c_1	n_2	n_1+n_2	c_2		n_1	c_1	n_2	n_1+n_2	c_2		n_1	c_1	n_2	n_1+n_2	c_2	
1–10	All	0	–	–	–	–	All	0	–	–	–	–	All	0	–	–	–	–
11–50	10	0	–	–	–	19.0	10	0	–	–	–	19.0	10	0	–	–	–	19.0
51–100	17	0	17	34	2	15.8	17	0	17	34	2	15.8	17	0	17	34	2	15.8
101–200	20	0	21	41	2	13.7	22	0	33	55	3	12.4	22	0	33	55	3	12.4
201–300	23	0	37	60	3	12.0	23	0	37	60	3	12.0	24	0	51	75	4	11.1
301–400	23	0	37	60	3	12.0	25	0	55	80	4	10.8	42	1	63	105	6	10.4
401–500	24	0	36	60	3	11.7	25	0	55	80	4	10.8	46	1	79	125	7	9.7
501–600	26	0	54	80	4	10.7	46	1	69	115	6	9.7	48	1	97	145	8	9.2
601–800	26	0	54	80	4	10.7	49	1	81	130	7	9.4	50	1	115	165	9	8.9
801–1000	27	0	58	85	4	10.3	49	1	86	135	7	9.2	70	2	120	190	10	8.4
1001–2000	49	1	76	125	6	9.1	50	1	150	200	10	8.0	100	3	180	280	14	7.5
2001–3000	50	1	95	145	7	8.7	80	2	165	245	12	7.6	130	4	260	390	19	6.9
3001–4000	55	1	110	165	8	8.5	105	3	200	305	14	7.0	155	5	330	485	23	6.5
4001–5000	60	1	135	195	9	7.8	110	3	225	335	15	6.7	215	7	390	605	27	6.0
5001–7000	60	1	165	225	10	7.3	110	3	250	360	16	6.6	270	9	505	775	34	5.7
7001–10,000	85	2	160	245	11	7.2	115	3	290	405	18	6.5	285	9	680	965	41	5.4
10,001–20,000	85	2	180	265	12	7.2	140	4	315	455	20	6.3	315	10	805	1120	47	5.3
20,001–50,000	85	2	205	290	13	7.0	170	5	420	590	26	6.0	390	13	940	1330	56	5.2
50,001–100,000	90	2	245	335	15	6.8	200	6	505	705	30	5.7	445	15	1105	1550	65	5.1

Trial 1: n_1 = first sample size; c_1 = acceptance number for first sample
"All" indicates that each piece in the lot is to be inspected
Trial 2: n_2 = second sample size; c_2 = acceptance number for first and second samples combined
p_t = lot tolerance per cent defective with a Consumer's Risk (P_C) of 0.10

6. CRITICAL PATH
SCHEDULING

Definition:

Critical Path Scheduling (CPS) may be defined as the longest chain of successive individual events which makes the completion of these events critical to all others.

Example:

The HRW Company is interested in determining the *minimum cost* considering indirect costs of outage, etc. to complete project P. Figure 6.1 depicts the project in arrow diagram form. The CPS[1] technique will be used to determine the solution. Facts regarding the pro-

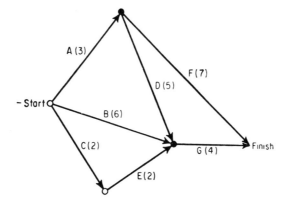

Figure 6.1   Arrow Diagram for Project P.

1. J. S. Sayer, J. E. Kelly, Jr., and Morgan R. Walker, "Critical Path Scheduling." By special permission of *Modern Manufacturing* (formerly *Factory*), July 1960. Copyright by McGraw-Hill Inc., New York, pp. 76, 77.

ject indicate that job E (figure 6.1) cannot be shortened regardless of costs. As shown in table 6.1, the "crash" column represents the minimum number of days to complete each respective job and the "cost slope" column lists the extra cost incurred for each day of time cut from the respective jobs. Outage loss cost by number of days has also been determined.

Solution:

From figure 6.1, the critical path is A—D—G or (3—5—4) days for a total of 12 days. Based upon data given in figure 6.1 and table 6.1, table 6.2 is developed to determine the project cost and outage loss cost. Examination of table 6.2 reveals that the total lowest cost to complete Project P is $1430, under schedule III—for 10 days.

COST TABLE

| JOB | NORMAL | | CRASH | | COST SLOPE |
	DAYS	DOLLARS	DAYS	DOLLARS	DOLLARS/DAY
A	3	$ 50	2	$ 100	$ 50
B	6	140	4	260	60
C	2	25	1	50	25
D	5	100	3	180	40
E	2	80	2	80	—
F	7	115	5	175	30
G	4	100	2	240	70
	Total	$610	Total	$1085	

TABLE 6.1   Cost Table

INDIRECT COSTS

Schedule	I	II	III	IV	V	VI
Days	12	11	10	9	8	7
Project Cost	$ 610	$ 650	$ 690	$ 770	$ 870	$1000
Outage Loss	$ 900	$ 820	$ 740	$ 700	$ 660	$ 620
Total Cost	$1510	$1470	$1430	$1470	$1530	$1620

TABLE 6.2   Indirect Costs

7. MANUFACTURING ECONOMICS

BUSINESS VOLUME BREAK-EVEN POINT

Formula:

$$BEP = \frac{F_c}{1 - \frac{V_c}{S}}$$

where

BEP = Break-even point (no profit-no loss point)
F_c = Fixed costs
V_c = Variable costs
S = Sales

Example:

For a sales volume (S) of \$1,250,000, fixed costs of \$210,000 (F_c) and variable costs of \$850,000 (V_c), calculate the break-even point (BEP).

Solution:

$$BEP = \frac{210,000}{1 - \frac{850,000}{1,250,000}} = \frac{210,000}{1 - 0.68} = \frac{210,000}{0.32} = \$656,250$$

CONTRIBUTION RATIO

Formula:

$$CR = S - V_c$$

$$\text{where}$$
$$\text{CR} = \text{Contribution ratio}$$
$$\text{and}$$
$$\text{S} = \text{F}_c + \text{V}_c + \text{P}$$
$$\text{where}$$
$$\text{S} = \text{Sales}$$
$$\text{F}_c = \text{Fixed costs}$$
$$\text{V}_c = \text{Variable costs}$$
$$\text{P} = \text{Profit}$$
$$\text{then,}$$
$$\text{CR} = \text{F}_c + \text{P}$$
$$\text{or}$$
$$\text{P} = \text{CR} - \text{F}_c$$

Example:

For a fixed cost (F$_c$) of \$250,000 and a variable cost (V$_c$) of 60% of sales (S), compute the contribution ratio (CR). Also, compute the profit (P) for sales (S) of \$975,000.

Solutions:

$$\text{CR} = \text{S} - 0.60\ (\text{S})$$
$$= \text{S}\ (1 - 0.60)$$
$$= \text{S}\ (0.40)$$
$$\text{or}$$
$$\text{CR} = 40\% \times \text{S}$$
$$\text{P} = 0.40 \times \$975{,}000 - \$250{,}000$$
$$= \$390{,}000 - \$250{,}000$$
$$\text{P} = \$140{,}000$$

ECONOMIC REPLACEMENT STUDY — ANNUAL COST METHOD

Formula:

$$\text{ACCR} = (\text{P} - \text{L}) \times (\text{C} - \text{RF}) + \text{L} \times \text{i}$$

$$\text{where}$$
$$\text{ACCR} = \text{Annual cost of capital recovery}$$
$$\text{P} = \text{First cost}$$
$$\text{L} = \text{Salvage value}$$
$$\text{C} - \text{RF} = \text{Capital} - \text{recovery factor} \quad \text{(Table VII.)}$$
$$\text{i} = \text{Interest rate}$$
$$\text{n} = \text{Life years}$$

Example:

Considering the alternate of replacing the present unit rather than a major maintenance overhaul of the present unit, the following data have been gathered for analyses:

The present unit may be repaired for a cost of $40,000 (P) for an estimated extended usage of 5 years (n), at which time its salvage value (L) is estimated to be $5000. Its present salvage value is $16,000.

For the proposed major maintenance overhaul of the present unit, it is estimated that subsequent inspection and maintenance costs would be about the same as the alternate method of a new unit. A new unit may be purchased and installed for $100,000 (P), an estimated life period of 20 years (n) with a salvage value (L) of $15,000.

An interest rate (i) of 10% is to be used.

Compute the respective alternate costs using the annual cost method.

Solution:

Present unit:

$$\begin{aligned} ACCR &= (\$40{,}000 + 16{,}000 - 5000) \times 0.26380 + \$5000 \times 0.10 \\ &= 51{,}000 \times 0.26380 + 500 \\ &= 13{,}454 + 500 \\ ACCR &= \$13{,}954 \end{aligned}$$

Proposed new unit:

$$\begin{aligned} ACCR &= (\$100{,}000 - 15{,}000) \times 0.11746 + \$15{,}000 \times 0.10 \\ &= 85{,}000 \times 0.11746 + 1500 \\ &= 9984 + 1500 \\ ACCR &= \$11{,}484 \end{aligned}$$

8. *BUDGETARY COST CONTROL*

CONVENTIONAL BUDGETING

Formula:

$$PSR = P/S$$

where

PSR = Profit to sales ratio
P = Profit
S = Sales
and
M = Material costs
E = Labor and other expense
or
$$P = S - (M + E)$$

Example 1:

Compute the profit (P) and profit to sales ratio (PSR) for the RLM Company for a period where the sales (S) amounted to $1,000,000, material costs (M) were $600,000 and labor and other expenses (E) were $200,000.

Solution:

$$P = \$1,000,000 - (\$600,000 + \$200,000)$$
$$= 1,000,000 - 800,000$$
$$P = \$200,000$$
$$PSR = \$200,000/\$1,000,000$$

$$PSR = 20\%$$

Example 2:

The URB Company, for a period where sales (S) amounted to $2,000,000, had material costs (M) of $700,000 and labor and other expenses (E) of $800,000.

Compute the profit (P) and profit to sales ratio (PSR).

Solution:

$$P = \$2,000,000 - (\$700,000 + \$800,000)$$
$$= 2,000,000 - 1,500,000$$
$$P = \$500,000$$
$$PSR = \$500,000/\$2,000,000$$
$$PSR = 25\%$$

FLEXIBLE BUDGETING—LOW—HIGH METHOD

Formula:

$$VC = VC_U \times U$$

where

VC = Variable cost
VC_U = Variable cost/Unit
U = Units
and
$VC_U = D_C/D_U$

where

$D_C = H_C - L_C$
$D_U = H_U - L_U$
and
H_C = Total high costs
L_C = Total low costs
H_U = High units
L_U = Low units
F_C = Fixed costs = $H_C - (VC_U \times H_U)$ or $L_C - (VC_U \times L_U)$

Example:

The WAL Company has determined the following data for two levels of activity and expenses for one of its product lines within a normal operating range for a given period of time; the number of units varies from a low (L_U) of 1000 to a high (H_U) of 2000 with respective total costs (L_C) of $340 and $455 ($H_C$). Determine the variable cost (V_C) for the low and high levels of activity, respectively, as well as the fixed costs (F_C) for each level of activity.

Solution:

$$D_U = 2000 - 1000 = 1000$$
$$D_C = 455 - 340 = 115$$
$$VC_U = 115/1000 = 0.115$$

For low level activity, $VC_L = 0.115 \times 1000 = \$115$

For high level activity, $VC_H = 0.115 \times 2000 = \$230$

$$F_C = 340 - 115 = \$225, \text{ or } 455 - 230 = \$225$$

Part II

PLANT FACILITIES

9. PLANT
LOCATION

CHOICE OF PLANT SITE—INTANGIBLE COST FACTORS

Formula: (See table 9.1)

Example:

After considerable analysis, the JRM Company has made a decision to relocate its facilities and is now confronted with the problem as to where. Tangible cost factors such as distribution costs of incoming and outgoing materials, labor, taxes, etc. have been carefully weighed for various potential locations. The result of this study would indicate a choice between the SLR, CCL and AMM locations based upon *tangible cost factors* only. The company realizes, however, that there are a number of other important *intangible cost factors* to be evaluated. As an *aid* in judging these factors, the JRM Company has designed table 9.1—Intangible Plant Site Cost Factors—using a *weighted degree point* approach to evaluate intangible cost figures on a basis designed for its *own use* in terms of *relative values*.

Solution:

Potential plant location sites at SLR, CCL and AMM have been evaluated using table 9.1 with the results shown in table 9.2— Intangible Plant Site Cost Factors for SLR, CCL and AMM locations.

PLANT LOCATION SURVEY

In a 1968 survey of industrial site selection,[1] the following questions and responses

---

1. Joseph L. Mazel, "Industrial Site Selection Today." By special permission of *Modern Manufacturing* (formerly *Factory*), May 1968. Copyright by McGraw-Hill Inc., New York, pp. 88–93.

| Factor | Degree—Points | | | % Max. Points |
	Degree 1. (Minimum Acceptable)	2. (Acceptable)	3. (Above Acceptable)	
I. Nearness to market	30 Points	90 Points	150 Points	30%
II. Nearness to incoming materials	25	75	125	25
III. Community attitude	15	45	75	15
IV. State & local ordinances	10	30	50	10
V. Educational facilities	5	15	25	5
VI. Capital availability	5	15	25	5
VII. Fire and police services	4	12	20	4
VIII. Recreational facilities	4	12	20	4
IX. Miscellaneous	2	6	10	2
Totals:	100 Points	300 Points	500 Points	100%

TABLE 9.1 The JRM Company—Intangible Plant Site Cost Factors

| | Location | | | | | |
| | SLR | | CCL | | AMM | |
	Degree	Points	Degree	Points	Degree	Points
I. Nearness to market	3	150	2	90	3	150
II. Nearness to incoming materials	2	75	2	75	3	125
III. Community attitude	3	75	2	45	3	75
IV. State & local ordinances	3	50	3	50	2	30
V. Educational facilities	2	15	1	5	3	25
VI. Capital availability	2	15	3	25	2	15
VII. Fire and police services	1	4	2	12	3	20
VIII. Recreational facilities	3	20	1	4	1	4
IX. Miscellaneous	1	2	3	10	2	6
Totals:		406 Points		316 Points		450 Points
% Max. Points		81%		63%		90%

TABLE 9.2 Intangible Plant Site Cost Factors for SLR, CCL and AMM Locations

were tabulated as follows:

1. Primary factors of site selection (in rank order)—those relating directly to the potential site (figure 9.1).
2. Secondary factors of site selection (in rank order)—those relating to area and employees (figure 9.2).
3. Shifting site selection factors (figure 9.3).
4. Reason for constructing new plant (figure 9.4).
5. Setting of new plants (figure 9.5).
6. Rating of company's planning methods (figure 9.6).

Figure 9.1   Primary Factors of Site Selection (in rank order)

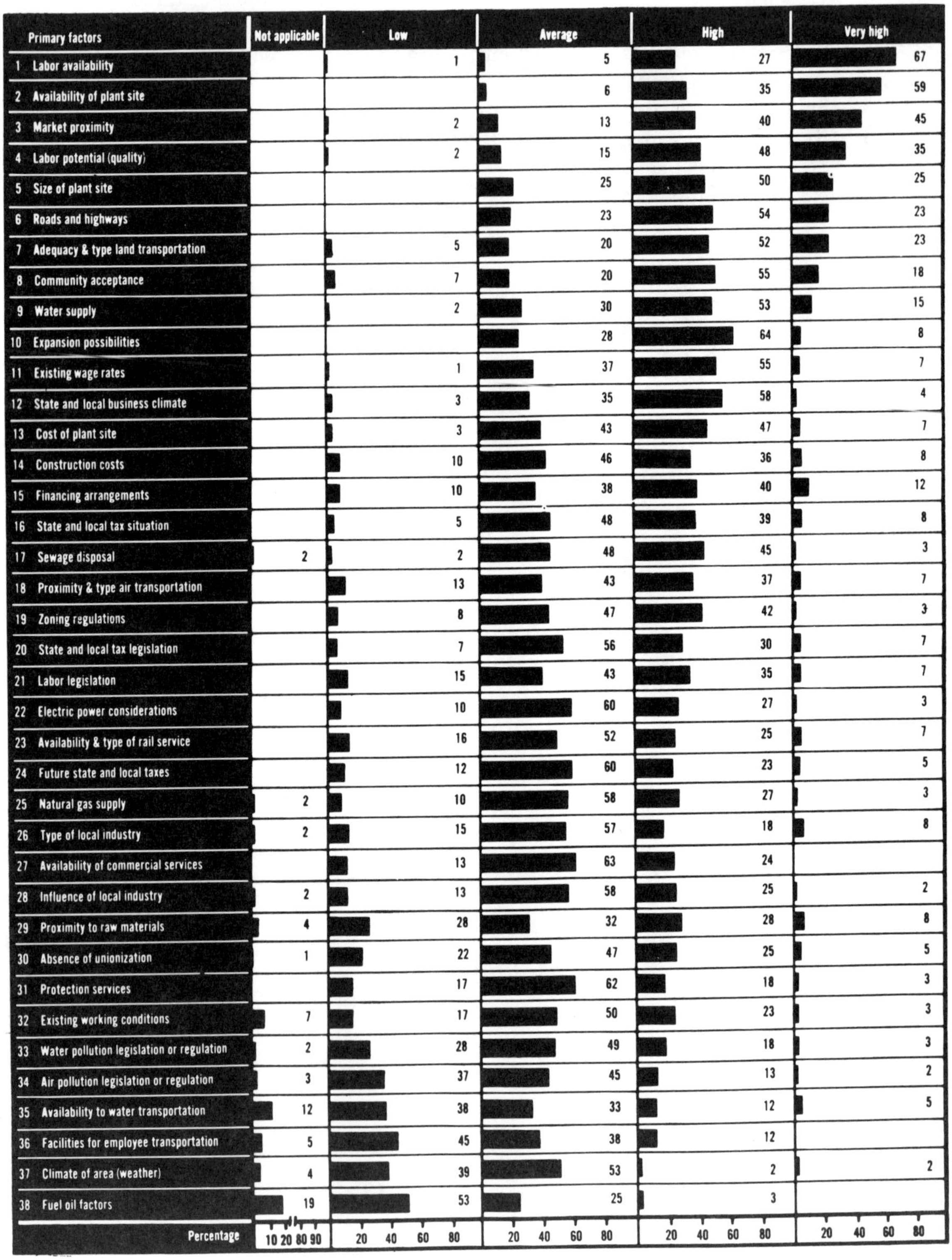

Primary factors	Not applicable	Low	Average	High	Very high
1 Labor availability		1	5	27	67
2 Availability of plant site			6	35	59
3 Market proximity		2	13	40	45
4 Labor potential (quality)		2	15	48	35
5 Size of plant site			25	50	25
6 Roads and highways			23	54	23
7 Adequacy & type land transportation		5	20	52	23
8 Community acceptance		7	20	55	18
9 Water supply		2	30	53	15
10 Expansion possibilities			28	64	8
11 Existing wage rates		1	37	55	7
12 State and local business climate		3	35	58	4
13 Cost of plant site		3	43	47	7
14 Construction costs		10	46	36	8
15 Financing arrangements		10	38	40	12
16 State and local tax situation		5	48	39	8
17 Sewage disposal	2	2	48	45	3
18 Proximity & type air transportation		13	43	37	7
19 Zoning regulations		8	47	42	3
20 State and local tax legislation		7	56	30	7
21 Labor legislation		15	43	35	7
22 Electric power considerations		10	60	27	3
23 Availability & type of rail service		16	52	25	7
24 Future state and local taxes		12	60	23	5
25 Natural gas supply	2	10	58	27	3
26 Type of local industry	2	15	57	18	8
27 Availability of commercial services		13	63	24	
28 Influence of local industry	2	13	58	25	2
29 Proximity to raw materials	4	28	32	28	8
30 Absence of unionization	1	22	47	25	5
31 Protection services		17	62	18	3
32 Existing working conditions	7	17	50	23	3
33 Water pollution legislation or regulation	2	28	49	18	3
34 Air pollution legislation or regulation	3	37	45	13	2
35 Availability to water transportation	12	38	33	12	5
36 Facilities for employee transportation	5	45	38	12	
37 Climate of area (weather)	4	39	53	2	2
38 Fuel oil factors	19	53	25	3	
Percentage	10 20 80 90	20 40 60 80	20 40 60 80	20 40 60 80	20 40 60 80

Figure 9.2   Secondary Factors of Site Selection (in rank order)

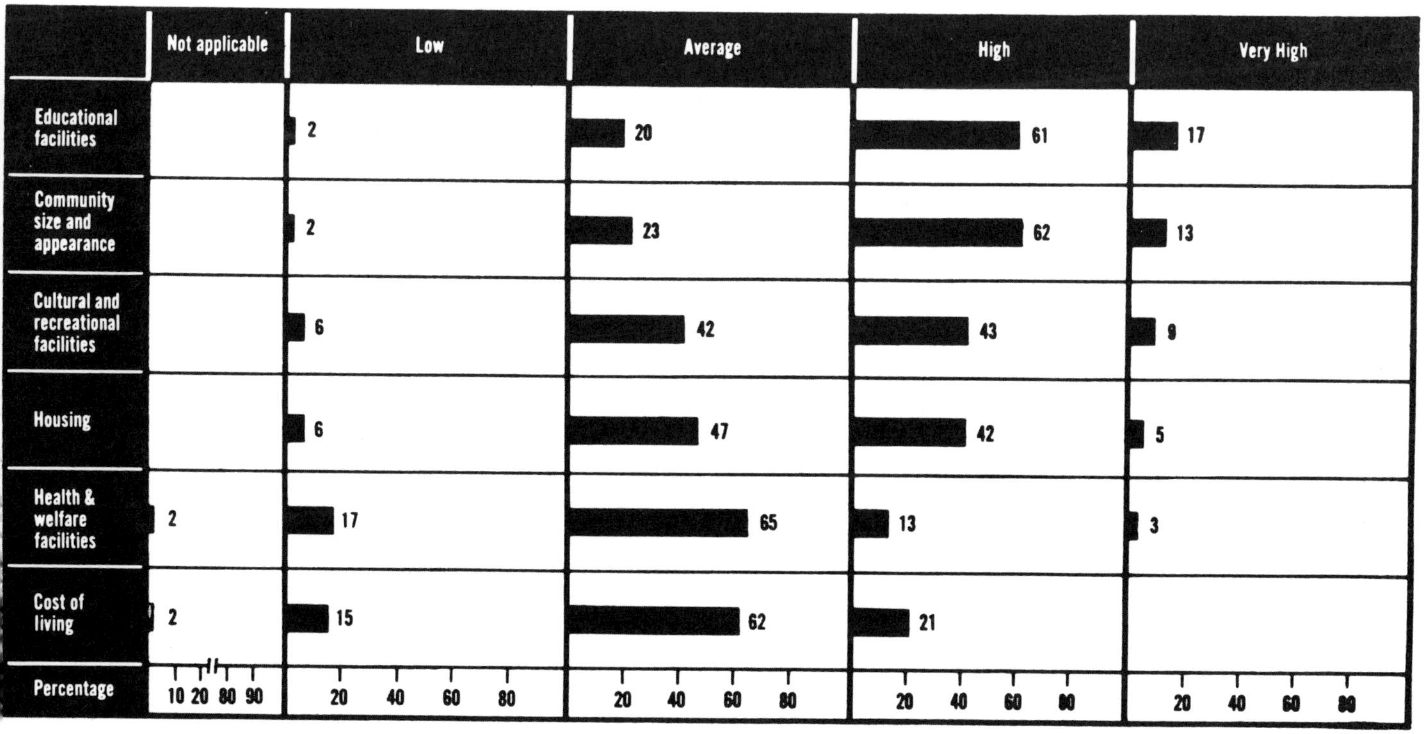

	Not applicable	Low	Average	High	Very High
Educational facilities		2	20	61	17
Community size and appearance		2	23	62	13
Cultural and recreational facilities		6	42	43	9
Housing		6	47	42	5
Health & welfare facilities	2	17	65	13	3
Cost of living	2	15	62	21	
Percentage	10 20 80 90	20 40 60 80	20 40 60 80	20 40 60 80	20 40 60 80

Figure 9.3   Shifting Site Selection Factors

Have factors shifted in your area in the last 5-10 years?	Yes	86.5%
	No	13.5%

These have Increased	These have Decreased
Labor considerations	Taxation
Financing arrangements	Raw material access
Educational facilities	Rail service
Marketing considerations	Unionization

Figure 9.4   Reason for Constructing New Plant

	Primary	Secondary
Expansion of production	79%	14%
Replacement of obsolete facility	11%	33%
Market considerations	8%	37%
Consolidation of production facilities	2%	16%

Figure 9.5   Setting of New Plants

Suburban	78%
Rural	12%
Urban	10%

Figure 9.6   Rating of Company's Planning Methods

	Excellent	Good	Average	Poor
Planning site requirements	25%	42%	19%	14%
Planning production requirements	28%	52%	20%	0%
Planning move into plant	20%	50%	25%	5%
Planning start-up	19%	56%	21%	4%

10. PLANT LAYOUT
AND MATERIALS HANDLING

ESTIMATING DEPARTMENT FLOOR AREA REQUIREMENTS

Formula:

$$D_A = N_{WC} \times W_{CA}$$

where

D_A = Dept. floor Area requirements

N_{WC} = Number of work centers required

W_{CA} = Work center area requirement

and

$$W_{CA} = P_{CA} + S_A + A_A + B_A$$

where

P_{CA} = Prime center area (machine, storage rack, etc.)

S_A = Supporting area (personnel working area, incoming–outgoing temporary materials area, operating supplies area, etc.)

A_A = Aisle area (main and service aisles)

B_B = Building structural areas (columns, etc.)

Example:

The EBB Company has decided to expand its facilities to meet increasing needs for its products. In order to determine total physical plant floor area requirements, the company is approaching the problem by estimating the individual function (department) floor area requirements—the sum of which with other factors of possible future expansion to be included in the final solution to the problem. For a particular department (P), the prime center area (P_{CA}) is estimated at 60 sq. ft., the supporting area (S_A) is 3 sq. ft. (5% of the prime center area P_{CA}), the aisle area (A_A) is 21 sq. ft. (35% of the prime center area P_{CA}) and the building structural area (B_B) is 0.60 sq. ft. (1% of the prime center area P_{CA}). The number of work centers (N_{WC}) is 150 for department P. Calculate the total floor area required.

Solution:

$$W_{CA} = P_{CA} + S_A + A_A + B_A$$
$$= 60 + 3 + 21 + 0.60$$
$$W_{CA} = 84.60 \text{ sq. ft.}$$
$$D_A = N_{WC} \times W_{CA}$$
$$= 150 \times 84.60$$
$$D_A = 12,690 \text{ sq. ft.}$$

BUILDING CONSTRUCTION OUTLINE

Formula:

$$B_{WL} = 2L + 2W \quad \text{(rectangular shape)}$$

where

B_{WL} = Total linear length of outside building wall
$\quad L$ = Longer wall of outside building
$\quad W$ = Shorter wall of outside building
$\qquad$ and

$$A = L \times W$$

where

$\quad A$ = Enclosed area of $L \times W$ building

Example:

The WFM Company, after determining plant floor space requirements, is considering various overall building plant construction shapes. From an overall viewpoint considering advantages and disadvantages, a rectangular plot plan has been decided upon. Following this decision, consideration will be given to materials handling warehousing, storage and handling costs. For a required enclosed area of 360,000 sq. ft. (A), the following combinations of length (L) and width (W) are under consideration:

	$\underline{L}$ (ft.)	$\times \underline{W}$ (ft.)	$= \underline{A}$ (sq. ft.)
1.	600	$\times$ 600	= 360,000
2.	900	$\times$ 400	= 360,000
3.	1200	$\times$ 300	= 360,000

Solution:

$$B_{WL} = 2L + 2W$$

1. $B_{WL} = 2 \times 600 + 2 \times 600$
    $\qquad = 1200 + 1200$
   $B_{WL} = 2400$ ft.
2. $B_{WL} = 2 \times 900 + 2 \times 400$
    $\qquad = 1800 + 800$
   $B_{WL} = 2600$ ft.
3. $B_{WL} = 2 \times 1200 + 2 \times 300$
    $\qquad = 2400 + 600$
   $B_{WL} = 3000$ ft.

CROSS DEPARTMENT RELATIONSHIPS

Priorities:

Degree factors of distance-movement frequency between departments. Nearness is:

1. essential
2. very important
3. important
4. unimportant

Example:

The RHM Company, after analyses of floor space and building construction outlines, described previously in this chapter, is now evaluating the importance of department cross relationships in terms of frequency of physical movements, communications, etc. When these data are determined, the company will develop alternate plans of plant layout and materials handling using templates and/or 3-D models to determine optimum solutions. The company has established the above priorities for evaluation of cross department relationships. Determine the cross relationships between departments A through H.

Solution:

	To:	A	B	C	D	E	F	G	H
From:	A	0	1	2	3	4	3	4	4
	B	1	0	2	3	4	4	3	4
	C	3	2	0	1	3	3	3	3
	D	3	3	1	0	1	3	4	4
	E	4	3	3	1	0	4	4	3
	F	4	4	4	3	2	0	4	2
	G	4	4	4	4	4	1	0	3
	H	4	4	3	2	3	2	2	0

Department

SPACE UTILIZATION

Formula:

   a) $V = A/P \times 100$

      where

      V = Vertical space utilization effectiveness (%)
      A = Actual average storage height (ft.)
      P = Potential average storage height (ft.)

   b) $R = S \times V$

      where

      R = Floor space requirements (sq. ft.)
      S = Actual floor area occupied (sq. ft.)
      V = Vertical space utilization effectiveness (%)

Example:

The ASC Company has conducted a survey of its warehouse space utilization. The findings indicate that although the physical characteristics of the building allow stacking material to an average height of 14 feet (P), the actual average storage height is 12 feet (A). Also, the actual floor area occupied is 183,000 sq. ft. (S). Compute a) the vertical space utilization effectiveness (V), and b) the floor space requirements (R).

Solution:

   a) $V = A/P$
        $= 12/14$
    $V = 86\%$

   b) $R = S \times V$
        $= 183{,}000 \times 86\%$
    $R = 157{,}380$ sq. ft.

Part III

STATISTICAL MATHEMATICS

11. GRAPHS
AND NOMOGRAPHS

DEFINITIONS—GRAPHS

When plotting graphs, the horizontal axis may be referred to as the abscissa axis. The vertical axis may be referred to as the ordinate axis. The intersection of the horizontal and vertical axes is termed the origin and may be designated by the letter "O". The magnitude of the value of the origin may not be equal to zero depending upon the scale graduation used.

Temperature Equivalents—Degrees Fahrenheit vs. *Centigrade*

Formula:

$$F = 1.8 \times C + 32$$

where

F = Degrees Fahrenheit
C = Degrees Centigrade

Example:

Plot a graph of the above formula, illustrating the relationship of temperatures of the English system (degrees Fahrenheit—F) and the Metric system (degrees Centigrade—C). From the graph, a) determine the value of C for an F value of 212 (boiling point of water), and b) determine the value of C for an F value of 32 (freezing point of water).

Solution:

Plot uniform scale values gradations for the Fahrenheit (F) and Centigrade scales, both plus (+) and minus values (−), as shown in figure 11.1 (Temperature Degrees—

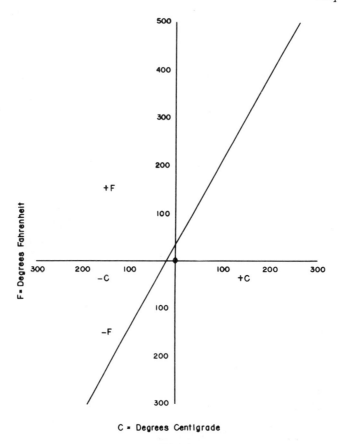

Figure 11.1   Temperature Degrees—Fahrenheit vs. Centigrade

Fahrenheit *vs.* Centigrade). Determine the relationships of the F and C values by use of the formula. The following relationships may be determined:

if C = −90,
   F = 1.8 × (−90) + 32 = −162 + 32 = −130
if C = 10,
   F = 1.8 × (10) + 32 = 18 + 32 = 50
if C = 60,
   F = 1.8 × (60) + 32 = 108 + 32 = 140

—tabulated as follows:

C	F
−90	−130
10	50
60	140

The above data are plotted in graphical form in figure 11.1, resulting in a straight line curve. From the graph, a plus value 212 (F) on the ordinate (vertical) scale gives a reading of 100 on the abscissa scale (C), at the point of intersection.

From the graph, a plus value of 32 (F) on the ordinate scale at its intersection with the curve gives a corresponding value on the abscissa scale (C) of 0 (zero).

Circumference of Circle

Formula:

$$C = \pi n$$

    where

C = Circumference of circle in units
π = Constant (3.141593)
n = Diameter of circle in units

Example:

For diameters of circles ranging from 1 through 10 units, establish the circumferences and plot the results in graphical form. From the graph, determine the circumferences of circles with diameters (n) of a) 3.25 units, and b) 7.40 units.

Solution:

n	C	(Table III.)
0	0.000000	
1	3.141593	
2	6.283185	
3	9.424778	

Figure 11.2—Circumference of Circle—shows the above data plotted in graphical form. The curve is in a straight line form as established by plotting the values of n (di-

Figure 11.2 Circumference of Circle

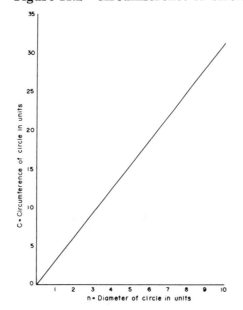

ameter of circle in units) on the abscissa axis and the values of C (circumference of circle in units) on the ordinate axis from the above data. From the graph, the circumference of a circle in units (C) for a diameter (n) of a) 3.25 units, is found to be 10.21 units, and of b) 7.40 units, is found to be 23.25 units.

Area of Circle

Formula:

$$A = \frac{\pi n^2}{4}$$

where

A = Area of circle in square units
π = Constant (3.141593)
n = Diameter of circle in units

Example:

For diameters of circles ranging from 1 through 10 units, establish the area and plot the results in graphical form. From the graph, determine the areas of circles with diameters (n) of a) 3.25 units, and b) 7.40 units.

Solution:

n	A	
0	0.0000000	(Table III.)
1	0.7853982	
2	3.141593	
3	7.068583	
4	12.56637	
5	19.63495	
6	28.27433	
7	38.48451	
8	50.26548	
9	63.61725	
10	78.53982	

Figure 11.3—Area of Circle—shows the above data plotted in graphical form. The curve relationship is established by plotting the corresponding values of n (diameter of circle in units) on the abscissa axis and the values of A (area of circle in square units) on the ordinate axis from the above data. From the graph, the area of a circle in square units (A) with a diameter (n) of a) 3.25 units, is found to be 8.294 square units, and of b) 7.40 units, is found to be 43.008 square units.

DEFINITIONS—NOMOGRAPHS

Nomographs, or alignment charts, are graphical means of depicting useful relationships of data involved in additions, subtractions, multiplications and divisions.

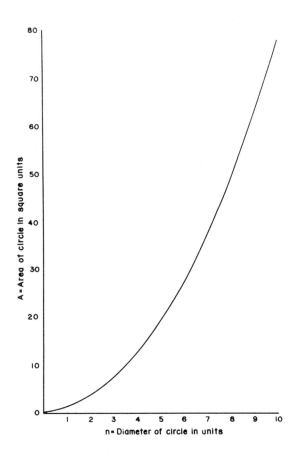

Figure 11.3   Area of Circle

NOMOGRAPH—ADDITION AND SUBTRACTION

Example:

Construct a nomograph for addition and subtraction of values from 0 to 30. From the chart, determine a) the sum (addition) of 3 plus 12, and b) the difference (subtraction) of 27 less 7.

Solution:

Plot scale values of 0 through 10 on a vertical scale (g) and scale values of 0 through 20 on a vertical scale (m) in a direct relationship on equal square graph paper as shown in figure 11.4—Addition and Subtraction. Align 1 on the (g) scale with 8 on the (m) scale to locate the intersection point on the horizontal line connecting 3 (g) scale and 6 (m) scale. Draw a vertical axis through this point to represent the (s) scale.   a) On figure 11.4, locate 3 on the (g) scale and align it with 12 on the (s) scale and find the intersection on the (m) scale, which gives a solution of 15.   b) On figure 11.4, align 27 on the (s) scale with 7 on the (g) scale and find the intersection on the (m) scale, which gives a solution of 20.

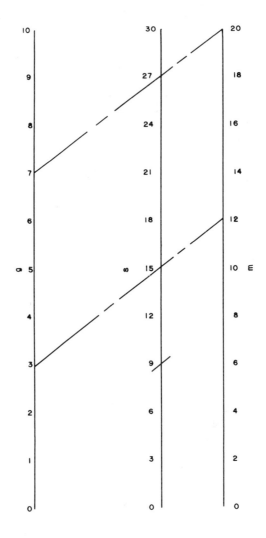

Figure 11.4   Addition and Subtraction

NOMOGRAPH—MULTIPLICATION AND DIVISION

Example:

Construct a nomograph for multiplication and division of values from 1 through 100. From the nomograph, determine a) the multiplication of 9 by 4, and b) 16 divided by 2.

Solution:

Plot scale values of w and 1 in scale gradations on semi-logarithm graph paper as shown in figure 11.5—Multiplication and Division. Draw a scale (a) midway between scales w and 1 with vertical scale gradations equal to multiples of scales w and 1.   a) On figure 11.5, locate 9 on the (w) scale and align it with 4 on the (1) scale, which gives the

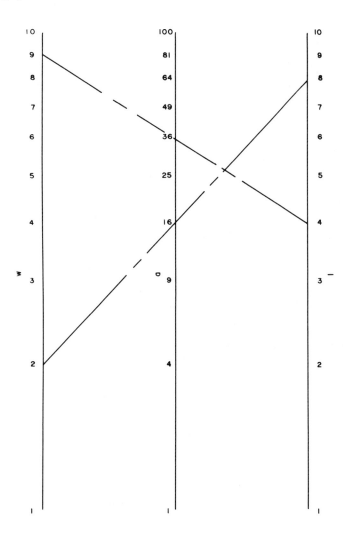

Figure 11.5 Multiplication and Division

solution of 36 on the (a) scale.   b) On figure 11.5, locate 16 on the (a) scale and align it with 2 on the (w) scale to find the intersection on the (1) scale, which gives the solution of 8.

NOMOGRAPH—EXCESS AIR IN A FURNACE

Formula:[1]

$$E = \frac{100(O_2 - 0.5CO)}{0.264N_2 - (O_2 - 0.5CO)}$$

---

1. Bill Sisson, "Nomogram Helps Find Excess Air," *Oil, Gas & Petrochem Equipment,* August, 1970, Tulsa, Okla. Copyright 1970, by The Petroleum Publishing Co., p. 26.

where

E = Excess air (percentage)
O_2 = Volumetric percentage of oxygen
CO = Volumetric percentage of carbon monoxide
N_2 = Volumetric percentage of nitrogen

Example:

Determine the excess air (percentage) (E) for an analysis of flue gas from a furnace where $CO = 1.5\%$, O_2 is 9.2% and N_2 is 78%.

Solution:

Plot the above formula in nomograph form as shown in figure 11.6 (Nomograph—Excess Air in a Furnace). Step 1: from 1.5 on the CO scale, draw a line to 9.2 on the O_2 scale, extend the line to the reference line, and mark the intersection. Step 2: from the marked point, draw a line to 78 on the N_2 scale, and read 69.6% where the line intersects the excess air scale (E).

Figure 11.6   Nomograph—Excess Air in a Furnace

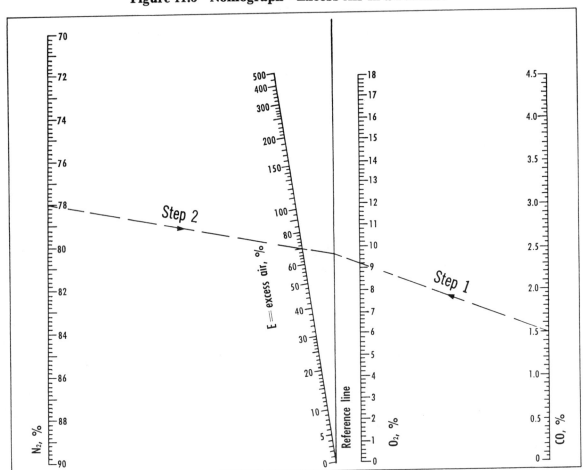

12. PROBABILITY—
POISSON DISTRIBUTION

FORMULA DEFINITIONS

The symbols and their definitions as used in this chapter are depicted and described in table 12.1—Symbols and Definitions.

POISSON DISTRIBUTION

For a known average number of arrivals in a given unit of time, a Poisson formula predicts the probability of detemining various numbers of arrivals in the same unit of time.

Formula:

$$P(c, a) = \sum_{x=c}^{\infty} \frac{e^{-a}a^x}{x!}$$

—which is the Cumulative Poisson Distribution tabulated in table VIII. This table is depicted graphically in figure 12.1.

Example:

For an average arrival rate of 3 (a), determine the probability that "c" or more arrivals occur per unit time P(c, a) for average arrival rates of 1 through 5, and 10.

Solution:

Using figure 12.1—Cumulative Poisson Distribution—locate a = 3 on the abscissa axis and its vertical intersections on curves c = 1 through 5, respectively, and c = 10, and read P(c, a) on the ordinate axis as follows:

a = 3

c	P(c, a)
1	0.950
2	0.801
3	0.577
4	0.353
5	0.185
10	0.001

Table 12.1

Symbols and Definitions

Symbol		*Definition*
a	=	average number of arrivals per unit of time
c	=	number of more arrivals per unit time
h	=	mean servicing or holding time
i	=	ordinal term number in a series of mathematical terms
m	=	mean waiting time of delayed arrivals
s	=	number of servers
u	=	servers occupancy ratio = ah/s
w	=	mean waiting time of all arrivals
x	=	multiples of average value of variable intervals between successive arrivals
e	=	natural logarithm base
P(c, a)	=	probability that "c" or more arrivals occur per unit time
F(x)	=	probability of exceeding a specified value of x
p > 0	=	probability of any delay greater than 0
∞	=	infinity
$\sum\limits_{i=f}^{k}$	=	summation of f th through k th terms in a series

---

Formula:

$$F(x) = e^{-x}$$

—which is a Poisson type arrival pattern characterized by an exponential cumulative probability distribution of intervals between successive arrivals. The values of e^{-x} are given in table IX. and shown graphically in figure 12.2—The Exponential Distribution.

Example:

Determine the probability—F(x)—of exceeding a value of x = 2.

Solution:

Using figure 12.2, locate x = 2 on the abscissa axis and its vertical intersection on the straight line curve and read a value of 0.135 for F(x) on the vertical axis.

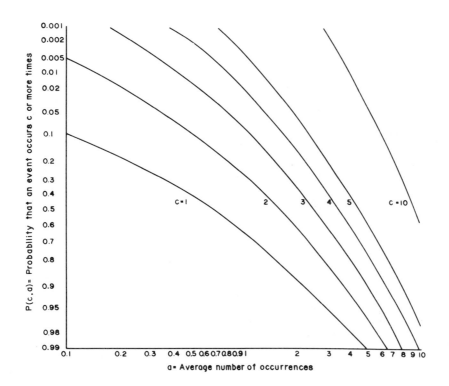

Figure 12.1   Cumulative Poisson Distribution

Figure 12.2   The Exponential Distribution

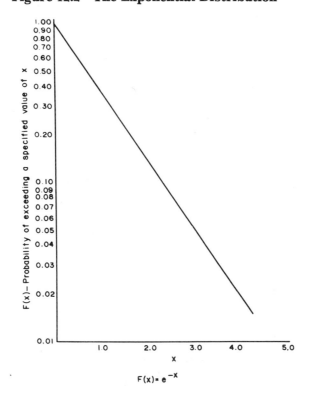

Formula:[1]

Formula 12.1 $P > 0 = \dfrac{\dfrac{(su)^s}{s!(1-u)}}{\displaystyle\sum_{i=0}^{s-1}\dfrac{(su)^i}{i!} + \dfrac{(su)^s}{s!(1-u)}}$

Formula 12.2 $w = \dfrac{h(P>0)}{s(1-u)}$

The above formulas are depicted in table 12.2. Formula 12.2 is plotted graphically in figure 12.3. This figure is a queuing system type with the following properties:

    a. Poisson arrivals (unlimited)
    b. Exponential service times
    c. Service in random or arrival order

Example 1:

For a Poisson arrival (unlimited) with a = 5.4 per minute and mean servicing or holding time h = 20 seconds with two servers (s), determine the mean waiting time of all arrivals (w) and the probability of delay ($P > 0$).

Solution:

$$w = \frac{ah}{s} = \frac{5.4(20/60)}{2} = 0.9$$

Using figure 12.3, the u = 0.90 curve intersects the s = 2 vertical line at the horizontal line (w/h) = 4.25. Therefore, w = 4.25h = 4.25 × 20 = 85 seconds. $P > 0$ may be calculated from formula 12.1 as 0.8526 (shown in table 12.2). As a check on the use of figure 12.3, formula 12.2 may be used as follows:

$$w = \frac{h(P>0)}{s(1-u)} = \frac{20(0.8526)}{2(1-0.9)} = 85 \text{ seconds}$$

Example 2:

For a Poisson arrival (unlimited) of a = 216 per hour, and mean servicing or holding time h = 120 seconds and a mean waiting time for all arrivals of w = 45 seconds, determine the number of servers (s) required to maintain the specified delay limits.

Solution:

Using figure 12.3—

$$ah = su = \frac{(216)}{(60)} \times \frac{(120)}{(60)} = 7.2$$

$$\text{and } w/h = \frac{45}{120} = 0.375,$$

---

1. "Solution Methods for Waiting Line Problems," The Port of New York Authority, September 1958, p. 18.

w/h = AVERAGE DELAY IN MULTIPLES OF AVERAGE SERVICING TIMES

u/s	1 P>0	1 w/h	2 P>0	2 w/h	3 P>0	3 w/h	4 P>0	4 w/h	5 P>0	5 w/h
0.1	0.0991	0.1101	0.0180	0.0100	0.0041	0.0015	0.0008	0.0002	0.0002	0.0001
0.2	0.2000	0.2500	0.0667	0.0417	0.0247	0.0103	0.0077	0.0024	0.0028	0.0007
0.3	0.3007	0.4296	0.1551	0.1108	0.0708	0.0337	0.0369	0.0132	0.0180	0.0051
0.4	0.4012	0.6687	0.2285	0.1738	0.1412	0.0784	0.1032	0.0430	0.0597	0.0199
0.5	0.5000	1.0000	0.3330	0.3330	0.2368	0.1579	0.1740	0.0870	0.1181	0.0472
0.6	0.6000	1.5000	0.4500	0.5625	0.3538	0.2948	0.2883	0.1800	0.2364	0.1182
0.7	0.6700	2.2333	0.5765	0.9608	0.4919	0.5466	0.4291	0.3577	0.3744	0.2496
0.8	0.8000	4.0000	0.7111	1.7778	0.6444	1.0740	0.5951	0.7439	0.5527	0.5527
0.9	0.9000	9.0000	0.8526	4.2630	0.8171	2.7237	0.7874	1.9683	0.7614	1.5228

u/s	6 P>0	6 w/h	7 P>0	7 w/h	8 P>0	8 w/h	9 P>0	9 w/h	10 P>0	10 w/h
0.1	0.0000	0.0000	0.0000	0.0000	0.0000	0.0000	0.0000	0.0000	0.0000	0.0000
0.2	0.0016	0.0003	0.0006	0.0001	0.0003	0.0000	0.0001	0.0000	0.0000	0.0000
0.3	0.0115	0.0027	0.0062	0.0012	0.0035	0.0006	0.0020	0.0003	0.0012	0.0002
0.4	0.0398	0.0111	0.0248	0.0059	0.0185	0.0039	0.0128	0.0024	0.0088	0.0015
0.5	0.0990	0.0330	0.0764	0.0219	0.0588	0.0147	0.0462	0.0103	0.0363	0.0073
0.6	0.1964	0.0818	0.1649	0.0589	0.1400	0.0438	0.1191	0.0330	0.1015	0.0254
0.7	0.3253	0.1807	0.3007	0.1432	0.2689	0.1120	0.2436	0.0902	0.2214	0.0738
0.8	0.5159	0.4283	0.4884	0.3489	0.4584	0.2865	0.4307	0.2393	0.4072	0.2038
0.9	0.7398	1.2360	0.7183	1.0262	0.7020	0.8775	0.6838	0.7598	0.6643	0.6643

TABLE 12.2 Probability of Delay

—the following values of u for various numbers of servers are obtained:

s	u	su
7	0.81	5.7
8	0.83	6.6
9	0.84	7.6
10	0.86	8.6

An analysis of the above data indicates that 9 servers (s) are the minimum number to maintain the condition of w = 45 seconds.

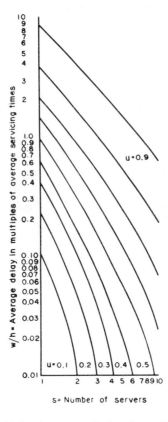

Figure 12.3   Average Delay in a Waiting Line System

13. MATHEMATICAL AVERAGES

ARITHMETIC AVERAGE

Definition:

An arithmetic or mean average is a single number representative of a set of figures and is determined by dividing the sum total of a set of figures by the number of figures.

Formula:

$$A_a = \frac{\sum\limits_{i=1}^{n} A_i}{n}$$

where

A_a = Arithmetic average

$A_i \ldots A_n$ = Value of respective figures

$\sum\limits_{i=1}^{n} A_i$ = Sum of respective figures $A_i \ldots A_n$

n = Number of figures

Example:

Determine the arithmetic average (A_a), to the nearest decimal place, for the following eleven figures (n), namely, $A_1 = 77$, $A_2 = 79$, $A_3 = 79$, $A_4 = 79$, $A_5 = 79$, $A_6 = 80$, $A_7 = 81$, $A_8 = 81$, $A_9 = 83$, $A_{10} = 84$ and $A_{11} = 85$.

Solution:

$$\sum\limits_{i=1}^{n=11} A_i = 77 + 79 + 79 + 79 + 79 + 80 + 81 + 81 + 83 + 84 + 85 = 887$$

or

$$\sum_{i=1}^{n=11} A_i = 77 + 4(79) + 80 + 2(81) + 83 + 84 + 85 = 887$$
$$A_a = 887/11 = 80.6$$

MEDIAN AVERAGE

Definition:

In a set of figures arranged in order of magnitude, a median average is that value that represents the point at which there are as many figures above it as there are figures below it.

Example:

Determine the median average for the figures used in the example of arithmetic average above, namely, 77, 79, 79, 79, 79, 80, 81, 81, 83, 84 and 85.

Solution:

Examination of the series in order of magnitude—77, 79, 79, 79, 79, 80, 81, 81, 83, 84, 85—reveals that 80 is the median average in that there are five figures above it (81, 81, 83, 84, 85), and, five figures below it (77, 79, 79, 79, 79).

MODE AVERAGE

Definition:

In a set of figures, the mode average is the value that occurs more often than any other.

Example:

Determine the mode average for the figures used in the examples of arithmetic average and median average above, namely, 77, 79, 79, 79, 79, 80, 81, 81, 83, 84, 85.

Solution:

Examination of these figures reveals that 79 is the mode value in that it occurs four times which is the most frequent occurrence.

Part IV

WORK MEASUREMENT

14. FLOW
PROCESS CHARTS

DEFINITION AND PURPOSE[1]:

A flow process chart is a graphic representation of the sequence of all operations, transportations, inspections, delays, and storages occurring during a process or procedure, and includes information considered desirable for analysis, such as time required and distance moved. An analysis of the chart may result in improvement of the process or procedure charted. There are two types:

a) The material type presents the process in terms of the events which occur to the material.

b) The man type presents the process in terms of the activities of the man.

The following are pertinent symbols and their definitions:

Symbol   *Definition*

Operation. An operation occurs when an object is intentionally changed in any of its physical or chemical characteristics, is assembled or disassembled from another object, or is arranged or prepared for another operation, transportation, inspection, or storage. An operation also occurs when information is given or received or when planning or calculating takes place.

Transportation. A transportation occurs when an object is moved from one place to another, except when such movements are a part of the operation or are caused by the operator at the work station during an operation or an inspection.

---

1. Extracted from ASME Standard Operation and Flow Process Charts, ASME Standard No. 101, with the permission of the publisher, The American Society of Mechanical Engineers, United Engineering Center, 345 East 47th Street, New York, N.Y. 10017, p. 3.

FLOW PROCESS CHART

Name of Part or Product   Requisition For Supplies - Rush Job

Chart Begins at Machine Shop Foreman's Desk, Ends on Typist's Desk in Pur. Dept.	Chart No.
Order No. Lot Size Dept.	Sheet 1
Charted by C.H.H. Date Charted Bldg.	of 1 Sheets

Travel in ft.	Time in min.	Symbol	Operations	Remarks
		1	Written longhand by foreman	
		1	On foreman's desk (awaiting messenger)	
1000		1	By messenger to secretary of head of department	
		2	On secretary's desk (awaiting typing)	
		2	Typed	
15		2	By messenger to head of department	
		3	On head of department's desk (awaiting approval)	
		O-3 INS-1	Examined, approved and coded (signed and code stamped)	
		4	On head of department's desk (awaiting messenger)	
2000		3	To Purchasing Department	
		5	On purchasing agent's desk (awaiting approval)	
		2	Examined and approved	
		6	On purchasing agent's desk (awaiting messenger)	
25		4	To typist's desk	
		7	On typist's desk (awaiting typing of purchase order)	

SUMMARY

Number of Operations   3
Number of Delays   7
Number of Inspections   2
Number of Transportations   4
Total Travel in Feet   3040

Figure 14.1   Flow Process Chart of the Material Type (Office Form) With Preprinted Headings and Captions

Inspection. An inspection occurs when an object is examined for identification or is verified for quality or quantity in any of its characteristics.
Delay. A delay occurs to an object when conditions except those which intentionally change the physical or chemical characteristics of the object, do not permit or require immediate performance of the next planned action.

FLOW PROCESS CHART

IDENTIFICATION

SUBJECT CHARTED	RELIEF VALVE BODY
DRAWING NO.	A-520652
PART NO.	16150
POINT AT WHICH CHART BEGINS	RECEIPT OF REQUISITION FROM MACHINE SHOP
LOCATION	NEAR STOCK STOREROOM
POINT AT WHICH CHART ENDS	STORING OF FINISHED PRODUCT IN STOREROOM
LOCATION	ASSEMBLY STOREROOM
CHART NO.	1021
TYPE OF CHART	
SHEET NO.	1 OF 1 SHEETS
CHARTED BY	
DATE	
APPROVED BY	
DATE	
YEARLY PRODUCTION	100,000 PIECES
COST UNIT	1 VALVE BODY

QUANTITY INFORMATION
1 BAR MAKES 50 BODIES
1 PACKING BOX HOLDS 25 BODIES

SUMMARY

INSTALLATION COST OF PROPOSED METHOD $150.00
TOTAL YEARLY SAVING–DIRECT LABOR $2200.00
ESTIMATED NET SAVING–FIRST YEAR $2050.00

	PRESENT METHOD		PROPOSED METHOD		DIFFERENCE	
	NO.	TIME IN HRS.	NO.	TIME IN HRS.	NO.	TIME IN HRS.
UNIT COST–DIRECT LABOR & INSP.	.300		.278		.022	
DISTANCE TRAVELED IN FEET	7383		1843		5540	
OPERATIONS	8	.3250	7	.3064	1	.0175
TRANSPORTATIONS	9	.0085	6	.00222	3	.00626
INSPECTIONS	2	.0085	1		1	.0085
DELAYS	12	31.5	8	25.5	4	6.0
STORAGES	2	110.0	1	60.0	1	38.0

PRESENT METHOD

QUANTITY UNIT CHARTED	DESCRIPTION OF EVENT	DIST. MOVED IN FEET	UNIT OPER. TIME	UNIT TRANSP. TIME	UNIT INSPECT. TIME	DELAY TIME	STORAGE TIME
1 BAR	Bar loaded on truck upon receipt of requisition from machine shop (2 men)	10	.0002				
20 BARS	Moved to #301 machine	210		.0002			
1 BAR	Bar unloaded to bar stock rack near #301 machine (2 men)	10	.0002				
1 VALVE BODY	Bar waits for operation to begin					4.00	
	Drill, bore, tap, seat, file and cut off		.0550				
	Await moveman					4.00	
300 VALVE BODIES	Moved to burring department	300		.0011			
	Await burring operator					1.50	
1 VALVE BODY	Burr	6	.0075				
	Await moveman					2.00	
100 VALVE BODIES	Moved to drill press in machine shop	320		.0011			
	Await drilling operator					2.00	
	Drill and countersink 8 holes		.0480				
	Await moveman					4.00	
100 VALVE BODIES	Moved to inspection department	100		.0001			
	Await inspector					1.50	
1 VALVE BODY	Inspect				.0085		
	Await moveman					2.00	
100 VALVE BODIES	Moved to storeroom #2	3009		.0030			
	Stored until requisitioned by detail department						48.0
300 VALVE BODIES	Moved to seat lapping machine	2550		.0025			
	Await operator					1.50	
1 VALVE BODY	Lap, seat, and test	6	.1700				
	Await moveman					2.00	
100 VALVE BODIES	Moved to paint booth	400		.0004			
	Await painter					4.00	
1 VALVE BODY	Mask, prime, paint, dry and unmask	10	.0350				
100 VALVE BODIES	Carried to packer by painter	125		.0001			
	Await packing					3.00	
1 VALVE BODY BOX 25	Pack in box	12	.0100				
VALVE BODY	Sent by conveyor to assembly department storeroom	300					
	Storage until requisitioned						60.0

PROPOSED METHOD

QUANTITY UNIT CHARTED	DESCRIPTION OF EVENT	DIST. MOVED IN FEET	UNIT OPER. TIME	UNIT TRANSP. TIME	UNIT INSPECT. TIME	DELAY TIME	STORAGE TIME
1 BAR	Bar loaded on truck upon receipt of requisition from machine shop (2 men)	10	.0002				
20 BARS	Moved to #301 machine	210		.0002			
1 BAR	Bar unloaded to bar stock rack near 301 machine	10	.0002				
1 VALVE BODY	Bar waits for operation to begin					4.00	
	Drill, bore, tap, seat, file and cut off		.0550				
	Await drill press operator					2.00	
	Moved to drill press by operator	20		.0002			
300 VALVE BODIES	Drill 8 holes		.0350				
	Await moveman					2.00	
	Moved to burring department	300		.0011			
	Await burring operator					1.50	
1 VALVE BODY	Burr	6	.0100				
	Await moveman					2.00	
300 VALVE BODIES	Moved to seat lapping machine–detail department	550		.0005			
	Await operator					6.00	
1 VALVE BODY	Lap, seat, test and inspect		.1700				
	Await moveman					2.00	
100 VALVE BODIES	Moved to paint booth	400		.0004			
	Await painter					6.00	
1 VALVE BODY	Mask, prime, paint, dry, unmask, pack in box	15	.0380				
BOX 25 VALVE BODY	Sent by conveyor to assembly department storeroom	425					
							60.0

Figure 14.2 Flow Process Chart With Preprinted Symbols Showing Present and Proposed Methods With a Summary of Cost Savings

Storage. A storage occurs when an object is kept and protected against unauthorized removal.

Combined Activity. When it is desired to show activities performed either concurrently or by the same operator at the same work station, the symbols for those activities are combined, as shown by the circle placed within the square to represent a combined operation and inspection.

Examples:

Figures 14.1[2] and 14.2[3] are sample Flow Process Charts. Figure 14.2 brings out clearly the differences between the present and the proposed methods as shown in the summary form in the upper right-hand corner. The summary provides space for recording the total yearly direct labor saving, the cost of putting the proposed method into effect, and the estimated net savings for the first year. This information is supported by a summary of the detailed changes in terms of unit cost, distance traveled in feet, and number and time of operations, transportations, inspections, delays, and storages under the present and proposed methods.

---

2. *Ibid.,* p. 13.
3. *Ibid.,* p. 16.

15. WORK
STANDARD

DEFINITION:

A work standard is a reasonable measure of work output expected from a worker with all due respect for working conditions, and the like.

Formula:

$$WS = ST + BTA$$

where

WS = Work standard

ST = Standard basic time values

BTA = Base time allowances

ST Definition:

Standard basic time values (ST) are distinguishable as logical, different basic functions that have to be performed in order to complete an operation. Their time values are considered to be reasonable and fair values of expected performance.

BTA Definition:

Base time allowances (BTA) are established values for personal needs, fatigue, etc. and are expressed as minutes/shift (S).

Expressed as a percentage figure,

$$BTA = \frac{B_T}{S - B_T} \times 100\%$$

where

BTA $= \%$ Base time allowance
$B_T =$ Sum of individual allowances in minutes/man/shift
$S =$ Shift time in minutes

Example 1:

The SAF Company is analyzing an assembly operation for the purpose of establishing a work standard. The one man operation has been broken down into three standard basic time values (ST) of: 1. Get parts (4 minutes/assembly); 2. assemble parts (10 minutes/ assembly); and 3. remove finished assembly (1 minute/assembly).

The Company works an 8 hour (480 minute) shift (S) and has two scheduled coffee breaks of 15 minutes each, figures an additional 20 minutes per shift for personal needs, etc. and 5 minutes per shift for miscellaneous (minor interruptions, etc.). Calculate the work standard (WS) for this operation.

Solution:

ST — Standard basic time values

1. Get parts	4 minutes/assembly
2. Assemble parts	10 " "
3. Remove finished assembly	2 " "
Total	16 minutes/assembly

BTA — Base time allowances

1. Coffee breaks	2 @ 15 minutes/shift =	30 minutes/shift
2. Personal, etc.	=	20 " "
3. Miscellaneous	=	5 " "
	Total	55 minutes/shift

BTA $= \%$ Base time allowances $=$

$$\frac{55 \text{ minutes}}{480 \text{ minutes} - 55 \text{ minutes}} \times 100\% = 12.9\%$$

WS $=$ ST $+$ BTA $=$ ST $+ \%$ BTA (ST)
   $=$ 16 minutes/assembly $+ 12.9\% \times$ 16 minutes/assembly
   $= 1.129 \times$ 16 minutes/assembly
WS $= 18.06$ minutes/assembly or 0.301 hours/assembly or
   3.32 assemblies/hour (man hour or clock hour)

Example 2:

The FMR Company is analyzing a lift truck operator-helper (2 man) operation for the purpose of establishing a work standard. The operation has been broken down into two standard basic time values (ST) of: 1. Driving time (3 clock minutes/order) and 2. pick order (10 minutes/order). The company figures a one hour "lost time" per 8 hour shift for cognizance of personal needs, fatigue, etc. Calculate the work standard (WS) for this operation.

Solution:

ST = Standard basic time values

1. Driving time   3 clock minutes/order
2. Pick order   10   〃   〃   〃

       Total   13 clock minutes/order

BTA = % Basic time allowances =

$$\frac{60 \text{ minutes}}{480 \text{ minutes} - 60 \text{ minutes}} \times 100\% = 14.3\%$$

WS = ST + BTA = ST + % BTA (ST)

     = 13 clock minutes/order + 14.3% × 13 clock minutes/order

     = 1.143 × 13 clock minutes/order

WS = 14.86 clock minutes/order or 0.25 clock hours/order or 4 orders/clock hour

$$0.25 \text{ clock hours/order} \times \frac{2 \text{ man hours}}{\text{clock hour}} = 0.50 \text{ man hours/order}$$

or 2 orders/man hour

16. MANPOWER
PLANNING*

FORMULA

ESMR = PSH/APH

      where

ESMR = Equivalent standard manpower requirements

  PSH = Productive standard hours

  APH = Available productive hours

Example:

The EHW Company has forecast a requirement of 38,000 productive standard hours (PSH) for its operation in the forthcoming year. The operation is basically repetitive such that productive requirements are basically the same from month to month. Based on past records, the company estimates that the average employee has 1880 available productive hours (APH) (after time off for vacation, etc.). Compute the equivalent standard manpower requirements for the forthcoming year.

Solution:

ESMR = 38,000 hours/1880 hours

ESMR = 20.2 men

Example:

The WJR Company which has a "peak and valley" operation is planning its forth-

---

* This chapter is adapted from *Industrial Management*, August 1969, "Manpower Planning for Fluctuating Operations," by Harry W. Muni, courtesy of Industrial Management Society, Chicago, Illinois, pp. 4–7.

coming year's manpower requirements. The compnay wants to avoid a hire and layoff practice and, instead, to have a practical fixed work force at "optimum" cost with respect to overtime requirements, fatigue, etc. Develop a procedure to establish a basis for determining its manpower requirements.

Solution:

The company first tabulates its past year's history as to productive standard-hour requirements (PSH) by month and makes any foreseeable changes for the forthcoming year. In this particular case, the WJR Company anticipates that the previous year's history will be adequate for immediate future planning purposes. Table 16.1—Total Required Standard Productive Man Hours Per Month—depicts this data. This data is also plotted graphically in figure 16.1. Table 16.2—Available Straight Time Productive Hours Per Month—is also developed from available data. Table 16.3—Total Available Straight Time Productive Hours—is based upon the available productive hours per man as derived from table 16.2, extended by different numbers of men. Table 16.4—Overtime Hour Requirements—represents the difference in number of hours between tables 16.1 and 16.3.

Table 16.5—Standard Manpower Requirements—also shown graphically in figure 16.2, indicates that the optimum cost or "break-even" number is 26 men. The final decision as to planned manpower would involve consideration of incremental payroll cost figures for number of men greater than 26, fatigue, job rotation, etc.

THE WJR COMPANY

TOTAL REQUIRED STANDARD PRODUCTIVE MAN HOURS PER MONTH

	Unit A Req'd. Std. Man Hours	Unit B Req'd. Std. Man Hours	Unit C Req'd. Std. Man Hours	Total Req'd. Std. Man Hours
19__				
Jan.	1,531	1,961	872	4,364
Feb.	1,592	2,044	908	4,544
Mar.	1,670	2,148	954	4,772
Apr.	1,830	2,334	1,064	5,228
May	2,012	2,566	1,146	5,724
June	1,763	2,257	1,004	5,024
July	1,455	1,870	831	4,156
Aug.	1,499	1,927	842	4,268
Sept.	1,572	2,019	897	4,488
Oct.	1,785	2,297	1,020	5,102
Nov.	1,893	2,435	1,082	5,410
Dec.	1,496	1,927	843	4,266
TOTALS	20,098	25,785	11,463	57,346

TABLE 16.1

THE WJR COMPANY

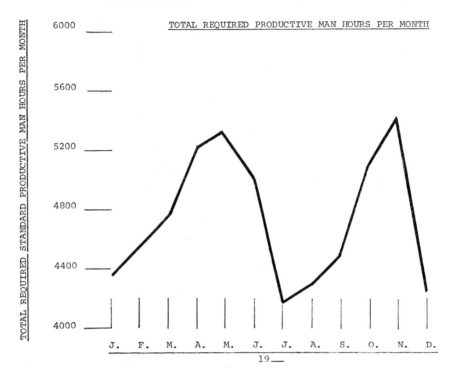

Figure 16.1

THE WJR COMPANY

**AVAILABLE STRAIGHT TIME PRODUCTIVE
HOURS PER MONTH**

Annual Holidays ....10 Days/Man
Annual Vacation ....11 Days/Man
Annual Sickness and Other Time Off ...... 8 Days/Man

	Holiday Hours/ Man	Vacation Hours/ Man	Sickness & Other Hours/ Man	Total Unavail- able Hours/ Man	Avail- able Produc- tive Hours/ Man
19__					
Jan.	8	—	7	15	169
Feb.	8	—	14	22	146
Mar.	—	—	6	6	162
Apr.	8	—	2	10	164
May	8	—	4	12	172
June	—	—	2	2	158
July	8	38	4	50	134
Aug.	—	37	3	40	136
Sept.	8	14	1	23	145
Oct.	8	—	5	13	171
Nov.	16	—	5	21	147
Dec.	8	—	7	15	161

TOTAL 1,865

TABLE 16.2

THE WJR COMPANY

TOTAL AVAILABLE STRAIGHT TIME PRODUCTIVE HOURS

19__	Avail.* Prod. Hours	25	26	27	28	Number of Men in Department 29	30	31	32	33	34	35
Jan.	169	4,225	4,394	4,563	4,732	4,901	5,070	5,239	5,408	5,577	5,746	5,915
Feb.	146	3,650	3,796	3,942	4,088	4,234	4,380	4,526	4,672	4,818	4,964	5,110
March	162	4,050	4,212	4,374	4,536	4,698	4,860	5,022	5,184	5,346	5,508	5,670
April	164	4,100	4,264	4,428	4,592	4,756	4,920	5,084	5,248	5,412	5,576	5,740
May	172	4,300	4,472	4,644	4,816	4,988	5,160	5,332	5,504	5,676	5,848	6,020
June	158	3,950	4,108	4,266	4,424	4,582	4,740	4,898	5,056	5,214	5,372	5,530
July	134	3,350	3,484	3,618	3,752	3,886	4,020	4,154	4,288	4,422	4,556	4,690
Aug.	136	3,400	3,536	3,672	3,808	3,944	4,080	4,216	4,352	4,488	4,624	4,760
Sept.	145	3,625	3,770	3,915	4,060	4,205	4,350	4,495	4,640	4,785	4,930	5,075
Oct.	171	4,275	4,446	4,617	4,788	4,959	5,130	5,301	5,472	5,643	5,814	5,985
Nov.	147	3,675	3,822	3,969	4,116	4,263	4,410	4,557	4,704	4,851	4,998	5,145
Dec.	161	4,025	4,186	4,347	4,508	4,669	4,830	4,991	5,152	5,313	5,474	5,635
TOTALS	1,865	46,625	48,490	50,355	52,220	54,085	55,950	57,815	59,680	61,545	63,410	65,275

*AVAILABLE PRODUCTIVE HOURS PER MAN AS DERIVED IN TABLE 2.

TABLE 16.3

THE WJR COMPANY

OVERTIME HOUR REQUIREMENTS

19__	Total* Req. Prod. Man-Hours	25	26	27	28	Number of Men 29	30	31	32	33	34	35
Jan.	4,364	139	—	—	—	—	—	—	—	—	—	—
Feb.	4,544	894	748	602	456	310	164	—	—	—	—	—
March	4,772	722	560	398	236	74	—	—	—	—	—	—
April	5,228	1,128	964	800	636	472	308	144	—	—	—	—
May	5,724	1,424	1,252	1,080	908	736	564	392	220	48	—	—
June	5,024	1,074	916	758	600	442	284	126	—	—	—	—
July	4,156	806	672	538	404	270	136	2	—	—	—	—
Aug.	4,268	868	732	596	460	324	188	52	—	—	—	—
Sept.	4,488	863	718	573	428	283	138	—	—	—	—	—
Oct.	5,102	827	656	485	314	143	—	—	—	—	—	—
Nov.	5,410	1,735	1,588	1,441	1,294	1,147	1,000	853	706	559	412	265
Dec.	4,266	241	80	—	—	—	—	—	—	—	—	—
TOTALS	57,346	10,721	8,886	7,271	5,736	4,201	2,782	1,569	926	607	412	265

*TOTAL REQUIRED PRODUCTIVE MAN-HOURS AS DERIVED IN TABLE 1.

TABLE 16.4

THE WJR COMPANY

STANDARD MANPOWER REQUIREMENTS

Number	Straight Time*		Overtime**		Annual Total
Of Men	Hours/Yr.	$/Yr.	Hours/Yr.	$/Yr.	Payroll
25	52,000	177,840	10,721	44,707	$222,547
26	54,080	184,954	8,886	37,055	222,009
27	56,160	192,067	7,271	30,319	222,236
28	58,240	199,181	5,736	23,919	223,100
29	60,320	206,294	4,201	17,518	223,812
30	62,400	213,408	2,782	11,600	225,008
31	64,480	220,522	1,569	6,543	227,065
32	66,560	227,635	926	3,852	231,487
33	68,640	234,749	607	2,531	237,280
34	70,720	241,862	412	1,718	243,580
35	72,800	248,976	265	1,105	250,081

*STRAIGHT TIME HOURLY RATE IS $2.78 BASE RATE PLUS $0.64 FRINGE EQUALS $3.42/HOUR.

**OVERTIME HOURLY RATE IS $4.17/HOUR. (1.5 × $2.78 EQUALS $4.17/HOUR).

TABLE 16.5

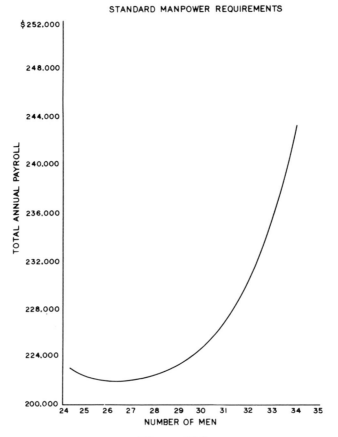

THE WJR COMPANY

STANDARD MANPOWER REQUIREMENTS

Figure 16.2

17. WORK
SAMPLING

DEFINITION:

Work sampling is a statistical technique based on the law of probability employing the use of a random number of observations to determine the magnitude of an activity. Over a sufficiently large number of observations, the number of times a particular condition is observed tends to give an accurate percentage for time in that state. The technique may be employed to determine machine down time, relationships of personnel work time activities, etc.

Formula:

$A_n/N \times 100 =$ Percentage of total observations during which

the condition is percent.
where
$A_n =$ Number of observations of activity
$N =$ Total number of observations

Example 1:

The JJP Company is using the work sampling technique to analyze one of its important man–machine operations in terms of productive and non-productive time. A total of 500 random observations (N) were made of which 370 observations (A_n) found the job in operation. Observations were not made during two 15 minute authorized rest periods in the 8 hour (480 minute) work shift. Determine the productive and non-productive category percentages and equate these to minutes per shift.

Solution:

	No. of Observations	Per cent of Total	Minutes/shift (480 − 30 = 450) Per cent of Total × 450
Productive (A$_n$)	370	74%	333 minutes
Non-Productive (A$_n$)	130	26	117
Total (N)	500	100%	450
Authorized rest periods (2 at 15 minutes = 30 minutes)			30
		Total shift	480 minutes

Example 2:

The JJP Company, after studying the above data, feels that the magnitude of the non-productive time (A$_n$), namely, 26%, requires analysis as to its make-up with a view, if possible, to reduce this figure. To insure greater accuracy on the next work sampling study, it has been decided to use 1000 random unit observations (N) and to classify the job into: productive time categories of set-up and man–machine working; and non-productive time categories of machine down–operator adjusting, machine down–maintenance repairing, machine down–waiting for incoming material, and machine down–operator away. Conduct the study and determine the percentage of each of these categories. Determine the productive and non-productive category percentages and equate these to minutes per shift.

Solution:

The following data were observed and tabulated:

	No. of Observations	Per cent of Total	Minutes/Shift (480 − 30 = 450) Per cent of Total × 450
Productive time			
Set-up time	39	3.9%	18 minutes
Working time	689	68.9	310
Sub-Total	728	72.8%	328 minutes
Non-productive time			
Machine down–operator adjusting	11	1.1%	5 minutes
Machine down–maintenance repairing	92	9.2%	41 minutes
Machine down–waiting for incoming material	22	2.2%	10 minutes
Machine down–operator away	147	14.7	66 minutes
Sub-Total	272	27.2%	122 minutes

	Total	1000	100%	450 minutes

Authorized rest periods
(2 at 15 minutes = 30 minutes)   30
                                                                      Total Shift   480 minutes

Example 3:

The JAN Company is analyzing its clerical activities through the use of the work sampling technique. A secretary's job is being studied to determine the magnitude of the different activities being handled and to explore the possibilities of re-assigning activities to lower clerical levels. The job will be divided into these categories: greeting and escorting visitors, answering telephone calls, taking dictation, typing, filing, using duplicating machine, collating, and away from desk. The work day consists of 7 hours (420 minutes) and there are no scheduled rest periods. Determine the percentages of each category of activity and equate these to minutes per shift. The study employed 600 random observations.

Solution:

	No. of Observations	Per cent of Total	(Minutes/Shift) (420 minutes) Per cent of Total × 420
Visitors	5	0.8%	3
Telephone	47	7.8	33
Dictation	27	4.5	19
Typing	363	60.5	254
Filing	19	3.2	13
Duplicating	34	5.7	24
Collate	31	5.2	22
Away from desk	74	12.3	52
Total	600	100%	420 minutes

USE OF RANDOM NUMBERS:

Table X.[1]—Table of Random Numbers—is presented to offer many thousands of combinations of numbers for random sampling. Establish some arbitrary plan of selecting numbers from the table. Examples could be an arbitrary plan of every other column down, some relationship between line and column values such as an entry on line 1 corresponding to its value in column 1; an entry on line 2 corresponding to its value in column 2, etc. Any number of arbitrary plans of selecting random numbers may be devised. Establish a plan of definitions for the digits in the random numbers chosen for the work sampling study.

1. "Table of 105,000 Random Decimal Digits," Interstate Commerce Commission, Bureau of Transport Economics and Statistics, Washington, D.C., 1949.

Example 1:

Two separate operations, A and B, are to be observed for a one-day work sampling study. It is desired to have 15 random observations for each. The work day is from 9:00 A.M. to 5:00 P.M. with a one hour lunch period from 12:00 Noon to 1:00 P.M. There are no scheduled rest periods. Develop a plan and establish a schedule of times that observations should be made.

Solution:

Designate the hours and minutes of the day with four digits, starting with 9 A.M.: 0900 = 9 A.M., 1200 = Noon, 1300 = 1 P.M., 1400 = 2 P.M., 1500 = 3 P.M., etc. In the five digit numbers in table X., the first four digits would represent the hour and minute of the day when an observation is to be made. The fifth digit will represent the operation to be observed, with values of 0 through 4 representing operation A and 5 through 9 representing operation B. An arbitrary decision has been made to use the values in column 2. The first figure shown in column 2 is 15011. The first four digits, 1501, indicate 3:01P.M. and the last digit, 1, indicates that operation A is to be observed at this time. The next figure, 46573, is discarded because its first four digits do not fit the conditions of the situation, namely, the work period being from 9:00 A.M. to 12:00 Noon (0900 to 1200) and 1:00 P.M. to 5:00 P.M. (1300 to 1700). The following figures were taken from column 2 and listed under operations A and B, respectively. The second column under each operation is a rearrangement of the figures in sequential order, and the third column shows the conversion to the time the observations are to be taken. Readings are taken for 15 values of operations A and B, respectively.

Operation A			Operation B		
Number	*Sequence*	*Time*	*Number*	*Sequence*	*Time*
15011	09193	9:19 A.M.	14367	10339	10:33 A.M.
14342	09352	9:35	14229	10375	10:37
10493	09452	9:45	11128	10396	10:39
09193	10150	10:15	10396	10538	10:53
13162	10493	10:49	13537	10559	10:55
13263	10553	10:55	16508	11085	11:08
09352	13114	1:11 P.M.	11085	11128	11:12
10150	13162	1:16	10538	11206	11:20
15262	13172	1:17	10559	11409	11:40
13114	13263	1:26	10375	13537	1:53 P.M.
09452	14342	2:34	14199	14199	2:19
16503	15011	3:01	16228	14229	2:22
15382	15262	3:26	11409	14367	2:36
13172	15382	3:38	10339	16228	4:22
10553	16503	4:50	11206	16508	4:50

Example 2:

An operation that works "around the clock," that is, 24 hours a day, is to be analyzed for a two day period. There are no specified meal or rest periods; consequently, observations will be made at random intervals throughout the day. It is desired to have 35 readings per day. Develop a schedule of times that observations should be made.

Solution:

Of the five-digit numbers in table X., the first digit when 0 through 4 will designate the first day of the study and when 5 through 9 the second day. The remaining four digits will designate the hours and minutes of the day starting with 0001 (one minute past midnight) and extending to 2400 (midnight). Therefore, no number whose last four digits are greater than 2400 can be used.

Numbers have been selected diagonally from the 14 columns by testing the first number in the first column, the second number in the second column, and so on, to the fourteenth number in the fourteenth column—then, from the left again, the fifteenth number in the first column, the sixteenth in the second, and so on—until as many numbers as desired are obtained. Also, the selection is alternated by first finding one for the first day and then one for the second, then again one for the first, and so on. (The first is line 11, column 11; the second, line 12, column 12; the third, line 13, column 13; and the fourth, line 21, column 7.)

These figures are listed below for each of the two days of the study, followed by a column placing them in numerical sequence and a third column depicting the conversion to the time the observations are to be taken.

| | 1st Day | | | 2nd Day | |
Number	Sequence	Time	Number	Sequence	Time
01547	00033	0:33 A.M.	90511	90229	2:29 A.M.
30613	20103	1:03	90229	60249	2:49
41135	00147	1:47	80428	70408	4:08
20103	10251	2:51	81652	70418	4:18
31355	30337	3:37	70735	80428	4:28
00033	30504	5:04	50501	50501	5:01
00959	10538	5:38	91227	80504	5:04
01221	30613	6:13	90730	90511	5:11
12143	40839	8:39	70408	90730	7:30
01715	40937	9:37	61005	70735	7:35
21252	00959	9:59	61448	50904	9:04
11959	41034	10:34	91055	90937	9:37
30337	01120	11:20	51736	61005	10:05
10251	41135	11:35	92121	51032	10:32

Number	1st Day Sequence	Time	Number	2nd Day Sequence	Time
02114	01221	12:21 P.M.	71246	91055	10:55
40839	31231	12:31	92006	91227	12:27 P.M.
31606	21252	12:52	61401	71246	12:46
32121	31355	1:55	51424	91326	1:26
40937	01509	3:09	61527	61401	2:01
01120	01547	3:47	91623	51424	2:24
31231	01604	4:04	50904	61448	2:48
31612	31606	4:06	70418	81454	2:54
42116	31612	4:12	80504	81456	2:56
01604	11623	4:23	81454	91507	3:07
42157	01715	5:15	81456	51514	3:14
01921	21828	6:28	91507	61527	3:27
11930	01921	7:21	91326	71533	3:33
00147	11930	7:30	60249	51547	3:47
10538	11959	7:59	92012	91623	4:23
11623	02114	9:14	51514	81652	4:52
41034	42116	9:16	82356	51736	5:36
21828	32121	9:21	51447	92006	8:06
30504	12143	9:43	90937	92012	8:12
01509	42157	9:57	51032	92121	9:21
42319	42319	11:19	71533	82356	11:56

18. LEARNING CURVES

DEFINITION:

A learning curve is a mathematical prediction of rate of learning—applicable to new products and new operators.

Formula:

    a) $P_1 \approx T_1$
          $2 P_1 \approx P_2 \approx p\% \times T_1 \approx T_2$
          $2 P_2 \approx P_3 \approx p\% \times T_2 \approx T_3$, etc.

        where

        $P_1 =$ Initial production unit
        $T_1 =$ Time/unit P_1
        $P_2 =$ Double initial production P_1 unit
        $p\% =$ Uniform percent-drop in time
        $T_2 =$ Time/unit P_2, etc.

Example:

The RLT Company is planning production of a new product (A). Product "A" is similar to an existing line of products which has experienced a 90% (p%) learning curve. The company will, then, make analyses on the prediction of an average 90% improvement curve. Determine and plot graphically (rectangular coordinates) the relationship of production units and production time/unit on the basis of the initial production unit (P_1) requiring 1 hour (T_1). Determine the unit time (T) for the 10th production unit (P).

Solution:

Production units	Time per unit (hours)
1	1
2	90% × 1 = 0.90
4	// × 0.90 = 0.81
8	// × 0.81 = 0.729
16	// × 0.729 = 0.656
32	// × 0.656 = 0.590
64	// × 0.590 = 0.531

The above data are plotted graphically (rectangular coordinates) in figure 18.1—90% Learning Curve. From this graph, a value of 0.705 hours is determined for the 10th production unit (P).

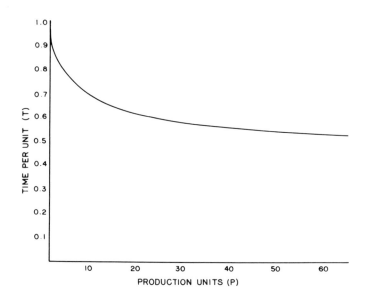

Figure 18.1   90% Learning Curve

Formula:

b) $T = T_1 \times P^s$

where

T = Time per unit
T_1 = Time per first unit
P = production unit
S = Slope (straight line curve on log-log graph paper)

Example:

From the data in the above example, plot a graph on log-log graph paper. Determine the unit time (T) for the 10th production unit (P).

Solution:

For a 90% learning curve, the slope (S) $= -0.152$, and the curve is a straight line as depicted in figure 18.2—Learning Curve—Log-Log Graph. From this graph, a value of 0.705 hours is determined for the 10th production unit (P).

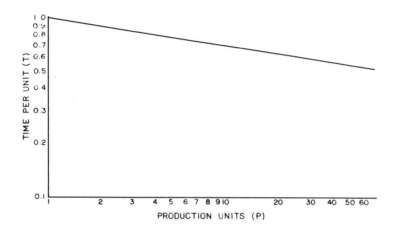

Figure 18.2 90% Learning Curve—Log-Log Graph

Example:

The RRH Company's line of products has an experience of an 85% learning rate improvement curve. For a new product (B), the prototype unit took 150 hours (T). Project the estimated unit times for succeeding units and plot these values graphically using rectangular coordinates and logarithm graph paper. The slope of the straight line curve on log-log paper for an 85% learning curve is -0.234. Determine the estimated time for the 50th unit.

Solution:

Production units	Time per unit (hours)
1	$= 150$
2	$85\% \times 150 = 128$
4	$\prime\prime \times 128 = 109$
8	$\prime\prime \times 109 = 92$
16	$\prime\prime \times 92 = 78$
32	$\prime\prime \times 78 = 66$
64	$\prime\prime \times 66 = 56$

These figures are plotted in rectangular coordinate form in figure 18.3—85% Learning Curve—and also in logarithm form in figure 18.4—Learning Curve—Log-Log Graph. For a production unit of 50 (P), a corresponding value of 60 hours (T) may be found from each graph.

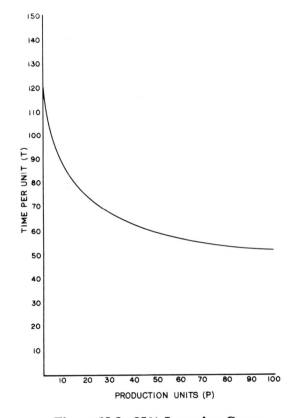

Figure 18.3   85% Learning Curve

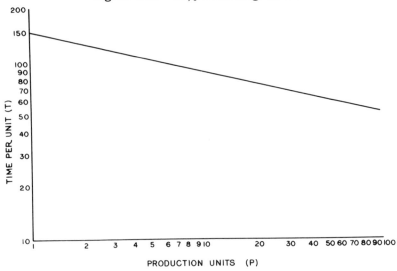

Figure 18.4   85% Learning Curve—Log-Log Graph

19. WAGE
INCENTIVES

PIECE WORK PLAN

Formula:

$$E_{PW} = R_u \times N$$

where

E_{PW} = Piece work earnings
R_u = Rate per piece (unit of work)
N = Number of units
and

$$E_G = H_R \times A_T$$

where

E_G = Guaranteed earnings
H_R = Hourly base rate
A_T = Total actual hours worked

Example:

The WER Company uses a piece work incentive plan for wage payment of its hourly rated direct labor force. The company guarantees earnings at its hourly base rate (H_R) of \$2.75/hour. For a particular operation in a given 8 hour work day (A_T), operator A made 56 units (N), operator B made 44 units (N), and operator C made 40 units (N). The piece work rate (R_u) is \$0.50/unit. Compute the earnings for operators A, B and C, respectively, for the given day.

Solution:

    Operator A

$$E_{PW} = \$0.50/\text{unit} \times 56 \text{ units}$$
$$E_{PW} = \$28.00$$

    Operator B

$$E_{PW} = \$0.50/\text{unit} \times 44 \text{ units}$$
$$= \$22.00$$
$$E_{PW} = \$2.75/\text{hour} \times 8 \text{ hours} = \$22.00$$

    Operator C

$$E_{PW} = \$.50/\text{unit} \times 40 \text{ units}$$
$$= \$20.00$$
$$E_G = \$2.75/\text{hour} \times 8 \text{ hours} = \$22.00$$

        Operator C is paid \$22.00 for the day.

100 PER CENT PREMIUM PLAN

Formula:

$$E_P = H_R \times A_T + H_R (SH_n \times n - A_T)$$

    where

E_P = Total earnings
H_R = Hourly base rate
A_T = Actual hours worked
SH_n = Standard hour rate per unit
n = Number of units

Example:

The SLJ Company uses a 100 per cent premium incentive plan for wage payment of its hourly rated direct labor force. The company guarantees earnings at its hourly base rate (H_R) of \$2.60/hour. For a particular operation in a given 8 hour work day (A_T), operator A made 36 units (n), operator B made 32 units, and operator C made 30 units. The standard hour rate per unit (SH_n) is 0.25 hours per unit. Compute the earnings for operators A, B and C, respectively, for the given day.

Solution:

    Operator A

$$E_P = \$2.60/\text{hour} \times 8 \text{ hours} + \$2.60/\text{hour} \times$$
$$(0.25 \text{ hours/unit} \times 36 \text{ units} - 8 \text{ hours})$$
$$= \$20.80 + \$2.60/\text{hour} (9 \text{ hours} - 8 \text{ hours})$$
$$= \$20.80 + 2.60/\text{hour} (1 \text{ hour})$$
$$E_P = \$23.40$$

Operator B

$$E_P = \$2.60/\text{hour} \times 8 \text{ hours} + \$2.60/\text{hour} \times$$
$$(0.25 \text{ hours/unit} \times 32 \text{ units} - 8 \text{ hours})$$
$$= \$20.80 + \$2.60/\text{hour} (8 \text{ hours} - 8 \text{ hours})$$
$$= \$20.80 + \$2.60/\text{hour} (0)$$
$$E_P = \$20.80$$

Operator C

$$E_P = \$2.60/\text{hour} \times 8 \text{ hours} + \$2.60/\text{hour} \times$$
$$(0.25 \text{ hours/unit} \times 30 \text{ units} - 8 \text{ hours})$$
$$= \$20.80 + \$2.60/\text{hour} (7\tfrac{1}{2} \text{ hours} - 8 \text{ hours})$$
$$= \$20.80 + \$2.60/\text{hours} (-\tfrac{1}{2} \text{ hour})$$
$$= \$20.80 \text{ (guaranteed rate of pay)}$$

HALSEY PLAN

Formula:

$$E_H = H_R \times A_T + 0.50 \times H_R (SH_n \times n - A_T)$$

where

E_P = Total earnings
H_R = Hourly base rate
A_T = Actual hours worked
SH_n = Standard hour rate per unit
n = Number of units

Example:

The GRS Company uses a Halsey (50–50 employee–employer saving share) incentive plan for wage payment of its hourly rated direct labor force. The company guarantees earnings at its hourly base rate (H_R) of \$2.50/hour. For a particular operation in a given 8 hour work day (A_T), operator A made 11 units (n), operator B made 8 units, and operator C made 7 units. The standard hour rate per unit (SH_n) is 1 hour per unit. Compute the earnings for operators A, B and C, respectively, for the given day.

Solution:

Operator A

$$E_H = \$2.50/\text{hour} \times 8 \text{ hours} + 0.50 \times \$2.50/\text{hours} \times$$
$$(1 \text{ hour/unit} \times 11 \text{ units} - 8 \text{ hours})$$
$$= \$20.00 + \$1.25 \text{ hour} (11 \text{ hours} - 8 \text{ hours})$$
$$= \$20.00 + \$1.25/\text{hour} (3 \text{ hours})$$
$$= \$20.00 + \$3.75$$
$$E_H = \$23.75$$

Operator B

$$E_H = \$2.50/\text{hour} \times 8 \text{ hours} + 0.50 \times \$2.50/\text{hours} \times$$
$$(1 \text{ hour/unit} \times 8 \text{ units} - 8 \text{ hours})$$
$$= \$20.00 + \$1.25/\text{hour} \ (8 \text{ hours} - 8 \text{ hours})$$
$$= \$20.00 + \$1.25/\text{hour} \ (0)$$
$$E_H = \$20.00$$

Operator C

$$E_H = \$2.50/\text{hour} \times 8 \text{ hours} + 0.50 \times \$2.50/\text{hour} \times$$
$$(1 \text{ hour/unit} \times 7 \text{ units} - 8 \text{ hours})$$
$$= \$20.00 + \$1.25/\text{hour} \ (7 \text{ hours} - 8 \text{ hours})$$
$$= \$20.00 + \$1.25/\text{hour} \ (-1 \text{ hour})$$
$$E_H = \$20.00 \ (\text{guaranteed rate of pay})$$

Part V

PERSONNEL

20. JOB
EVALUATION

HOURLY RATED JOBS

Formula: (Table 20.1)

Factor		(Degree)—Points (1)	(2)	(3)	% Max. Points
1. Skill					
a. Education		30	60	90	
b. Experience		90	135	180	
	Sub-Total	120	195	270	45%
2 Effort					
a. Mental		20	25	30	
b. Physical		70	80	90	
	Sub-Total	90	105	120	20
3. Responsibility					
a. Process equipment		70	80	90	
b. Materials		20	25	30	
	Sub-Total	90	105	120	20
4. Job Conditions					
a. Working conditions		60	65	70	
b. Safety hazards		10	15	20	
	Sub-Total	70	80	90	15
	Total	370	485	600 Points	100%

TABLE 20.1   The UDP Company—Job Rating Plan for Hourly Rated Jobs

Example:

The UDP Company has designed its own modified version of a job rating "point plan" for hourly rated jobs. The plan, as shown in mathematical form in table 20.1 (The UDP Company—Job Rating Plan for Hourly Rated Jobs) has descriptive definitions for each factor and degree point values assigned to each. Determine the point value rating for jobs A., B. and C., respectively, from the ratings as described below.

Solution:

		(Degree)—Points					
		Job A.		*Job B.*		*Job C.*	
1. Skill							
	a. Education	(2)	60	(1)	30	(1)	30
	b. Experience	(3)	180	(1)	90	(2)	135
	Sub-Total		240		120		165
2. Effort							
	a. Mental	(3)	30	(2)	25	(2)	25
	b. Physical	(2)	80	(3)	90	(2)	80
	Sub-Total		110		115		105
3. Responsibility							
	a. Process equipment	(3)	90	(1)	70	(2)	80
	b. Materials	(1)	20	(3)	30	(3)	30
	Sub-Total		110		100		110
4. Job Conditions							
	a. Working conditions	(1)	60	(3)	70	(2)	65
	b. Safety hazards	(1)	15	(3)	20	(2)	15
	Sub-Total		75		90		80
	Total		535		425		460

Formula:

$$PR/G = \frac{P_{max.} - P_{min.}}{N}$$

where

PR/G = Point range per grade
$P_{max.}$ = Maximum number of points possible in job rating plan
$P_{min.}$ = Minimum number of points possible in job rating plan
N = Number of job grades

Example:

The UDP Company, following its above job rating plan, wants to slot its hourly rated jobs into ten (10) job grades (N). Determine the point ranges for each of these ten job grades and, then, determine the job grade for jobs A., B. and C., respectively, determined above.

Solution:

From the job rating plan described, the maximum number of points possible ($P_{max.}$) is 600 and the minimum number of points possible ($P_{min.}$) is 370.

$$\text{PR/G} = \frac{600 - 370}{10} = \frac{230}{10} = 23$$

Job Grade	Point Range
1	370–392
2	393–415
3	416–438
4	439–461
5	462–484
6	485–507
7	508–530
8	531–553
9	554–576
10	577–600

From the previous data, the job grades for jobs A., B. and C., respectively, are determined as follows:

Job	Points	Point Range	Job Grade
A	535	531–553	8
B	425	416–438	3
C	465	462–484	5

21. MERIT
RATING

CLERICAL JOBS

Formula: (Table 21.1)

| | | | *(Degree)—Points* | | |
	Factor	(1)	(2)	(3)	*% Max. Points*
1.	Dependability	70	85	100	20%
2.	Quality	70	85	100	20
3.	Quantity	70	85	100	20
4.	Cooperation	60	65	70	14
5.	Personality	40	50	60	12
6.	Initiative	30	35	40	8
7.	Punctuality & Attendance	20	25	30	6
	Total	360	430	500	100%

TABLE 21.1   The UMB Company—Merit Rating Plan for Clerical Jobs

Example:

The UMB Company has designed its own merit rating "point plan" for the company's clerical rated jobs. The plan, as depicted in mathematical form in table 21.1 (The UMB Company—Merit Rating Plan for Clerical Jobs), has descriptive definitions for each factor and degree point values assigned to each. Determine the point rating for clerk A., B. and C., respectively, from the ratings as described below.

Solution:

(Degree)—Points

	Factor		Clerk A.		Clerk B.		Clerk C.
1.	Dependability	(2)	85	(1)	70	(3)	100
2.	Quality	(3)	100	(1)	70	(3)	100
3.	Quantity	(1)	70	(3)	100	(2)	85
4.	Cooperation	(2)	65	(1)	60	(2)	65
5.	Personality	(3)	60	(1)	40	(2)	50
6.	Initiative	(1)	30	(1)	30	(2)	35
7.	Punctuality & Attendance	(2)	25	(1)	20	(3)	30
	Total		435		390		465

MANAGEMENT JOBS

Formula: (Table 21.2)

(Degree)—Points

	Factor	(1)	(2)	(3)	% Max. Points
1.	Leadership	160	180	200	20%
2.	Initiative	120	140	160	16
3.	Quality	110	115	120	12
4.	Quantity	110	115	120	12
5.	Follow-up	100	105	110	11
6.	Communications	90	95	100	10
7.	Knowledge of job	90	95	100	10
8.	Cooperation	80	85	90	9
	Total	860	930	1000	100%

TABLE 21.2   The ASM Company—Merit Rating Plan for Management Jobs

Example:

The ASM Company has designed its own merit rating "point plan" for the company's management level jobs. The plan, as depicted in mathematical Form in table 21.2 (The ASM Company—Merit Rating Plan for Management Jobs), has descriptive definitions for each factor, and degree point values assigned to each. Determine the point rating for Managers A. and B., respectively, from the ratings as described below.

Solution:

	Factor	(Degree)—Points	
		Manager A.	*Manager B.*
1.	Leadership	(3) 200	(2) 180
2.	Initiative	(2) 140	(1) 120
3.	Quality	(3) 120	(2) 115
4.	Quantity	(2) 115	(2) 115
5.	Follow-up	(3) 110	(3) 110
6.	Communications	(2) 95	(1) 90
7.	Knowledge of job	(1) 90	(3) 100
8.	Cooperation	(3) 90	(2) 85
	Total	960	905

Part VI

BUILDING SERVICES

22. ELECTRICITY

OHM'S LAW

Formula:

$$I = E/R$$

where

I = Current in units of amperes
E = Electromotive force in units of volts
R = Resistance in units of ohms

Example 1:

Determine the current (I) flowing in a conductor where the electromotive force (E) is 110 volts and the resistance (R) is 20 ohms.

Solution:

$$I = E/R$$
$$= 110 \text{ volts}/20 \text{ ohms}$$
$$I = 5.5 \text{ amperes}$$

Example 2:

A circuit has a resistance (R) of 150 ohms with a current (I) of 0.3 amperes flowing through it. What is the necessary voltage (E) required for this?

Solution:

$$I = E/R$$
$$0.3 \text{ amperes} = E/150 \text{ ohms}$$

or

$$E = 0.3 \text{ amperes} \times 150 \text{ ohms}$$
$$E = 45 \text{ volts}$$

SERIES CONNECTIONS

Formula:

(Current passes through a series of units in a circuit):

$$I = E/\sum R_s$$

where

I = Current in units of amperes
E = Electromotive force in units of volts
$\sum R_s$ = Total resistance in units of ohms $(R_1 + R_2 + \ldots R_s)$

Example:

A 110 volt (E) circuit has a resistance R_1 of 6 ohms, a resistance R_2 of 3 ohms and a resistance R_3 of 2 ohms connected in series. What is the current (I) that will flow through this circuit?

Solution:

$$\sum R_s = R_1 + R_2 + R_3$$
$$= 6 \text{ ohms} + 3 \text{ ohms} + 2 \text{ ohms}$$
$$\sum R_3 = 11 \text{ ohms}$$
$$I = E/\sum R_s$$
$$= 110 \text{ volts}/11 \text{ ohms}$$
$$I = 10 \text{ amperes}$$

PARALLEL CONNECTIONS

Formula:

(Current divides between units connected in a circuit.)

$$I = E/\sum R_p$$

where

I = Current in units of amperes
E = Electromotive force in units of volts
$\sum R_p$ = Total resistance in units of ohms

where

$$\frac{1}{\sum R_p} = \frac{1}{R_1} + \frac{1}{R_2} + \cdots \frac{1}{R_p}$$

Example:

Conductors of a resistance of 3 ohms (R_1) and 6 ohms (R_2) are connected in parallel to a 1.5 volt (E) battery. Determine the total resistance (R_p) in ohms and the current (I) in amperes.

Solution:

$$\frac{1}{\sum R_p} = \frac{1}{R_1} + \frac{1}{R_2}$$

$$\frac{1}{\sum R_p} = \frac{1}{3 \text{ ohms}} + \frac{1}{6 \text{ ohms}}$$

$$= \frac{2}{6 \text{ ohms}} + \frac{1}{6 \text{ ohms}}$$

$$= \frac{3}{6 \text{ ohms}}$$

$$\frac{1}{\sum R_p} = \frac{1}{2 \text{ ohms}}$$

or

$$\sum R_p = 2 \text{ ohms}$$
$$I = E/\sum R_p$$
$$= 1.5 \text{ volts}/2 \text{ ohms}$$
$$I = 0.75 \text{ amperes}$$

ELECTRICAL POWER

Formula:

W = E × I

where

W = Watts expended
KW = Kilowatts = W/1000
E = Electromotive forces in units of volts
I = Current in units of amperes

Example 1:

For a circuit requiring 115 volts (E) and 12 amperes (I) of current, compute the electrical power required (W) in watts.

Solution:

   W = E × I
      = 115 volts × 12 amperes
   W = 1380 watts

Example 2:

   For a bulb that is marked 75 watts (W) with an electromotive voltage (E) force of 115 volts, compute the current (I) in units of amperes required to operate the bulb.

Solution:

   W = E × I
      75 watts = 115 volts × I
                or
      $\frac{75 \text{ watts}}{115 \text{ volts}} = I$
      0.652 amperes = I

Example 3:

   A generator carries a load of 300 amperes (I) and an electromotive force (E) of 220 volts. Compute the kilowatts (KW) of power required.

Solution:

    W = E × I
       = 220 volts × 300 amperes
    W = 66,000 watts
   KW = W/1000
       = 66,000 watts/1000
   KW = 66 kilowatts

ELECTRICAL HORSEPOWER

Formula:

   H.P. = W/746
            where
   H.P. = Electrical horsepower
     W = Watts
            and
     W = E × I
            where
     E = Electromotive force in units of volts
     I = Current in units of amperes

Example:

A generator maintains an electromotive force (E) of 220 volts with a current usage (I) of 85 amperes. Compute the electrical horsepower being generated.

Solution:

$$W = E \times I$$
$$= 220 \text{ volts} \times 85 \text{ amperes}$$
$$W = 18,700 \text{ watts}$$
$$\text{H.P.} = W/746$$
$$= 18,700 \text{ watts} /746$$
$$\text{H.P.} = 25.1 \text{ electrical horsepower}$$

23. HEAT,
WATER AND AIR

HEATING AND COOLING

A British thermal unit (BTU) is the quantity of heat required to raise the temperature of 1 lb. of water by 1 Fahrenheit degree. The specific heat of a substance is the number of British thermal units required to raise the temperature of 1 lb. of that substance 1 Fahrenheit degree.

Formula:

$$BTU = sh \times W \times t_d$$

where

BTU = Heat in British thermal units

sh = Specific heat of substance

W = Weight of the substance in lbs.

t_d = Temperature difference of state of substance (degrees Fahrenheit —°F.)

Example 1:

Compute the heat required in British thermal units (BTU's) to raise the temperature of 60 lbs. (W) of water from 60°F. to 190°F. The specific heat (sh) of water is 1.0.

Solution:

$$BTU's = 1.0 \times 60 \times (190 - 60)$$
$$= 60 \times 130$$
$$BTU's = 7800 \text{ British thermal units}$$

122

Example 2:

Compute the heat given off in British thermal units (BTU's) in cooling 100 lbs. (W) of water from 170°F. to 80°F. The specific heat (sh) of water is 1.0.

Solution:

$$BTU's = 1.0 \times 100 \times (170 - 80)$$
$$= 100 \times 90$$
$$BTU's = 9000 \text{ British thermal units}$$

Example 3:

Compute the heat given off in British thermal units (BTU's) in cooling 35 lbs. (W) of copper from 200°F. to 40°F. The specific heat (sh) of copper is 0.093.

Solution:

$$BTU's = 0.093 \times 35 \times (200 - 40)$$
$$= 3.255(160)$$
$$BTU's = 521 \text{ British thermal units}$$

FUEL CONSUMPTION

Formula:

$$A_c = C/E$$

where

A_c = Annual consumption expressed in fuel units
C = Constant for a given set of conditions in units of BTU's
E = Effective fuel value utilized expressed in BTU's/hour

where

$$C = \frac{H \times 24 \times D}{(65° - T_o)}$$

and

H = Heating requirements in units of BTU's/hour
D = Degree-days—temperature degree differences Fahrenheit (°F.) for any one day between 65°F. and the mean temperature of the day (°F.)

Example:

A given building is designed for an outside temperture (T_o) of $-10°F$.; the heating requirements (H) are 160,000 BTU's /hour; and it is located in a city where the average annual degree-days (D) are 6000. Compute the annual consumption of fuel units for the following possible heat-producing agents: a. Fuel Oil—Effective fuel value utilized (E_o) of 100,000 BTU's/gal.; b. Coal—Effective fuel value utilized (E_c) of 7500 BTU's/lb.;

c. Electricity—Effective fuel value utilized (E_e) of 3412 BTU's/kw-hr.; and d. Natural Gas —Effective fuel value utilized (E_{ng}) of 750,000 BTU's/therm. (A therm is equal to 100,000 BTU's in the gas.)

Solution:

$$C = \frac{H \times 24 \times D}{(65° - T_o)}$$

$$= \frac{160,000 \times 24 \times 6000}{65 - (-10)}$$

$$= \frac{23,040,000,000}{75}$$

$$C = 307,200,000 \text{ BTU's}$$

a. Fuel Oil
$$A_c = C/E_o$$
$$= 307,200,000 \text{ BTU's}/100,000 \text{ BTU's/gal.}$$
$$A_c = 3072 \text{ gals.}$$

b. Coal
$$A_c = C/E_c$$
$$= 307,200,000 \text{ BTU's}/7500 \text{ BTU's/1b.}$$
$$A_c = 40,960 \text{ 1bs.}$$

c. Electricity
$$A_c = C/E_e$$
$$= 307,200,000 \text{ BTU's}/3412 \text{ BTU's/kw-hr.)}$$
$$A_c = 90,035 \text{ kw-hrs.}$$

d. Natural Gas
$$A_c = C/E_{ng}$$
$$= 307,200,000 \text{ BTU's}/75,000 \text{ BTU's/therm}$$
$$= 4096 \text{ Therms}$$

APPENDIX

I. ALGEBRA

Algebra is that branch of mathematics in which letters representing numbers are combined according to the rules of arithmetic. The treatment of the subject in this chapter will be limited to its fundamentals.

BASIC LAWS AND DEFINITIONS:

Addition: $a + b = b + a$; $a + (b + c) = a + (c + b) = (c + b) + a$

Multiplication: $ab = ba$; $a \times b \times c = (a \times b) \times c = a \times (b \times c)$

Distributive: $a(b + c) = ab + ac$

Monomial: An expression containing only one term, e.g. $2ab$, xy^2

Polynomial: An expression containing two or more terms, e.g. $x + y$, $3x^2 + a + cy + (bd)^2$

Exponents:

    a) $n^a \times n^b = n^{a+b}$

    b) $n^{-x} = \dfrac{1}{n^x}$

    c) $n^\circ = 1$

Radicals:

    $\sqrt[n]{a} = a^{1/n}$

Factorials: $n! = n \times (n - 1) \times (n - 2) \times \ldots \times 1$

POLYNOMIALS:

Addition Formula:

Place like terms in columns and find the algebraic sum of the like terms.

Example:

Add the terms $(5x^2y + 2xy - 6x + 4y - 6)$ and $(2x^3y - xy + 2x - y)$.

Solution:

$$5x^2y + 2xy - 6x + 4y - 6$$
$$2x^3y \qquad\quad - \;\; xy + 2x - \;\; y$$
$$\overline{2x^3y + 5x^2y \; + \;\; xy - 4x + 3y - 6}$$

Subtraction Formula:

Place like terms in columns and, then, by mentally changing the algebraic signs of the subtrahend, proceed as in the addition process.

Example:

From the term $(4xy^3 - 2xy^2 + x - 2y^2 - 7)$ subtract $(xy^2 - 4x - 4xy - 3y^2 + 3)$.

Solution:

$$4xy^3 - 2xy^2 + \;\; x - 2y^2 \qquad\quad - 7$$
$$xy^2 - 4x - 3y^2 - 4xy + 3$$
$$\overline{4xy^3 - 3xy^2 + 5x + \;\; y^2 + 4xy - 10}$$

Multiplication Formula:

1. $(a + b)(m + n) = am + an + bm + bn$
2. If there are more than two terms in either expression, place the polynomial with the fewer terms beneath the other polynomial and multiply term by term beginning at the left. Like terms of the partial multiplications are placed beneath each other in columns to faciliate addition.

Example:

Find the product of the expressions $(3x - y)$ and $(x + 2y)$.

Solution:

$$(a + b)(m + n) = am + an + bm + bn$$
$$(3x - y)(x + 2z) = 3x^2 + 6xz - xy - 2yz$$

Example:

Find the product of the expressions $(2x^3 - x^2 + 4x + 6)$ and $(x - 4)$.

Solution:

$$
\begin{array}{l}
2x^3 - x^2 + 4x + 6 \\
\underline{x - 4} \\
2x^4 - x^3 + 4x^2 + 6x \\
 \underline{- 8x^3 + 4x^2 - 16x - 24} \\
2x^4 - 9x^3 + 8x^2 - 10x - 24
\end{array}
$$

Division Formula:

1. Arrange both the dividend and divisor in either descending or ascending powers of the same letter. 2. Divide the first term of the dividend by the first term of the divisor and write the result as the first term of the quotient. 3. Multiply the complete divisor by the first term of the quotient and subtract the product from the dividend. 4. If there is a remainder, consider it as a new dividend and repeat steps 1, 2 and 3.

Example:

Divide the expression $(10x^3 - 7x^2y - 16xy^2 + 12y^3)$ by $(2x^2 + xy - 2y^2)$.

Solution:

$$
\begin{array}{r}
5x - 6y \\
2x^2 + xy - 2y^2 \overline{\smash{)}10x^3 - 7x^2y - 16xy^2 + 12y^3} \\
\underline{10x^3 + 5x^2y - 10xy^2} \\
- 12x^2y - 6xy^2 + 12y^3 \\
\underline{- 12x^2y - 6xy^2 + 12y^3} \\
0
\end{array}
$$

Solution is $5x - 6y$

FACTORING POLYNOMIALS:

The factors of a polynomial are two or more expressions which when multiplied together equal the polynomial. As an example, for the polynomial $x^3y^2 - xy^2$, the factors (x), (y^2) and $x^2 - 1$ when multiplied together $= x^3y^2 - xy^2$. The factors x and y^2 are common to both terms or the polynomial $x^3y^2 - xy^2$ and are therefore referred to as *common factors*. An algebraic expression is a *prime factor* if it has no other factors except itself and 1. An expression is said to be *factored* completely when it has been separated into its prime factors.

Example:

Completely factor the polynomial $2b^3 - 8b^2 - 6b$.

Solution:

$$
\begin{array}{l}
2b^3 - 8b^2 - 6b \\
2(b^3 - 4b^2 - 3b) = 2b^3 - 8b^2 - 6b \\
2b(b^2 - 4b - 3) = 2b^3 - 8b^2 - 6b
\end{array}
$$

Example:

Completely factor the polynomial $3x^3 - 12x$.

Solution:

$3x^3 - 12x$

$(3)\ (x)\ (x^2 - 4) = 3x^3 - 12x$

$(3)\ (x)\ (x + 2)\ (x - 2) - 3x^3 - 12x$

AXIOMS OF EQUALITY:

Given equation $y = mx + b$:

1. $y + a = mx + b + a$
2. $y - a = mx + b - a$
3. $a(y) = a(mx + b)$
4. $\dfrac{y}{a} = \dfrac{mx + b}{a}$

Example:

Solve the equation $y = 3x + 4$ for x.

Solution:

$$y = 3x + 4$$

from (2) $\quad y - 4 = 3x + 4 - 4$

$$y - 4 = 3x$$

from (4) $\quad \dfrac{y - 4}{3} = \dfrac{3x}{3}$

$$\dfrac{y - 4}{3} = x$$

Example:

Solve the equation $\dfrac{a}{\sin A} = \dfrac{b}{\sin B}$ for b.

Solution:

$$\dfrac{a}{\sin A} = \dfrac{b}{\sin B}$$

from (3) $\sin B \times \dfrac{a}{\sin A} = \sin B \times \dfrac{b}{\sin B}$

$$\sin B \times \dfrac{a}{\sin A} = b$$

Example:

Solve the equation $10x - 4 = 4x + 32$ for x.

Solution:

$$10x - 4 = 4x + 32$$
from (1) $\quad 10x - 4 + 4 = 4x + 32 + 4$
$$10x = 4x + 36$$
from (2) $\quad 10x - 4x = 4x - 4x + 36$
$$6x - 46$$
from (4) $\quad \dfrac{6x}{6} = \dfrac{36}{6}$
$$x = 6$$

SIMULTANEOUS EQUATIONS:

Formula:

To solve two simultaneous equations of two unknowns, eliminate temporarily one of the unknowns through the use of Axioms of Equality. When one of the original unknowns is determined, substitute it in either of the two equations and solve for the other unknown.

Example:

Solve the simultaneous equations 1) $x - 2y = -1$ and 2) $2x + 3y = 12$ for x and y, respectively.

Solution:

$$1) \quad x - 2y = -1$$
$$2) \quad 2x + 3y = 12$$
From 1) $\quad x - 2y = -1$
$$x = 2y - 1$$
Substituting $\quad x = 2y - 1$ in 2), $2(2y - 1) + 3y = 12$
$$4y - 2 + 3y = 12$$
$$7y = 14$$
$$y = 2$$
Substituting $\quad y = 2$ in 1), $x - 2(2) = -1$
$$x - 4 = -1$$
$$x = 3$$

Example:

Solve the simultaneous equations 1) $2x + y = 0$, and 2) $2x - y = 1$.

Solution:

$$1)\quad 2x + y = 0$$
$$2)\quad 2x - y = 1$$
$$\text{From 1)}\quad 2x + y = 0$$
$$y = -2x$$

Substituting $y = -2x$ in 2), $2x - (-2x) = 1$
$$2x + 2x = 1$$
$$4x = 1$$
$$x = \frac{1}{4}$$

Substituting $x = \frac{1}{4}$ in 1), $2 \times \frac{1}{4} + y = 0$
$$\frac{1}{2} + y = 0$$
$$y = -\frac{1}{2}$$

QUADRATIC EQUATION FORMULA:

The roots of a quadratic equation ($ax^2 + bx + c = 0$) are expressed by the formula—

$$x = \frac{-b \pm \sqrt{b^2 - 4ac}}{2a}$$

Example:

Find the roots of the quadratic equation $x^2 + 7x - 8 = 0$ by use of the quadratic equation formula. Verify the results.

Solution:

$$x^2 + 7x - 8 = 0$$
$$x = \frac{-b \pm \sqrt{b^2 - 4ax}}{2a}$$
$$a = 1, \ b = 7, \ c = -8$$
$$x = \frac{-7 \pm \sqrt{7^2 - 4(1)(-8)}}{2(1)}$$
$$= \frac{-7 \pm \sqrt{49 + 32}}{2}$$
$$= \frac{-7 \pm \sqrt{81}}{2} \quad \text{(From table II.)}$$
$$x_1 = \frac{-7 \pm 9}{2}$$
$$x_1 = \frac{-7 + 9}{2}$$

$$x_1 = \frac{2}{2}$$

$$x_1 = 1$$

$$x_2 = \frac{-7 - 9}{2}$$

$$x_2 = \frac{-16}{2}$$

$$x_2 = -8$$

Verify for $x_1 = 1$

$$x^2 + 7x - 8 = 0$$
$$(1)^2 + 7(1) - 8 = 0$$
$$1 + 7 - 8 = 0$$
$$8 - 8 = 0$$
$$0 = 0$$

for $x_2 = -8$

$$x^2 + 7x - 8 = 0$$
$$(-8)^2 + 7(-8) - 8 = 0$$
$$64 - 56 - 8 = 0$$
$$64 - 64 = 0$$
$$0 = 0$$

Example:

Find the roots of the quadratic equation $2x^2 - 4x + 1 = 0$ by use of the quadratic equation formula. Verify the results.

Solution:

$$2x^2 - 4x + 1 = 0$$

$$x = \frac{-b \pm \sqrt{b^2 - 4ac}}{2a}$$

$$a = 2, \ b = -4, \ c = 1$$

$$x = \frac{-(-4) \pm \sqrt{(-4)^2 - 4(2)(1)}}{2(2)}$$

$$= \frac{4 \pm \sqrt{16 - 8}}{4}$$

$$= \frac{4 \pm \sqrt{8}}{4}$$

$$= \frac{4 \pm \sqrt{4 \times 2}}{4}$$

$$= \frac{4 \pm 2\sqrt{2}}{4}$$

$$= \frac{2 \pm \sqrt{2}}{2}$$

$$x_1 = 1 + \frac{\sqrt{2}}{2}$$

$$x_2 = 1 - \frac{\sqrt{2}}{2}$$

Verify for $\quad x_1 = 1 + \frac{\sqrt{2}}{2}$

$$2x^2 - 4x + 1 = 0$$

$$2(1 + \frac{\sqrt{2}}{2})^2 - 4(1 + \frac{\sqrt{2}}{2}) + 1 = 0$$

$$2(1 + \frac{\sqrt{2}}{2} + \frac{\sqrt{2}}{2} + \frac{\sqrt{2}^2}{4}) - 4 - 2\sqrt{2} + 1 = 0$$

$$2(1 + \frac{2\sqrt{2}}{2} + \frac{2}{4}) - 3 - 2\sqrt{2} = 0$$

$$2 + 2\sqrt{2} + 1 - 3 - 2\sqrt{2} = 0$$

$$2 + 1 - 3 = 0$$

$$x_2 = 1 - \frac{\sqrt{2}}{2}$$

$$2x^2 - 4x + 1 = 0$$

$$2(1 - \frac{\sqrt{2}}{2})^2 - 4(1 - \frac{\sqrt{2}}{2}) + 1 = 0$$

$$2(1 - \frac{\sqrt{2}}{2} - \frac{\sqrt{2}}{2} + \frac{\sqrt{2}^2}{4}) - 4 + 2\sqrt{2} + 1 = 0$$

$$2(1 - \frac{2\sqrt{2}}{2} + \frac{2}{4}) - 3 + 2\sqrt{2} = 0$$

$$2 - 2\sqrt{2} + 1 - 3 + 2\sqrt{2} = 0$$

$$2 + 1 - 3 = 0$$

II. FUNDAMENTALS OF LOGARITHMS

DEFINITION:

A logarithm is defined as the power to which a fixed number must be raised in order to produce a given number. Expressed algebraically,

$$b^x = N$$

or

$$x = \log_b N$$

where

x = logarithm of N

b = base

N = number N

Example:

Determine the number (N) when the base (b) is 2 and the logarithm (x) is 3.

Solution:

$$b^x = N$$
$$2^3 = N$$
$$8 = N$$

Example:

For a number (N) of 16, determine its logarithm (x) when the base (b) is 2.

Solution:

$$x = \log_b N$$
$$x = \log_2 16$$
$$x = 4$$

NATURAL AND COMMON LOGARITHMS

Natural or Napierian logarithms use a base (b) of e which is equal approximately to 2.71828. To distinguish natural logarithms from other logarithmic systems the symbol l_n is employed. (See table VI., which shows the values of this logarithmic system.)

Common logarithms use as their base (b) the value *10*. This logarithmic system is the most accepted one. When the common logarithmic system is used, it is not necessary to indicate the base (b) is 10 as this is understood.

If a base (b) other than 10 is used in a logarithmic system, it must be indicated by a subscript to the right and below the abbreviation "log." The exception, of course, as mentioned in conjunction with the natural logarithmic system, is the subscript l_n to represent the base (b) of e = 2.71828 approximately.

The relationship between the natural and common logarithms is such that by multiplying the common logarithm of a number by 2.3026 the natural logarithm is determined. Conversely, multiplying the natural logarithm by 0.4343 (the reciprocal of 2.3026 or 1/2.3026) gives the common logarithm of a number.

For the remainder of this chapter, *only* the *common logarithmic* system will be discussed.

EXPONENTIAL AND LOGARITHM RELATIONS:

The following table shows the exponential and corresponding logarithmic relations:

$$10^{-3} = 0.001 \qquad \log 0.001 = -3$$
$$10^{-2} = 0.01 \qquad \log 0.01 = -2$$
$$10^{-1} = 0.01 \qquad \log 0.1 = -1$$
$$10^{0} = 1 \qquad \log 1 = 0$$
$$10^{1} = 10 \qquad \log 10 = 1$$
$$10^{2} = 100 \qquad \log 100 = 2$$
$$10^{3} = 1000 \qquad \log 1000 = 3$$

COMPONENTS OF LOGARITHMS

A logarithm consists of two parts—the number to the left of the decimal point, designated the *characteristic,* and the digits to the right, designated the *mantissa.* The rules for determining the characteristic of a number are as follows:

1. For a number greater than 1, the characteristic is positive and equal to one less than the number of digits to the left of the decimal point.
2. For a positive number less than 1, the characteristic is negative and has an absolute value one more than the number of zeroes between the decimal point and the first nonzero digit.

The *mantissa* is the same for any particular sequence of digits regardless of the position of the decimal point. Table V gives the mantissas of numbers to *ten decimal places,* which gives more than sufficient accuracy for the solution of most problems. For illustrative purposes, however, only five decimal places will be used—the fifth obtained by rounding.

The following table is compiled for the number series 1079, showing the logarithm for each different position of the decimal point. (See table V, column x = 1079, for the mantissa as shown below.)

Number	Characteristic	Mantissa	Logarithm
1,079	3	0.03302	3.03302
107.9	2	0.03302	2.03302
10.79	1	0.03302	1.03302
1.079	0	0.03302	0.03302
0.1079	-1	0.03302	9.03302 -10
0.01079	-2	0.03302	8.03302 -10
0.001079	-3	0.03302	7.03302 -10

To find the *anti-logarithm* (*number corresponding* to a *logarithm*), locate the mantissa in table V and read the corresponding value of x. Then, use the characteristic of the logarithm to establish the location of the decimal point. As an example, in the case of the logarithm 1.30936, the mantissa of 0.30936 in table V. corresponds to a value of 204 for x. The characteristic of the logarithm, namely, 1, indicates that the number has two digits to the left of the decimal point. Therefore, the anti-logarithm, or the number corresponding to the logarithm 1.30936, is 20.4.

INTERPOLATION OF TABLES:

Where a number series or a mantissa may not be read directly from the tables because it lies between two successive number series or between two successive mantissas, it is necessary to interpolate to determine the desired value. By this is meant that proportionate parts are needed. This may be best explained through the use of an example, as follows:

It is desired to find the mantissa of the number series 11374. The table does not show five consecutive figures. Instead, however, the following number series and their corresponding mantissas are given as follows:

x	Mantissa
11380	.05614
11370	.05576

The figure 11374 lies $\frac{4}{10}$ or 0.4 of the way between 11370 and 11380. The difference between the mantissas of these two figures is .05614—.05576 or .00038. For the same proportion of 0.4 between the number series, 0.4 $\times$ .00038 = .00015 is the difference in the mantissas, so that .05576 + .00015 gives .05591 as the mantissa for 11374.

USES OF LOGARITHMS:

1. *Multiplication—*

To multiply, add the logarithms of the numbers to obtain the logarithm of the product. Determine the anti-logarithm for the solution.

Example:

Find the product of 1247 $\times$ 78.3.

Solution:

Determine the sum of the logarithms of 1247 and 78.3.

The logarithm of 1247 = 3.09587   (Table V.)

The logarithm of 78.3 = 1.89376

Total   4.98963

The product (anti-logarithm of 4.98963) is equal to 97,640.

2. *Division—*

To divide, subtract the logarithm of the divisor from the logarithm of the dividend to obtain the logarithm of the quotient. Determine the anti-logarithm for the solution.

Example:

Divide 543 by 91.4.

Solution:

Determine the difference of the logarithm of 91.4 from the logarithm of 543.

The logarithm of 543 = 2.73480   (Table V.)

The logarithm of 91.4 = 1.96095

Difference   = 0.77385

The quotient (anti-logarithm of 0.77385) is 5.9409.

3. *Powers—*

To determine the solution of a number to a power, first multiply the logarithm of the number by the power. The anti-logarithm of the product is the answer.

Example:

Determine the solution of the expression 12^4.

Solution:

Determine the logarithm of 12 and multiply by 4.

The logarithm of 12 = 1.07918   (Table V.)

$4 \times 1.07918 = 4.31672$

The answer (anti-logarithm of 4.31672) = 20,736.

4. *Roots of a Number—*

To find the root of a number, divide the logarithm of the number by the index of the root indicated. The anti-logarithm of the quotient is the solution.

Example:

Determine the root of the expression $\sqrt[5]{27}$.

Solution:

Determine the logarithm of 27 and divide by 5

The logarithm of 27 = 1.43136   (Table V.)

$1.43136/5 = 0.28627$

The answer (anti-logarithm of 0.28627) = 1.9332.

III. GEOMETRY

DEFINITION AND SCOPE:

Geometry is that branch of mathematics that deals with space and its relations, particularly with respect to the properties and measurement of points, lines, angles, surfaces and solids. The scope of this chapter will be limited to the more important fundamental aspects of the subject.

A *point* is a location and is represented by a dot. Mathematically, a point has no physical dimensions at all.

A *line* is a connection of a series of points and may be either straight or curved. A line has length but no width or thickness. A straight line is the shortest distance between two points.

An *angle* is the space between two intersecting lines.

A *surface* has length and breadth, but not thickness. A surface is determined by three points which do not lie on the same line.

A *solid* is an object with dimensions of length, breadth and thickness.

A *plane* is a surface such that a straight line joining any two of its points lies wholly within the surface.

TRIANGLE:

A plane *triangle* is a three-sided closed figure, as shown in figure III a.

The *perimeter* of a triangle is the sum of the lengths of its three sides—expressed as a formula in $P = a + b + c$, where P is the perimeter and a, b and c are the three respective sides.

Example:

In figure III a. sides a, b and c are 8″, 10″ and 7″, respectively. Determine its perimeter (P).

Solution:

P = a + b + c
a = 8″, b = 10″, and c = 7″
P = 8 + 10 + 7
P = 25″

The area of a triangle (A) = $\sqrt{s(s-a)(s-b)(s-c)}$

where s = $\frac{1}{2}$ (a + b + c)

and

a, b and c are the three respective sides

Example:

In figure III a. sides a, b and c are 8″, 10″ and 7″, respectively (same as in previous example). Determine its area (A).

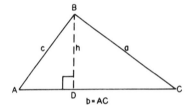

Figure III a.   Triangle

Solution:

A = $\sqrt{s(s-a)(s-b)(s-c)}$

s = $\frac{1}{2}$ (a + b + c)

= $\frac{1}{2}$ (8 + 10 + 7)

= $\frac{1}{2}$ (25)

s = 12.5 in.

A = $\sqrt{12.5(12.5-8)(12.5-10)(12.5-7)}$

= $\sqrt{12.5(4.5)(2.5)(5.5)}$

= $\sqrt{773.4}$

A = 27.8 in.²   (from table II.)

The area of a triangle may also be determined by the formula A = $\frac{1}{2}$ hb, where A is the area, h is the height of the triangle and b is the base. In figure III a. the height (h) is shown as the dotted line h (the perpendicular distance from vertex B to side b with right angles at D), and the base (b) is represented by b or the distance of the line between A and C.

Example:

Compute the area of a triangle whose height (h) is 22′ and base (b) 34′.

Solution:

$$A = \frac{1}{2} \, hb$$

$$= \frac{1}{2} \times 22 \times 34$$

$$= 374 \text{ ft.}^2$$

QUADRILATERAL:

A *quadrilateral* is a plane four-sided closed figure, as shown in figure III b. AC and **BD** are designated as its two diagonals. The perimeter (P) is equal to the sum of the sides. In the figure, it is equal to the sum of e, f, g and h. The area of a quadrilateral may be determined by dividing the figure into triangles as shown in figure III b. and summing the areas of the triangles.

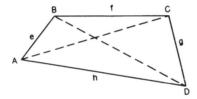

Figure III b. Quadrilateral

PARALLELOGRAM:

A *parallelogram* is a quadrilateral figure where the opposite sides are parallel, as shown in figure III c. A *rhombus* (figure III d.) is a parallelogram where all four sides are of the same length. If each of the four angles are right angles (90°), the figure is a *rectangle* (figure III e.). If the four sides are the same length and the four angles right angles, the figure is a *square* (figure III f.).

The area of a parallelogram is expressed by the formula A-bh, where A is the Area, b the base and h the height (see figure III c.).

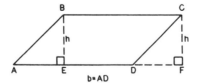

Figure III c. Parallelogram

Figure III d.   Rhombus   **Figure III e.   Rectangle**   **Figure III f.   Square**

The area of a rectangle is also expressed by the formula $A = hb$ (see figure III e.).

The area of a square is expressed by the formula $A = b^2$, where A is the area and b a side (see figure III f.). Thus, the area of any parallelogram is the product of the base and height.

Example:

The base (b) of a parallelogram is 20″ and the height is 14″. Determine the area (A).

Solution:

$$A = bh$$
$$= 20 \times 14$$
$$A = 280 \text{ in.}^2$$

TRAPEZOID:

A *trapezoid* is a quadrilateral figure where two sides are parallel and the other two are not. The area of a trapezoid is expressed by the formula $A = \frac{1}{2}h(b_1 + b_2)$, where A is the area, h is the height, and b_1 and b_2, respectively, are the lengths of the parallel bases (see figure III g.).

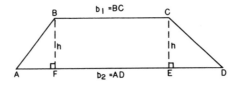

Figure III g.   Trapezoid

Example:

Find the area (A) of a trapezoid whose height (h) is 6″ and whose bases (b_1 and b_2) are 6″ and 4″, respectively.

Solution:

$$A = \frac{1}{2} h (b_1 + b_2)$$

$$= \frac{1}{2} \times 6 \, (6 + 4)$$
$$= 3 \, (10)$$
$$A = 30 \text{ in.}^2$$

CIRCLE:

A *circle* is a plane figure bounded by a curved line called the circumference every point of which is equally distant from the center, as shown in figure III h. The *radius* (r) of a circle is the straight line distance from the center to any point on the circumference. The *diameter* (d) of a circle is a straight line drawn through the center of the circle with its terminations on the circumference. The diameter (d) is equal to 2 times the radius (r). An *arc* (a) is a part of the circumference. A *chord* is the straight line connection between the ends of an arc (DB in figure III h.). The circumference of a circle is expressed by the formula $C_c = \pi d$, where C_c is the circumference, π is a constant of approximate value 3.1416 and d is the diameter. The area is expressed by the formula $A_c = \frac{\pi \, d^2}{4}$, where A_c is the area, π is a constant of approximate value 3.1416 and d is the diameter.

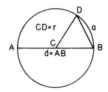

Figure III h.   Circle

Example:

For a circle whose diameter (d) is 8″, compute a) the circumference (C_c), and b) the area A_c.

Solution:

a) $C_c = \pi d$
        where d = 8″
     $C = 3.1416 \times 8$
     $C_c = 25.1″$

b) $A_c = \frac{\pi d^2}{4}$

        $= \frac{3.1416 \times (8)^2}{4}$

     $A_c = 50.3 \text{ in.}^2$

CYLINDER:

The surface area of a cylinder is expressed by the formula $S_c = \pi d(h + \frac{d}{2})$, where S_C is the surface area, π is a constant of approximate value 3.1416, d is the diameter and h is the height of the cylinder. The volume of a cylinder is expressed by the formula $V_c = \frac{\pi d^2}{4} \times h$, where V_c is the volume, π is a constant of approximately value 3.1416, d is the diameter and h is the height of the cylinder.

Example:

For a cylinder whose diameter (d) is 1′ and height (h) 3′, compute a) the surface S_c, and b) the volume V_c.

Solution:

a) $S_c = \pi d \ (h + \frac{d}{2})$

where $d = 1'$ and $h = 3'$

$S_c = 3.1416 \times 1 \ (3 + \frac{1}{2})$

$= 3.1416 \ (3.5)$

$S_c = 11 \ ft.^2$

b) $V_c = \frac{\pi d^2}{4} \times h$

$= 3.1416 \times \frac{(1)^2}{4} \times 3$

$V_c = 2.4 \ ft.^3$

SPHERE:

The surface area of a sphere is expressed by the formula $S_s = \pi d^2$, where S_s is the surface area, π is a constant of approximate value 3.1416 and d is the diameter. The volume of a sphere is expressed by the formula $V_s = \frac{1}{6} \pi d^3$, where V_s is the volume, π is a constant of approximate value 3.1416 and d is the diameter.

Example:

For a sphere whose diameter is 6′, compute a) the surface area S_s, and b) the volume V_s.

Solution:

a) $S_s = \pi d^2$
$= 3.1416 \times (6)^2$
$= 3.1416 \times 36$
$S_s = 113 \text{ ft.}^2$

b) $V_s = \frac{1}{6}\pi d^3$
$= \frac{1}{6} \times 3.1416 \times (6)^3$
$= \frac{1}{6} \times 3.1416 \times (6)^2 \times 6$
$= 3.1416 \times 6^2$
$= 3.1416 \times 36$
$= 113 \text{ ft.}^3$

IV. TRIGONOMETRY

DEFINITION AND SCOPE:

Trigonometry is that branch of mathematics which deals with the relationships of the angles and sides of triangles. The scope of this chapter is limited to an introduction of the more important formulas as they relate to measurement of lengths and angles of plane triangles.

An *angle* is the space between two intersecting lines and is measured in units of degrees (°). A degree is 1/360 of the circumference of a circle. An *acute* angle is less than 90°, a *right* angle is equal to 90°, and an *obtuse* angle is greater than 90°. The sum of the three angles of any triangle is 180°.

A *right triangle* as shown in figure IV a. is defined as any triangle which has a right angle. In this figure, angle A is opposite side a, angle B is opposite side b, and angle C (right angle—90°) is opposite side c, which is called the hypotenuse.

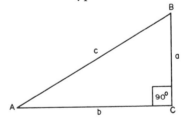

Figure IV a.   Right Triangle

PYTHAGOREAN THEOREM:

(Reference figure IV a.)

$$c^2 = a^2 + b^2$$

where
c = hypotenuse (side opposite right angle C)
a = side opposite angle A
b = side opposite angle B

Example:

Find the value of the hypotenuse (c) in a right triangle where side a is 3″ and side b is 4″ (figure IV a.).

Solution:

$$c^2 = a^2 + b^2$$
$$= 3^2 + 4^2$$
$$= 9 + 16$$
$$c^2 = 25$$
$$c = \sqrt{25}$$
$$c = 5''$$

Example:

In a right triangle where the hypotenuse (c) is 16′ and side a is 4′, determine the length of side b (figure IV a.).

Solution:

$$c^2 = a^2 + b^2$$
$$16^2 = 4^2 + b^2$$
$$16^2 - 4^2 = b^2$$
$$256 - 16 = b^2 \qquad \text{(Table II.)}$$
$$240 = b^2$$
$$\sqrt{240} = b \qquad \text{(Table II.)}$$
$$15.49' = b$$

TRIGONOMETRIC RATIOS—RIGHT TRIANGLE:

(Reference figure IV a.)

(1) sine A = sin A = $\dfrac{\text{side opposite A}}{\text{hypotenuse}} = \dfrac{a}{c}$

(2) cosine A = cos A = $\dfrac{\text{side adjacent to A}}{\text{hypotenuse}} = \dfrac{b}{c}$

(3) tangent A = tan A = $\dfrac{\text{side opposite A}}{\text{side adjacent to A}} = \dfrac{a}{b}$

(4) cosecant A = csc A = $\dfrac{\text{hypotenuse}}{\text{side opposite A}} = \dfrac{c}{a} = \dfrac{1}{\sin A}$

(5) secant A = sec A = $\dfrac{\text{hypotenuse}}{\text{side adjacent to A}} = \dfrac{c}{b} = \dfrac{1}{\cos A}$

(6) cotangent A = cot A = $\dfrac{\text{side adjacent to A}}{\text{side opposite A}} = \dfrac{b}{a} = \dfrac{1}{\tan A}$

(7) A + B = 90°
(8) sin A = cos (90° − A)
(9) cos A = sin (90° − A)
(10) tan A = cot (90° − A)

Example:

In a right triangle, angle A is 30° and side a equals 10′. Determine the value of the hypotenuse (c), side b and angle B, using the above trigonometric ratios. Verify the values of the sides by use of the Pythagorean Theorem.

Solution:

(1) $\sin A = \dfrac{a}{c}$

       $\sin 30° = 0.5000$ (Table XI.)

       $0.5 = \dfrac{10'}{c}$

       $c = \dfrac{10'}{0.5}$

       $c = 20'$

(2) $\cos A = \dfrac{b}{c}$

       $\cos 30° = 0.8660$ (Table XI.)

       $0.866 = \dfrac{b}{20'}$

       $0.866 \times 20' = b$

       $17.32' = b$

(3) $\tan A = \dfrac{a}{b}$

       $\tan 30° = 0.5774$

       $0.5774 = \dfrac{10'}{b}$

       $b = \dfrac{10}{0.5774}$

       $b = 17.32'$

(4) A + B = 90°   (the 7th formula)
    30° + B = 90°
         B = 90° − 30°
         B = 60°

Verification of sides by Pythagorean Theorem:

$$c^2 = a^2 + b^2$$
$$20^2 = 10^2 + 17.32^2 \qquad \text{(Table II.)}$$
$$400 = 100 + 300$$
$$400 = 400$$

TRIGONOMETRIC RELATIONSHIPS—ALL TRIANGLES:

(Reference figure IV b.)

(11) $A + B + C = 180°$

(12) $\sin (180° - A) = \sin A$

(13) $\cos (180° - A) = -\cos A$

(14) $\tan (180° - A) = -\tan A$

(15)   Law of sines: $\dfrac{a}{\sin A} = \dfrac{b}{\sin B} = \dfrac{c}{\sin C}$

(16)   Law of cosines: $a^2 = b^2 + c^2 - 2bc \cos A$

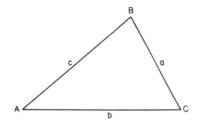

Figure IV b.   All Triangles

Example:

In figure IV b. angle A is 50°, angle C is 70° and side a is 20 units. Determine angle B and sides b and c.

Solution:

(11) $A + B + C = 180°$

$\qquad 50° + B + 70° = 180°$

$\qquad\qquad\quad B = 180° - 50° - 70°$

$\qquad\qquad\qquad = 180° - 120°$

$\qquad\qquad\quad B = 60°$

(15) $\dfrac{a}{\sin A} = \dfrac{b}{\sin B} = \dfrac{c}{\sin C}$

$\qquad \dfrac{20}{\sin 50°} = \dfrac{b}{\sin 60°} = \dfrac{c}{\sin 70°}$

$\sin 50° = 0.7660 \qquad \text{(Table XI.)}$

$\sin 60° = 0.8660 \qquad\quad // \quad\quad //$

$\sin 70° = 0.9397 \qquad\quad // \quad\quad //$

$$\frac{20}{0.7660} = \frac{b}{0.8660}$$

$$\frac{0.8660 \times 20}{0.7660} = b$$

$$22.6 \text{ units} = b$$

$$\frac{20}{0.7660} = \frac{c}{0.9397}$$

$$\frac{0.9397 \times 20}{0.7660} = c$$

$$24.6 \text{ units} = c$$

Example:

In figure IV b. angle A is 100° and angle C is 45°. Side b is 41 units. Determine the unit values of sides a and c.

Solution:

(12) $\sin (180° - A) = \sin A$
      $\sin (180° - 100°) = \sin A$
      $\sin 80° = \sin A$

(15) $\dfrac{a}{\sin A} = \dfrac{b}{\sin B} = \dfrac{c}{\sin C}$

$$\frac{a}{\sin 80°} = \frac{41}{\sin 35°} = \frac{c}{\sin 45°}$$

$\sin 80° = 0.9848$ (Table XI.)
$\sin 45° = 0.7071$ //   //

$$\frac{41}{0.5736} = \frac{c}{0.7071}$$

$$\frac{41 \times 0.7071}{0.5776} = c$$

$$50.4 \text{ units} = c$$

Example:

In figure IV b. side a is 7 units, side b is 9 units and side c is 10 units. Using the law of cosines, determine each angle to the nearest full degree. Verify the solutions.

Solution:

(16) $a^2 = b^2 + c^2 - 2bc \cos A$
      $2bc \cos A = b^2 + c^2 - a^2$

$$\cos A = \frac{b^2 + c^2 - a^2}{2bc}$$

$$\cos\ B = \frac{a^2 + c^2 - b^2}{2ac}$$

and, $\cos C = \dfrac{a^2 + b^2 - c^2}{2ab}$

$$\cos A = \frac{9^2 + 10^2 - 7^2}{2 \times 9 \times 10}$$

$$= \frac{81 + 100 - 49}{180}$$

$$= \frac{132}{180}$$

$\cos A = 0.7333$

$A = 43°$ (Table XI.)

$$\cos B = \frac{7^2 + 10^2 - 9^2}{2 \times 7 \times 10}$$

$$= \frac{49 + 100 - 81}{140}$$

$$= \frac{68}{140}$$

$\cos B = 0.4857$

$B = 61°$ (Table XI.)

$$\cos C = \frac{7^2 + 9^2 - 10^2}{2 \times 7 \times 9}$$

$$= \frac{49 + 81 - 100}{126}$$

$$= \frac{30}{126}$$

$\cos C = 0.2380$

$C = 76°$ (Table XI.)

Verify solutions

(11) $A + B + C = 180°$

$43° + 61° + 76° = 180°$

$180° = 180°$

V. SLIDE RULE

Operations and Scope:

A slide rule is a mechanical device designed to permit rapid mathematical computations with the exception of addition and subtraction. The most usual operations performed with the slide rule are multiplication, division and square and cube powers and roots of numbers. Trigonometric relationships are also frequently used. The slide rule depicted in figure V a. provides for the above mentioned operations as well as logarithm scales for additional operations.

The scope of this chapter is limited to describing the more important mathematical relationships readily ascertained through use of this instrument. This chapter is not intended to represent a manual containing complete instructions for the use of any particular slide rule.

Reading the Scales:

Figure V b. shows the reading on the C and D scales. The illustrative reading of 246 is independent of the location of a decimal point. Thus, 246 may represent 24.6, 2460, etc. As a general rule, to locate the placement of a decimal point for any operation, make a rough calculation as to the magnitude of the solution.

Multiplication and Division—C and D Scales:

As depicted in figure V c., to multiply 2 by 4, set index of C to 2 on D, push hairline to 4 on C, and at the hairline read 8 on D. The division process is just the opposite. To divide 8 by 4, align 8 on the D scale and 4 on the C scale with the hairline and read the solution under the index of the C scale on scale D, namely, 2.

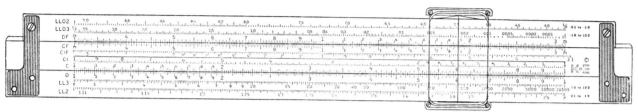

Front Face

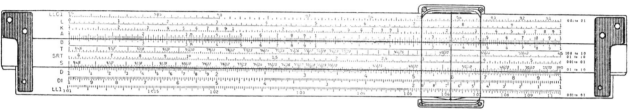

Reverse Face

Figure V a.   Slide Rule[1]

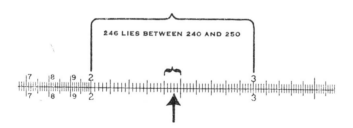

Figure V b.   Reading C and D Scales[2]

Figure V c.   Multiplication and Division—C and D Scales[3]

1. By permission of Keuffel & Esser Co., © 1947.
2. *Ibid.*, p. 5.
3. *Ibid.*, p. 7.

Trigonometry

Figure V d. shows the relations on the SRT scale (Sine, Radian, Tangent) and S (Sine) scale with scale D.

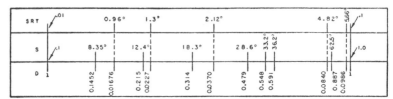

Figure V d.   Trigonometry—SRT, S and D Scales[4]

Powers of a Number:

Figure V e. shows the setting on a slide rule to find the values of 3^2 and 3^{-2}. The index of the C scale is aligned with 3 on the LL3 scale and the hairline set at 2 on the C scale. The value of 3^2 or 9 is read on the LL3 scale and the value of 3^{-2} or 0.1111 on the LL03 scale.

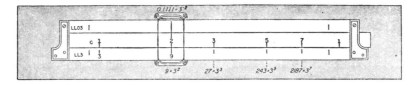

Figure V e.   Powers of a Number[5]

Logarithms:

The L scale and the D scale are used for determining logarithms and anti-logarithms for the common base or base 10. These scales are depicted in figure V f. The logarithm of 2 (mantissa) is found by aligning the hairline with 2 on the D scale to find the mantissa value of 0.301 on the L scale.

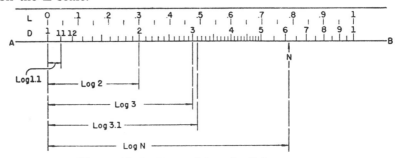

Figure V f.   Logarithm Scale[6]

4. *Ibid.*, p. 44.
5. *Ibid.*, p. 83.
6. *Ibid.*, p. 103.

VI. USE OF INTEREST FACTORS

SYMBOLS AND DEFINITIONS:

The following symbols and their definitions will be employed throughout this chapter in explanation of the use of the interest tables given in table VII[1]:

i = annual interest rate

n = number of interest periods

P = present sum or principal of money

S = sum of money n interest periods from present date that is equivalent to P with an annual interest rate of i

R = one in a series of n equal payments paid at the end of each of such n periods

SINGLE-PAYMENT–COMPOUND-AMOUNT FACTOR:

Formula:

$$S = P(1 + i)^n$$
$$= P(\text{C-af})$$

where $(1 + i)^n$ or (C-af) is defined as the single-payment–compound-amount factor.

Example:

If $1000 (P) is invested at an interest rate (i) of 6% compounded annually, what will be its worth (S) 10 years (n) from now?

---

1. H. G. Thuesen and W. J. Fabrycky, *Engineering Economy,* Third Edition, © 1964. Reprinted by Permission of Prentice-Hall, Inc., Englewood Cliffs, N. J., pp. 490–498.

Solution:

$S = P(\text{C-af})$

where (C-af) is 1.791 for i = 6%, n = 10   (Table VII.)

$= 1000(1.791)$

$S = \$1791$

SINGLE-PAYMENT–PRESENT-WORTH FACTOR:

Formula:

$$P = S\left[\frac{1}{(1+i)^n}\right]$$

$$= S\,(\text{P-wf})$$

where $\left[\dfrac{1}{(1+i)^n}\right]$ or (P-wf) is defined as the single-payment–present-worth factor.

Example:

How much money should be invested now (P) to realize \$1000 (S) at an interest rate (i) of 6%, 10 years (n) from now?

Solutiom:

$P = S(\text{P-wf})$

where (P-wf) is 0.55839 for i = 6%, n = 10

$P = 1000(0.55839)$ (Table VII.)

$P = \$558.39$

EQUAL-PAYMENT–COMPOUND-AMOUNT FACTOR:

Formula:

$$S = R\left[\frac{(1+i)^n - 1}{i}\right]$$

$$= R(\text{C-afe})$$

where $\left[\dfrac{(1+i)^n - 1}{i}\right]$ or (C-afe) is defined as the equal-payment–compound-amount factor.

Example:

By investing \$100 now and \$100 every year (R) from now, what will be the worth of this investment 10 years (n) from now at an interest rate (i) of 6%?

Solution:

$$S = R(\text{C-afe})$$

where (C-afe) is 13.181 for i $= 6\%$, n $= 10$

$$S = 100(13.181) \qquad \text{(Table VII.)}$$
$$S = \$1318.10$$

EQUAL-PAYMENT–SINKING-FUND FACTOR:

Formula:

$$R = S \left[\frac{i}{(1 + i)^n - 1} \right]$$
$$= S \, (\text{S-ffe})$$

where $\left[\dfrac{i}{(1 + i)^n - 1} \right]$ or (S-ffe) is defined as the equal-payment–sinking-fund factor.

Example:

It is desired to have a fund of $1000 (S) at the end of 10 years (n) by investing an equal amount of money (R) each year with an interest rate (i) of 6%. Determine the value of R to do this.

Solution:

$$R = S(\text{S-ffe})$$

where (S-ffe) is 0.07587 for i $= 6\%$, n $= 10$

$$R = 1000 \, (0.07587) \qquad \text{(Table VII.)}$$
$$R = \$75.87$$

EQUAL-PAYMENT–PRESENT-WORTH FACTOR:

Formula:

$$P = R \left[\frac{(1 + i)^n - 1}{i(1 + i)^n} \right]$$
$$= R(\text{P-wfe}) \text{ or (P-wfe) is defined as the equal-payment–present-worth factor.}$$

Example:

To realize an annual return of $100 (R) for each of the next 10 years (n), with an interest rate (i) of 6%, how much investment money is needed now (P)?

Solution:

$$P = R(P\text{-wfe})$$

where (P-wfe) is 7.36009 for i $= 6\%$, n $= 10$

$$P = 100(7.36009) \qquad \text{(Table VII.)}$$

$$P = \$736.01$$

EQUAL-PAYMENT–CAPITAL-RECOVERY FACTOR:

Formula:

$$R = P\left[\frac{i(1 + i)^n}{(1 + i)^n - 1}\right]$$

$$= P(C\text{-rfe})$$

where $\left[\dfrac{i(1 + i)^n}{(1 + i)^n - 1}\right]$ or (C-rfe) is defined as the equal-payment–capital-recovery factor.

Example:

For an initial investment of $1000 (P), what is the expected annual return (R) for a period of 10 years (n) with an interest rate (i) of 6%?

Solution:

$$R = P(C\text{-rfe})$$

where (C-rfe) is 0.13587 for i $= 6\%$, n $= 10$

$$R = 1000(0.13587) \qquad \text{(Table VII.)}$$

$$R = \$135.87$$

TABLES

Table   I 161

IMPORTANT CONSTANTS

Number	Log	Number	Log
$\pi = 3.14159265$	0.4971499	$\pi^2 = 9.86960440$	0.9942997
$2\pi = 6.28318531$	0.7981799	$\dfrac{1}{\pi^2} = 0.10132118$	9.0057003−10
$4\pi = 12.56637061$	1.0992099	$\sqrt{\pi} = 1.77245385$	0.2485749
$\dfrac{\pi}{2} = 1.57079633$	0.1961199	$\dfrac{1}{\sqrt{\pi}} = 0.56418958$	9.7514251−10
$\dfrac{\pi}{3} = 1.04719755$	0.0200286	$\sqrt{\dfrac{3}{\pi}} = 0.97720502$	9.9899857−10
$\dfrac{4\pi}{3} = 4.18879020$	0.6220886	$\sqrt{\dfrac{4}{\pi}} = 1.12837917$	0.0524551
$\dfrac{\pi}{4} = 0.78539816$	9.8950899−10	$\sqrt[3]{\pi} = 1.46459189$	0.1657166
$\dfrac{\pi}{6} = 0.52359878$	9.7189986−10	$\dfrac{1}{\sqrt[3]{\pi}} = 0.68278406$	9.8342834−10
$\dfrac{1}{\pi} = 0.31830989$	9.5028501−10	$\sqrt[3]{\pi^2} = 2.14502940$	0.3314332
$\dfrac{1}{2\pi} = 0.15915494$	9.2018201−10	$\sqrt[3]{\dfrac{3}{4\pi}} = 0.62035049$	9.7926371−10
$\dfrac{3}{\pi} = 0.95492966$	9.9799714−10	$\sqrt[3]{\dfrac{\pi}{6}} = 0.80599598$	9.9063329−10
$\dfrac{4}{\pi} = 1.27323954$	0.1049101		

		Log
e = Naperian Base....................	= 2.71828183	0.43429448
$M = \log_{10} e$.........................	= 0.43429448	9.63778431−10
$1 \div M = \log_e 10$.....................	= 2.30258509	0.36221569
$180 \div \pi$ = degrees in 1 radian.........	= 57.2957795	1.75812263
$\pi \div 180$ = radians in 1°..............	= 0.01745329	8.24187737−10
$\pi \div 10800$ = radians in 1'............	= 0.0002908882	6.46372612−10
$\pi \div 648000$ = radians in 1"...........	= 0.000004848136811095	4.68557487−10
sin 1".............................	= 0.000004848136811076	4.68557487−10
tan 1"..............................	= 0.000004848136811133	4.68557487−10
centimeters in 1 ft....................	= 30.480	1.4840150
feet in 1 cm.........................	= 0.032808	8.5159850−10
inches in 1 m........................	= 39.37	1.5951654
pounds in 1 kg.......................	= 2.20462	0.3433340
kilograms in 1 lb.....................	= 0.453593	9.6566660−10

$\pi = 3.14159$	26535	89793	23846	26433	83280
$e = 2.71828$	18284	59045	23536	02874	71353
$M = 0.43429$	44819	03251	82765	11289	18917
$1 \div M = 2.30258$	50929	94045	68401	79914	54684
$\log_{10}\pi = 0.49714$	98726	94133	85435	12682	88291
$\log_e \pi = 1.14472$	98858	49400	17414	34273	51353

TABLE I.

TABLE I.   IMPORTANT CONSTANTS
From *Mathematical Tables and Formulas* by Robert D. Carmichael and Edwin R. Smith, ©1931, Dover Publications, Inc., New York, p. 50.

POWERS AND ROOTS n^k

k		2	3	4
1	$n^1=$	2	3	4
2	$n^2=$	4	9	16
3	$n^3=$	8	27	64
4	$n^4=$	16	81	256
5	$n^5=$	32	243	1024
6	$n^6=$	64	729	4096
7	$n^7=$	128	2187	16384
8	$n^8=$	256	6561	65536
9	$n^9=$	512	19683	2 62144
10	$n^{10}=$	1024	59049	10 48576
24	$n^{24}=$	167 77216	(11)2.8242 95365	(14)2.8147 49767
1/2	$n^{1/2}=$	1.4142 13562	1.7320 50808	2.0000 00000
1/3	$n^{1/3}=$	1.2599 21050	1.4422 49570	1.5874 01052
1/4	$n^{1/4}=$	1.1892 07115	1.3160 74013	1.4142 13562
1/5	$n^{1/5}=$	1.1486 98355	1.2457 30940	1.3195 07911

k	5	6	7	8	9
1	5	6	7	8	9
2	25	36	49	64	81
3	125	216	343	512	729
4	625	1296	2401	4096	6561
5	3125	7776	16807	32768	59049
6	15625	46656	1 17649	2 62144	5 31441
7	78125	2 79936	8 23543	20 97152	47 82969
8	3 90625	16 79616	57 64801	167 77216	430 46721
9	19 53125	100 77696	403 53607	1342 17728	3874 20489
10	97 65625	604 66176	2824 75249	(9)1.0737 41824	(9)3.4867 84401
24	(16)5.9604 64478	(18)4.7383 81338	(20)1.9158 12314	(21)4.7223 66483	(22)7.9766 44308
1/2	2.2360 67977	2.4494 89743	2.6457 51311	2.8284 27125	3.0000 00000
1/3	1.7099 75947	1.8171 20593	1.9129 31183	2.0000 00000	2.0800 83823
1/4	1.4953 48781	1.5650 84580	1.6265 76562	1.6817 92831	1.7320 50808
1/5	1.3797 29662	1.4309 69081	1.4757 73162	1.5157 16567	1.5518 45574

k	10	11	12	13	14
1	10	11	12	13	14
2	100	121	144	169	196
3	1000	1331	1728	2197	2744
4	10000	14641	20736	28561	38416
5	1 00000	1 61051	2 48832	3 71293	5 37824
6	10 00000	17 71561	29 85984	48 26809	75 29536
7	100 00000	194 87171	358 31808	627 48517	1054 13504
8	1000 00000	2143 58881	4299 81696	8157 30721	(9)1.4757 89056
9	(9)1.0000 00000	(9)2.3579 47691	(9)5.1597 80352	(10)1.0604 49937	(10)2.0661 04678
10	(10)1.0000 00000	(10)2.5937 42460	(10)6.1917 36422	(11)1.3785 84918	(11)2.8925 46550
24	(24)1.0000 00000	(24)9.8497 32676	(25)7.9496 84720	(26)5.4280 07704	(27)3.2141 99700
1/2	3.1622 77660	3.3166 24790	3.4641 01615	3.6055 51275	3.7416 57387
1/3	2.1544 34690	2.2239 80091	2.2894 28485	2.3513 34688	2.4101 42264
1/4	1.7782 79410	1.8211 60287	1.8612 09718	1.8988 28922	1.9343 36420
1/5	1.5848 93192	1.6153 94266	1.6437 51830	1.6702 77652	1.6952 18203

k	15	16	17	18	19
1	15	16	17	18	19
2	225	256	289	324	361
3	3375	4096	4913	5832	6859
4	50625	65536	83521	1 04976	1 30321
5	7 59375	10 48576	14 19857	18 89568	24 76099
6	113 90625	167 77216	241 37569	340 12224	470 45881
7	1708 59375	2684 35456	4103 38673	6122 20032	8938 71739
8	(9)2.5628 90625	(9)4.2949 67296	(9)6.9757 57441	(10)1.1019 96058	(10)1.6983 56304
9	(10)3.8443 35938	(10)6.8719 47674	(11)1.1858 78765	(11)1.9835 92904	(11)3.2268 76978
10	(11)5.7665 03906	(12)1.0995 11628	(12)2.0159 93900	(12)3.5704 67227	(12)6.1310 66258
24	(28)1.6834 11220	(28)7.9228 16251	(29)3.3944 86713	(30)1.3382 58845	(30)4.8987 62931
1/2	3.8729 83346	4.0000 00000	4.1231 05626	4.2426 40687	4.3588 90744
1/3	2.4662 12074	2.5198 42100	2.5712 81591	2.6207 41394	2.6684 01649
1/4	1.9679 89671	2.0000 00000	2.0305 43185	2.0597 67144	2.0877 97630
1/5	1.7187 71928	1.7411 01127	1.7623 40348	1.7826 02458	1.8019 83127

k	20	21	22	23	24
1	20	21	22	23	24
2	400	441	484	529	576
3	8000	9261	10648	12167	13824
4	1 60000	1 94481	2 34256	2 79841	3 31776
5	32 00000	40 84101	51 53632	64 36343	79 62624
6	640 00000	857 66121	1133 79904	1480 35889	1911 02976
7	(9)1.2800 00000	(9)1.8010 88541	(9)2.4943 57888	(9)3.4048 25447	(9)4.5864 71424
8	(10)2.5600 00000	(10)3.7822 85936	(10)5.4875 87354	(10)7.8310 98528	(11)1.1007 53142
9	(11)5.1200 00000	(11)7.9428 00466	(12)1.2072 69218	(12)1.8011 52661	(12)2.6418 07540
10	(13)1.0240 00000	(13)1.6679 88098	(13)2.6559 92279	(13)4.1426 51121	(13)6.3403 38097
24	(31)1.6777 21600	(31)5.4108 19838	(32)1.6525 10926	(32)4.8025 07640	(33)1.3337 35777
1/2	4.4721 35955	4.5825 75695	4.6904 15760	4.7958 31523	4.8989 79486
1/3	2.7144 17617	2.7589 24176	2.8020 39331	2.8438 66980	2.8844 99141
1/4	2.1147 42527	2.1406 95143	2.1657 36771	2.1899 38703	2.2133 63839
1/5	1.8205 64203	1.8384 16287	1.8556 00736	1.8721 71231	1.8881 75023

TABLE II.

TABLE II.   POWERS AND ROOTS

From *Handbook of Mathematical Functions With Formulas, Graphs, and Mathematical Tables*. National Bureau of Standards Applied Mathematics Series—55, 1964, pp. 24–63.

Table II   163

POWERS AND ROOTS n^k

k	25	26	27	28	29
1	25	26	27	28	29
2	625	676	729	784	841
3	15625	17576	19683	21952	24389
4	3 90625	4 56976	5 31441	6 14656	7 07281
5	97 65625	118 81376	143 48907	172 10368	205 11149
6	2441 40625	3089 15776	3874 20489	4818 90304	5948 23321
7	(9)6.1035 15625	(9)8.0318 10176	(10)1.0460 35320	(10)1.3492 92851	(10)1.7249 87631
8	(11)1.5258 78906	(11)2.0882 70646	(11)2.8242 95365	(11)3.7780 19983	(11)5.0024 64130
9	(12)3.8146 97266	(12)5.4295 03679	(12)7.6255 97485	(13)1.0578 45595	(13)1.4507 14598
10	(13)9.5367 43164	(14)1.4116 70957	(14)2.0589 11321	(14)2.9619 67667	(14)4.2070 72333
24	(33)3.5527 13679	(33)9.1066 85770	(34)2.2528 39954	(34)5.3925 32264	(35)1.2518 49008
1/2	5.0000 00000	5.0990 19514	5.1961 52423	5.2915 02622	5.3851 64807
1/3	2.9240 17738	2.9624 96068	3.0000 00000	3.0365 88972	3.0723 16826
1/4	2.2360 67977	2.2581 00864	2.2795 07057	2.3003 26634	2.3205 95787
1/5	1.9036 53939	1.9186 45192	1.9331 82045	1.9472 94361	1.9610 09057

k	30	31	32	33	34
1	30	31	32	33	34
2	900	961	1024	1089	1156
3	27000	29791	32768	35937	39304
4	8 10000	9 23521	10 48576	11 85921	13 36336
5	243 00000	286 29151	335 54432	391 35393	454 35424
6	7290 00000	8875 03681	(9)1.0737 41824	(9)1.2914 67969	(9)1.5448 04416
7	(10)2.1870 00000	(10)2.7512 61411	(10)3.4359 73837	(10)4.2618 44298	(10)5.2523 35014
8	(11)6.5610 00000	(11)8.5289 10374	(12)1.0995 11628	(12)1.4064 08618	(12)1.7857 93905
9	(13)1.9683 00000	(13)2.6439 62216	(13)3.5184 37209	(13)4.6411 48440	(13)6.0716 99277
10	(14)5.9049 00000	(14)8.1962 82870	(15)1.1258 99907	(15)1.5315 78985	(15)2.0643 77754
24	(35)2.8242 95365	(35)6.2041 26610	(36)1.3292 27996	(36)2.7818 55434	(36)5.6950 03680
1/2	5.4772 25575	5.5677 64363	5.6568 54249	5.7445 62647	5.8309 51895
1/3	3.1072 32506	3.1413 80652	3.1748 02104	3.2075 34330	3.2396 11801
1/4	2.3403 47319	2.3596 11062	2.3784 14230	2.3967 81727	2.4147 36403
1/5	1.9743 50486	1.9873 40755	2.0000 00000	2.0123 46617	2.0243 97459

k	35	36	37	38	39
1	35	36	37	38	39
2	1225	1296	1369	1444	1521
3	42875	46656	50653	54872	59319
4	15 00625	16 79616	18 74161	20 85136	23 13441
5	525 21875	604 66176	693 43957	792 35168	902 24199
6	(9)1.8382 65625	(9)2.1767 82336	(9)2.5657 26409	(9)3.0109 36384	(9)3.5187 43761
7	(10)6.4339 29688	(10)7.8364 16410	(10)9.4931 87713	(11)1.1441 55826	(11)1.3723 10067
8	(12)2.2518 75391	(12)2.8211 09907	(12)3.5124 79454	(12)4.3477 92138	(12)5.3520 09260
9	(13)7.8815 63867	(14)1.0155 99567	(14)1.2996 17398	(14)1.6521 61013	(14)2.0872 83612
10	(15)2.7585 47354	(15)3.6561 58440	(15)4.8085 84372	(15)6.2782 11848	(15)8.1404 06085
24	(37)1.1419 13124	(37)2.2452 25771	(37)4.3335 25711	(37)8.2187 60383	(38)1.5330 29700
1/2	5.9160 79783	6.0000 00000	6.0827 62530	6.1644 14003	6.2449 97998
1/3	3.2710 66310	3.3019 27249	3.3322 21852	3.3619 75407	3.3912 11443
1/4	2.4322 99279	2.4494 89743	2.4663 25715	2.4828 23796	2.4989 99399
1/5	2.0361 68005	2.0476 72511	2.0589 24137	2.0699 35054	2.0807 16549

k	40	41	42	43	44
1	40	41	42	43	44
2	1600	1681	1764	1849	1936
3	64000	68921	74088	79507	85184
4	25 60000	28 25761	31 11696	34 18801	37 48096
5	1024 00000	1158 56201	1306 91232	1470 08443	1649 16224
6	(9)4.0960 00000	(9)4.7501 04241	(9)5.4890 31744	(9)6.3213 63049	(9)7.2563 13856
7	(11)1.6384 00000	(11)1.9475 42739	(11)2.3053 93332	(11)2.7181 86111	(11)3.1927 78097
8	(12)6.5536 00000	(12)7.9849 25229	(12)9.6826 51996	(13)1.1688 20028	(13)1.4048 22363
9	(14)2.6214 40000	(14)3.2738 19344	(14)4.0667 13838	(14)5.0259 26119	(14)6.1812 18395
10	(16)1.0485 76000	(16)1.3422 65931	(16)1.7080 19812	(16)2.1611 48231	(16)2.7197 36094
24	(38)2.8147 49767	(38)5.0911 10945	(38)9.0778 09315	(39)1.5967 72093	(39)2.7724 53276
1/2	6.3245 55320	6.4031 24237	6.4807 40698	6.5574 38524	6.6332 49581
1/3	3.4199 51893	3.4482 17240	3.4760 26645	3.5033 98060	3.5303 48335
1/4	2.5148 66859	2.5304 39534	2.5457 29895	2.5607 49602	2.5755 09577
1/5	2.0912 79105	2.1016 32478	2.1117 85765	2.1217 47461	2.1315 25513

k	45	46	47	48	49
1	45	46	47	48	49
2	2025	2116	2209	2304	2401
3	91125	97336	1 03823	1 10592	1 17649
4	41 00625	44 77456	48 79681	53 08416	57 64801
5	1845 28125	2059 62976	2293 45007	2548 03968	2824 75249
6	(9)8.3037 65625	(9)9.4742 96896	(10)1.0779 21533	(10)1.2230 59046	(10)1.3841 28720
7	(11)3.7366 94531	(11)4.3581 76572	(11)5.0662 31205	(11)5.8706 83423	(11)6.7822 30728
8	(13)1.6815 12539	(13)2.0047 61223	(13)2.3811 28666	(13)2.8179 28043	(13)3.3232 93057
9	(14)7.5668 06426	(14)9.2219 01627	(15)1.1191 30473	(15)1.3526 05461	(15)1.6284 13598
10	(16)3.4050 62892	(16)4.2420 74748	(16)5.2599 13224	(16)6.4925 06211	(16)7.9792 26630
24	(39)4.7544 50505	(39)8.0572 70802	(40)1.3500 46075	(40)2.2376 37322	(40)3.6703 36822
1/2	6.7082 03932	6.7823 29983	6.8556 54600	6.9282 03230	7.0000 00000
1/3	3.5568 93304	3.5830 47871	3.6088 26080	3.6342 41186	3.6593 05710
1/4	2.5900 20064	2.6042 90687	2.6183 30499	2.6321 48026	2.6457 51311
1/5	2.1411 27368	2.1505 60013	2.1598 30012	2.1689 43542	2.1779 06425

TABLE II.

Table II

POWERS AND ROOTS n^k

k					
1	50	51	52	53	54
2	2500	2601	2704	2809	2916
3	1 25000	1 32651	1 40608	1 48877	1 57464
4	62 50000	67 65201	73 11616	78 90481	85 03056
5	3125 00000	3450 25251	3802 04032	4181 95493	4591 65024
6	(10)1.5625 00000	(10)1.7596 28780	(10)1.9770 60966	(10)2.2164 36113	(10)2.4794 91130
7	(11)7.8125 00000	(11)8.9741 06779	(12)1.0280 71703	(12)1.1747 11140	(12)1.3389 25210
8	(13)3.9062 50000	(13)4.5767 94457	(13)5.3459 72853	(13)6.2259 69041	(13)7.2301 96134
9	(15)1.9531 25000	(15)2.3341 65173	(15)2.7799 05884	(15)3.2997 63592	(15)3.9043 05912
10	(16)9.7656 25000	(17)1.1904 24238	(17)1.4455 51059	(17)1.7488 74704	(17)2.1083 25193
24	(40)5.9604 64478	(40)9.5870 33090	(41)1.5278 48342	(41)2.4133 53110	(41)3.7796 38253
1/2	7.0710 67812	7.1414 28429	7.2111 02551	7.2801 09889	7.3484 69228
1/3	3.6840 31499	3.7084 29769	3.7325 11157	3.7562 85754	3.7797 63150
1/4	2.6591 47948	2.6723 45118	2.6853 49614	2.6981 67876	2.7108 06011
1/5	2.1867 24148	2.1954 01897	2.2039 44575	2.2123 56822	2.2206 43035

k					
1	55	56	57	58	59
2	3025	3136	3249	3364	3481
3	1 66375	1 75616	1 85193	1 95112	2 05379
4	91 50625	98 34496	105 56001	113 16496	121 17361
5	5032 84375	5507 31776	6016 92057	6563 56768	7149 24299
6	(10)2.7680 64063	(10)3.0840 97946	(10)3.4296 44725	(10)3.8068 69254	(10)4.2180 53364
7	(12)1.5224 35234	(12)1.7270 94850	(12)1.9548 97493	(12)2.2079 84168	(12)2.4886 51485
8	(13)8.3733 93789	(13)9.6717 31157	(14)1.1142 91571	(14)1.2806 30817	(14)1.4683 04376
9	(15)4.6053 66584	(15)5.4161 69448	(15)6.3514 61955	(15)7.4276 58740	(15)8.6629 95819
10	(17)2.5329 51621	(17)3.0330 54891	(17)3.6203 33315	(17)4.3080 42069	(17)5.1111 67533
24	(41)5.8708 98173	(41)9.0471 67858	(42)1.3835 55344	(42)2.1002 54121	(42)3.1655 43453
1/2	7.4161 98487	7.4833 14774	7.5498 34435	7.6157 73106	7.6811 45748
1/3	3.8029 52461	3.8258 62366	3.8485 01131	3.8708 76641	3.8929 96416
1/4	2.7232 69815	2.7355 64800	2.7476 96205	2.7596 69021	2.7714 88002
1/5	2.2288 07384	2.2368 53829	2.2447 86134	2.2526 07878	2.2603 22470

k					
1	60	61	62	63	64
2	3600	3721	3844	3969	4096
3	2 16000	2 26981	2 38328	2 50047	2 62144
4	129 60000	138 45841	147 76336	157 52961	167 77216
5	7776 00000	8445 96301	9161 32832	9924 36543	(9)1.0737 41824
6	(10)4.6656 00000	(10)5.1520 37436	(10)5.6800 23558	(10)6.2523 50221	(10)6.8719 47674
7	(12)2.7993 60000	(12)3.1427 42836	(12)3.5216 14606	(12)3.9389 80639	(12)4.3980 46511
8	(14)1.6796 16000	(14)1.9170 73130	(14)2.1834 01056	(14)2.4815 57803	(14)2.8147 49767
9	(16)1.0077 69600	(16)1.1694 14609	(16)1.3537 08655	(16)1.5633 81416	(16)1.8014 39851
10	(17)6.0466 17600	(17)7.1334 29117	(17)8.3929 93659	(17)9.8493 02919	(18)1.1529 21505
24	(42)4.7383 81338	(42)7.0455 68477	(43)1.0408 79722	(43)1.5281 75339	(43)2.2300 74520
1/2	7.7459 66692	7.8102 49676	7.8740 07874	7.9372 53933	8.0000 00000
1/3	3.9148 67641	3.9364 97183	3.9578 91610	3.9790 57208	4.0000 00000
1/4	2.7831 57684	2.7946 82393	2.8060 66263	2.8173 13247	2.8284 27125
1/5	2.2679 33155	2.2754 43032	2.2828 55056	2.2901 72049	2.2973 96710

k					
1	65	66	67	68	69
2	4225	4356	4489	4624	4761
3	2 74625	2 87496	3 00763	3 14432	3 28509
4	178 50625	189 74736	201 51121	213 81376	226 67121
5	(9)1.1602 90625	(9)1.2523 32576	(9)1.3501 25107	(9)1.4539 33568	(9)1.5640 31349
6	(10)7.5418 89063	(10)8.2653 95002	(10)9.0458 38217	(10)9.8867 48262	(11)1.0791 81631
7	(12)4.9022 27891	(12)5.4551 60701	(12)6.0607 11605	(12)6.7229 88818	(12)7.4463 53253
8	(14)3.1864 48129	(14)3.6004 06063	(14)4.0606 76776	(14)4.5716 32397	(14)5.1379 83744
9	(16)2.0711 91284	(16)2.3762 68001	(16)2.7206 53440	(16)3.1087 10030	(16)3.5452 68744
10	(18)1.3462 74334	(18)1.5683 36881	(18)1.8228 37805	(18)2.1139 22820	(18)2.4461 94061
24	(43)3.2353 44710	(43)4.6671 78950	(43)6.6956 88867	(43)9.5546 30685	(44)1.3563 70007
1/2	8.0622 57748	8.1240 38405	8.1853 52772	8.2462 11251	8.3066 23863
1/3	4.0207 25759	4.0412 40021	4.0615 48100	4.0816 55102	4.1015 65930
1/4	2.8394 11514	2.8502 69883	2.8610 05553	2.8716 21711	2.8821 21417
1/5	2.3045 31620	2.3115 79249	2.3185 41963	2.3254 22030	2.3322 21626

k					
1	70	71	72	73	74
2	4900	5041	5184	5329	5476
3	3 43000	3 57911	3 73248	3 89017	4 05224
4	240 10000	254 11681	268 73856	283 98241	299 86576
5	(9)1.6807 00000	(9)1.8042 29351	(9)1.9349 17632	(9)2.0730 71593	(9)2.2190 06624
6	(11)1.1764 90000	(11)1.2810 02839	(11)1.3931 40695	(11)1.5133 42263	(11)1.6420 64902
7	(12)8.2354 30000	(12)9.0951 20158	(13)1.0030 61300	(13)1.1047 39852	(13)1.2151 28027
8	(14)5.7648 01000	(14)6.4575 35312	(14)7.2220 41363	(14)8.0646 00919	(14)8.9919 47402
9	(16)4.0353 60700	(16)4.5848 50072	(16)5.1998 69781	(16)5.8871 58671	(16)6.6540 41078
10	(18)2.8247 52490	(18)3.2552 43551	(18)3.7439 06243	(18)4.2976 25830	(18)4.9239 90397
24	(44)1.9158 12314	(44)2.6927 76876	(44)3.7668 63772	(44)5.2450 38047	(44)7.2704 49690
1/2	8.3666 00265	8.4261 49773	8.4852 81374	8.5440 03745	8.6023 25267
1/3	4.1212 85300	4.1408 17749	4.1601 67646	4.1793 39196	4.1983 36454
1/4	2.8925 07608	2.9027 83108	2.9129 50630	2.9230 12786	2.9329 72088
1/5	2.3389 42837	2.3455 87669	2.3521 58045	2.3586 55818	2.3650 82769

TABLE II.

Table II   165

POWERS AND ROOTS n^k

k	75	76	77	78	79
1	75	76	77	78	79
2	5625	5776	5929	6084	6241
3	4 21875	4 38976	4 56533	4 74552	4 93039
4	316 40625	333 62176	351 53041	370 15056	389 50081
5	(9)2.3730 46875	(9)2.5355 25376	(9)2.7067 84157	(9)2.8871 74368	(9)3.0770 56399
6	(11)1.7797 85156	(11)1.9269 99286	(11)2.0842 23801	(11)2.2519 96007	(11)2.4308 74555
7	(13)1.3348 38867	(13)1.4645 19457	(13)1.6048 52327	(13)1.7565 56885	(13)1.9203 90899
8	(15)1.0011 29150	(15)1.1130 34787	(15)1.2357 36292	(15)1.3701 14371	(15)1.5171 08810
9	(16)7.5084 68628	(16)8.4590 64385	(16)9.5151 69445	(17)1.0686 89209	(17)1.1985 15960
10	(18)5.6313 51471	(18)6.4288 88932	(18)7.3266 80473	(18)8.3357 75831	(18)9.4682 76083
24	(45)1.0033 91278	(45)1.3788 79182	(45)1.8870 23915	(45)2.5719 97041	(45)3.4918 06676
1/2	8.6602 54038	8.7177 97887	8.7749 64387	8.8317 60866	8.8881 94417
1/3	4.2171 63326	4.2358 23584	4.2543 20865	4.2726 58682	4.2908 40427
1/4	2.9428 30956	2.9525 91724	2.9622 56638	2.9718 27866	2.9813 07501
1/5	2.3714 40610	2.3777 30992	2.3839 55503	2.3901 15677	2.3962 12991

k	80	81	82	83	84
1	80	81	82	83	84
2	6400	6561	6724	6889	7056
3	5 12000	5 31441	5 51368	5 71787	5 92704
4	409 60000	430 46721	452 12176	474 58321	497 87136
5	(9)3.2768 00000	(9)3.4867 84401	(9)3.7073 98432	(9)3.9390 40643	(9)4.1821 19424
6	(11)2.6214 40000	(11)2.8242 95365	(11)3.0400 66714	(11)3.2694 03734	(11)3.5129 80316
7	(13)2.0971 52000	(13)2.2876 79245	(13)2.4928 54706	(13)2.7136 05099	(13)2.9509 03466
8	(15)1.6777 21600	(15)1.8530 20189	(15)2.0441 40859	(15)2.2522 92232	(15)2.4787 58911
9	(17)1.3421 77280	(17)1.5009 46353	(17)1.6761 95504	(17)1.8694 02553	(17)2.0821 57485
10	(19)1.0737 41824	(19)1.2157 66546	(19)1.3744 80313	(19)1.5516 04119	(19)1.7490 12288
24	(45)4.7223 66483	(45)6.3626 85441	(45)8.5414 66801	(46)1.1425 47375	(46)1.5230 10388
1/2	8.9442 71910	9.0000 00000	9.0553 85138	9.1104 33579	9.1651 51390
1/3	4.3088 69380	4.3267 48711	4.3444 81486	4.3620 70671	4.3795 19140
1/4	2.9906 97562	3.0000 00000	3.0092 16698	3.0183 49479	3.0274 00104
1/5	2.4022 48868	2.4082 24685	2.4141 41771	2.4200 01407	2.4258 04834

k	85	86	87	88	89
1	85	86	87	88	89
2	7225	7396	7569	7744	7921
3	6 14125	6 36056	6 58503	6 81472	7 04969
4	522 00625	547 00816	572 89761	599 69536	627 42241
5	(9)4.4370 53125	(9)4.7042 70176	(9)4.9842 09207	(9)5.2773 19168	(9)5.5840 59449
6	(11)3.7714 95156	(11)4.0456 72351	(11)4.3362 62010	(11)4.6440 40868	(11)4.9698 12910
7	(13)3.2057 70883	(13)3.4792 78222	(13)3.7725 47949	(13)4.0867 55964	(13)4.4231 33490
8	(15)2.7249 05250	(15)2.9921 79271	(15)3.2821 16715	(15)3.5963 45248	(15)3.9365 88806
9	(17)2.3161 69463	(17)2.5732 74173	(17)2.8554 41542	(17)3.1647 83818	(17)3.5035 64037
10	(19)1.9687 44043	(19)2.2130 15789	(19)2.4842 34142	(19)2.7850 09760	(19)3.1181 71993
24	(46)2.0232 71747	(46)2.6789 39031	(46)3.5355 91351	(46)4.6514 04745	(46)6.1004 25945
1/2	9.2195 44457	9.2736 18495	9.3273 79053	9.3808 31520	9.4339 81132
1/3	4.3968 29672	4.4140 04962	4.4310 47622	4.4479 60181	4.4647 45096
1/4	3.0363 70277	3.0452 61646	3.0540 75810	3.0628 14314	3.0714 78656
1/5	2.4315 53252	2.4372 47818	2.4428 89656	2.4484 79851	2.4540 19455

k	90	91	92	93	94
1	90	91	92	93	94
2	8100	8281	8464	8649	8836
3	7 29000	7 53571	7 78688	8 04357	8 30584
4	656 10000	685 74961	716 39296	748 05201	780 74896
5	(9)5.9049 00000	(9)6.2403 21451	(9)6.5908 15232	(9)6.9568 83693	(9)7.3390 40224
6	(11)5.3144 10000	(11)5.6786 92520	(11)6.0635 50013	(11)6.4699 01834	(11)6.8986 97811
7	(13)4.7829 69000	(13)5.1676 10194	(13)5.5784 66012	(13)6.0170 08706	(13)6.4847 75942
8	(15)4.3046 72100	(15)4.7025 25276	(15)5.1321 88731	(15)5.5958 18097	(15)6.0956 89385
9	(17)3.8742 04890	(17)4.2792 98001	(17)4.7216 13633	(17)5.2041 10830	(17)5.7299 48022
10	(19)3.4867 84401	(19)3.8941 61181	(19)4.3438 84542	(19)4.8398 23072	(19)5.3861 51141
24	(46)7.9766 44308	(47)1.0399 04400	(47)1.3517 85726	(47)1.7522 28603	(47)2.2650 01461
1/2	9.4868 32981	9.5393 92014	9.5916 63047	9.6436 50761	9.6953 59715
1/3	4.4814 04747	4.4979 41445	4.5143 57435	4.5306 54896	4.5468 35944
1/4	3.0800 70288	3.0885 90619	3.0970 41015	3.1054 22799	3.1137 37258
1/5	2.4595 09486	2.4649 50932	2.4703 44749	2.4756 91866	2.4809 93182

k	95	96	97	98	99
1	95	96	97	98	99
2	9025	9216	9409	9604	9801
3	8 57375	8 84736	9 12673	9 41192	9 70299
4	814 50625	849 34656	885 29281	922 36816	960 59601
5	(9)7.7378 09375	(9)8.1537 26976	(9)8.5873 40257	(9)9.0392 07968	(9)9.5099 00499
6	(11)7.3509 18906	(11)7.8275 77897	(11)8.3297 20049	(11)8.8584 23809	(11)9.4148 01494
7	(13)6.9833 72961	(13)7.5144 74781	(13)8.0798 28448	(13)8.6812 55332	(13)9.3206 53479
8	(15)6.6342 04313	(15)7.2138 95790	(15)7.8374 33594	(15)8.5076 30226	(15)9.2274 46944
9	(17)6.3024 94097	(17)6.9253 39958	(17)7.6023 10587	(17)8.3374 77621	(17)9.1351 72475
10	(19)5.9873 69392	(19)6.6483 26360	(19)7.3742 41269	(19)8.1707 28069	(19)9.0438 20750
24	(47)2.9198 90243	(47)3.7541 32467	(47)4.8141 72219	(47)6.1578 03365	(47)7.8567 81408
1/2	9.7467 94345	9.7979 58971	9.8488 57802	9.8994 94937	9.9498 74371
1/3	4.5629 02635	4.5788 56970	4.5947 00892	4.6104 36292	4.6260 65009
1/4	3.1219 85641	3.1301 69160	3.1382 88993	3.1463 46284	3.1543 42146
1/5	2.4862 49570	2.4914 61879	2.4966 30932	2.5017 57527	2.5068 42442

TABLE II.

POWERS AND ROOTS n^k

Table II

k	100	101	102	103	104
1	100	101	102	103	104
2	10000	10201	10404	10609	10816
3	10 00000	10 30301	10 61208	10 92727	11 24864
4	1000 00000	1040 60401	1082 43216	1125 50881	1169 85856
5	(10)1.0000 00000	(10)1.0510 10050	(10)1.1040 80803	(10)1.1592 74074	(10)1.2166 52902
6	(12)1.0000 00000	(12)1.0615 20151	(12)1.1261 62419	(12)1.1940 52297	(12)1.2653 19018
7	(14)1.0000 00000	(14)1.0721 35352	(14)1.1486 85668	(14)1.2298 73865	(14)1.3159 31779
8	(16)1.0000 00000	(16)1.0828 56706	(16)1.1716 59381	(16)1.2667 70081	(16)1.3685 69050
9	(18)1.0000 00000	(18)1.0936 85273	(18)1.1950 92569	(18)1.3047 73184	(18)1.4233 11812
10	(20)1.0000 00000	(20)1.1046 22125	(20)1.2189 94420	(20)1.3439 16379	(20)1.4802 44285
24	(48)1.0000 00000	(48)1.2697 34649	(48)1.6084 37249	(48)2.0327 94106	(48)2.5633 04165
1/2	(1)1.0000 00000	(1)1.0049 87562	(1)1.0099 50494	(1)1.0148 89157	(1)1.0198 03903
1/3	4.6415 88834	4.6570 09508	4.6723 28728	4.6875 48148	4.7026 69375
1/4	3.1622 77660	3.1701 53880	3.1779 71828	3.1857 32501	3.1934 36868
1/5	2.5118 86432	2.5168 90229	2.5218 54548	2.5267 80083	2.5316 67508

k	105	106	107	108	109
1	105	106	107	108	109
2	11025	11236	11449	11664	11881
3	11 57625	11 91016	12 25043	12 59712	12 95029
4	1215 50625	1262 47696	1310 79601	1360 48896	1411 58161
5	(10)1.2762 81563	(10)1.3382 25578	(10)1.4025 51731	(10)1.4693 28077	(10)1.5386 23955
6	(12)1.3400 95641	(12)1.4185 19112	(12)1.5007 30352	(12)1.5868 74323	(12)1.6771 00111
7	(14)1.4071 00423	(14)1.5036 30259	(14)1.6057 81476	(14)1.7138 24269	(14)1.8280 39121
8	(16)1.4774 55444	(16)1.5938 48075	(16)1.7181 86180	(16)1.8509 30210	(16)1.9925 62642
9	(18)1.5513 28216	(18)1.6894 78959	(18)1.8384 59212	(18)1.9990 04627	(18)2.1718 93279
10	(20)1.6288 94627	(20)1.7908 47697	(20)1.9671 51357	(20)2.1589 24997	(20)2.3673 63675
24	(48)3.2250 99944	(48)4.0489 34641	(48)5.0723 66953	(48)6.3411 80737	(48)7.9110 83175
1/2	(1)1.0246 95077	(1)1.0295 63014	(1)1.0344 08043	(1)1.0392 30485	(1)1.0440 30651
1/3	4.7176 93980	4.7326 23491	4.7474 59398	4.7622 03156	4.7768 56181
1/4	3.2010 85873	3.2086 80436	3.2162 21453	3.2237 09795	3.2311 46315
1/5	2.5365 17482	2.5413 30642	2.5461 07613	2.5508 49001	2.5555 55397

k	110	111	112	113	114
1	110	111	112	113	114
2	12100	12321	12544	12769	12996
3	13 31000	13 67631	14 04928	14 42897	14 81544
4	1464 10000	1518 07041	1573 51936	1630 47361	1688 96016
5	(10)1.6105 10000	(10)1.6850 58155	(10)1.7623 41683	(10)1.8424 35179	(10)1.9254 14582
6	(12)1.7715 61000	(12)1.8704 14552	(12)1.9738 22685	(12)2.0819 51753	(12)2.1949 72624
7	(14)1.9487 17100	(14)2.0761 60153	(14)2.2106 81407	(14)2.3526 05480	(14)2.5022 68791
8	(16)2.1435 88810	(16)2.3045 37770	(16)2.4759 63176	(16)2.6584 44193	(16)2.8525 86422
9	(18)2.3579 47691	(18)2.5580 36924	(18)2.7730 78757	(18)3.0040 41938	(18)3.2519 48521
10	(20)2.5937 42460	(20)2.8394 20986	(20)3.1058 48208	(20)3.3945 67390	(20)3.7072 21314
24	(48)9.8497 32676	(49)1.2239 15658	(49)1.5178 62893	(49)1.8788 09051	(49)2.3212 20685
1/2	(1)1.0488 08848	(1)1.0535 65375	(1)1.0583 00524	(1)1.0630 14581	(1)1.0677 07825
1/3	4.7914 19857	4.8058 95534	4.8202 84528	4.8345 88127	4.8488 07586
1/4	3.2385 31840	3.2458 67180	3.2531 53123	3.2603 90439	3.2675 79877
1/5	2.5602 27376	2.5648 65499	2.5694 70314	2.5740 42354	2.5785 82140

k	115	116	117	118	119
1	115	116	117	118	119
2	13225	13456	13689	13924	14161
3	15 20875	15 60896	16 01613	16 43032	16 85159
4	1749 00625	1810 63936	1873 88721	1938 77776	2005 33921
5	(10)2.0113 57188	(10)2.1003 41658	(10)2.1924 48036	(10)2.2877 57757	(10)2.3863 53660
6	(12)2.3130 60766	(12)2.4363 96323	(12)2.5651 64202	(12)2.6995 54153	(12)2.8397 60855
7	(14)2.6600 19880	(14)2.8262 19734	(14)3.0012 42116	(14)3.1854 73901	(14)3.3793 15418
8	(16)3.0590 22863	(16)3.2784 14892	(16)3.5114 53276	(16)3.7588 59203	(16)4.0213 85347
9	(18)3.5178 76292	(18)3.8029 61275	(18)4.1084 00333	(18)4.4354 53859	(18)4.7854 48563
10	(20)4.0455 57736	(20)4.4114 35079	(20)4.8068 28389	(20)5.2338 35554	(20)5.6946 83790
24	(49)2.8625 17619	(49)3.5236 41704	(49)4.3297 28675	(49)5.3109 00627	(49)6.5031 99444
1/2	(1)1.0723 80529	(1)1.0770 32961	(1)1.0816 65383	(1)1.0862 78049	(1)1.0908 71211
1/3	4.8629 44131	4.8769 98961	4.8909 73246	4.9048 68131	4.9186 84734
1/4	3.2747 22171	3.2818 18035	3.2888 68168	3.2958 73252	3.3028 33952
1/5	2.5830 90178	2.5875 66964	2.5920 12982	2.5964 28703	2.6008 14587

k	120	121	122	123	124
1	120	121	122	123	124
2	14400	14641	14884	15129	15376
3	17 28000	17 71561	18 15848	18 60867	19 06624
4	2073 60000	2143 58881	2215 33456	2288 86641	2364 21376
5	(10)2.4883 20000	(10)2.5937 42460	(10)2.7027 08163	(10)2.8153 05684	(10)2.9316 25062
6	(12)2.9859 84000	(12)3.1384 28377	(12)3.2973 03959	(12)3.4628 25992	(12)3.6352 15077
7	(14)3.5831 80800	(14)3.7974 98336	(14)4.0227 10830	(14)4.2592 75970	(14)4.5076 66696
8	(16)4.2998 16960	(16)4.5949 72986	(16)4.9077 07213	(16)5.2389 09443	(16)5.5895 06703
9	(18)5.1597 80352	(18)5.5599 17313	(18)5.9874 02800	(18)6.4438 58615	(18)6.9309 88312
10	(20)6.1917 36422	(20)6.7274 99949	(20)7.3046 31415	(20)7.9259 46096	(20)8.5944 25506
24	(49)7.9496 84720	(49)9.7017 23378	(50)1.1820 50242	(50)1.4378 80104	(50)1.7463 06393
1/2	(1)1.0954 45115	(1)1.1000 00000	(1)1.1045 36102	(1)1.1090 53651	(1)1.1135 52873
1/3	4.9324 24149	4.9460 87443	4.9596 75664	4.9731 89833	4.9866 30952
1/4	3.3097 50920	3.3166 24790	3.3234 56186	3.3302 45713	3.3369 93965
1/5	2.6051 71085	2.6094 98635	2.6137 97668	2.6180 68602	2.6223 11847

TABLE II.

Table II POWERS AND ROOTS n^k 167

k	125	126	127	128	129
1	125	126	127	128	129
2	15625	15876	16129	16384	16641
3	19 53125	20 00376	20 48383	20 97152	21 46689
4	2441 40625	2520 47376	2601 44641	2684 35456	2769 22881
5	(10)3.0517 57813	(10)3.1757 96938	(10)3.3038 36941	(10)3.4359 73837	(10)3.5723 05165
6	(12)3.8146 97266	(12)4.0015 04141	(12)4.1958 72915	(12)4.3980 46511	(12)4.6082 73663
7	(14)4.7683 71582	(14)5.0418 95218	(14)5.3287 58602	(14)5.6294 99534	(14)5.9446 73025
8	(16)5.9604 64478	(16)6.3527 87975	(16)6.7675 23424	(16)7.2057 59404	(16)7.6686 28202
9	(18)7.4505 80597	(18)8.0045 12848	(18)8.5947 54749	(18)9.2233 72037	(18)9.8925 30381
10	(20)9.3132 25746	(21)1.0085 68619	(21)1.0915 33853	(21)1.1805 91621	(21)1.2761 36419
24	(50)2.1175 82368	(50)2.5638 52774	(50)3.0994 83316	(50)3.7414 44192	(50)4.5097 56022
1/2	(1)1.1180 33989	(1)1.1224 97216	(1)1.1269 42767	(1)1.1313 70850	(1)1.1357 81669
1/3	5.0000 00000	5.0132 97935	5.0265 25695	5.0396 84200	5.0527 74347
1/4	3.3437 01525	3.3503 68959	3.3569 96823	3.3635 85661	3.3701 36005
1/5	2.6265 27804	2.6307 16865	2.6348 79413	2.6390 15822	2.6431 26458

k	130	131	132	133	134
1	130	131	132	133	134
2	16900	17161	17424	17689	17956
3	21 97000	22 48091	22 99968	23 52637	24 06104
4	2856 10000	2944 99921	3035 95776	3129 00721	3224 17936
5	(10)3.7129 30000	(10)3.8579 48965	(10)4.0074 64243	(10)4.1615 79589	(10)4.3204 00342
6	(12)4.8268 09000	(12)5.0539 13144	(12)5.2898 52801	(12)5.5349 00854	(12)5.7893 36459
7	(14)6.2748 51700	(14)6.6206 26219	(14)6.9826 05697	(14)7.3614 18136	(14)7.7577 10855
8	(16)8.1573 07210	(16)8.6730 20347	(16)9.2170 39521	(16)9.7906 86120	(17)1.0395 33255
9	(19)1.0604 49937	(19)1.1361 65665	(19)1.2166 49217	(19)1.3021 61254	(19)1.3929 74561
10	(21)1.3785 84918	(21)1.4883 77022	(21)1.6059 76966	(21)1.7318 74468	(21)1.8665 85912
24	(50)5.4280 07704	(50)6.5239 57088	(50)7.8302 26935	(50)9.3851 10346	(51)1.1233 50184
1/2	(1)1.1401 75425	(1)1.1445 52314	(1)1.1489 12529	(1)1.1532 56259	(1)1.1575 83690
1/3	5.0657 97019	5.0787 53078	5.0916 43370	5.1044 68722	5.1172 29947
1/4	3.3766 48375	3.3831 23282	3.3895 61224	3.3959 62690	3.4023 28159
1/5	2.6472 11681	2.6512 71840	2.6553 07280	2.6593 18337	2.6633 05339

k	135	136	137	138	139
1	135	136	137	138	139
2	18225	18496	18769	19044	19321
3	24 60375	25 15456	25 71353	26 28072	26 85619
4	3321 50625	3421 02016	3522 75361	3626 73936	3733 01041
5	(10)4.4840 33438	(10)4.6525 87418	(10)4.8261 72446	(10)5.0049 00317	(10)5.1888 84470
6	(12)6.0534 45141	(12)6.3275 18888	(12)6.6118 56251	(12)6.9067 62437	(12)7.2125 49413
7	(14)8.1721 50940	(14)8.6054 25688	(14)9.0582 43063	(14)9.5313 32163	(15)1.0025 44368
8	(17)1.1032 40377	(17)1.1703 37894	(17)1.2409 79300	(17)1.3153 23839	(17)1.3935 36672
9	(19)1.4893 74509	(19)1.5916 59535	(19)1.7001 41641	(19)1.8151 46897	(19)1.9370 15974
10	(21)2.0106 55587	(21)2.1646 56968	(21)2.3291 94048	(21)2.5049 02718	(21)2.6924 52204
24	(51)1.3427 97252	(51)1.6030 01028	(51)1.9111 44882	(51)2.2756 11258	(51)2.7061 70815
1/2	(1)1.1618 95004	(1)1.1661 90379	(1)1.1704 69991	(1)1.1747 34012	(1)1.1789 82612
1/3	5.1299 27840	5.1425 63181	5.1551 36735	5.1676 49252	5.1801 01467
1/4	3.4086 58099	3.4149 52970	3.4212 13222	3.4274 39296	3.4336 31623
1/5	2.6672 68608	2.6712 08461	2.6751 25206	2.6790 19145	2.6828 90577

k	140	141	142	143	144
1	140	141	142	143	144
2	19600	19881	20164	20449	20736
3	27 44000	28 03221	28 63288	29 24207	29 85984
4	3841 60000	3952 54161	4065 86896	4181 61601	4299 81696
5	(10)5.3782 40000	(10)5.5730 83670	(10)5.7735 33923	(10)5.9797 10894	(10)6.1917 36422
6	(12)7.5295 36000	(12)7.8580 47975	(12)8.1984 18171	(12)8.5509 86579	(12)8.9161 00448
7	(15)1.0541 35040	(15)1.1079 84764	(15)1.1641 75380	(15)1.2227 91081	(15)1.2839 18465
8	(17)1.4757 89056	(17)1.5622 58518	(17)1.6531 29040	(17)1.7485 91246	(17)1.8488 42589
9	(19)2.0661 04678	(19)2.2027 84510	(19)2.3474 43237	(19)2.5004 85481	(19)2.6623 33328
10	(21)2.8925 46550	(21)3.1059 26159	(21)3.3333 69396	(21)3.5757 94238	(21)3.8337 59992
24	(51)3.2141 99700	(51)3.8129 28871	(51)4.5177 29930	(51)5.3464 42484	(51)6.3197 48715
1/2	(1)1.1832 15957	(1)1.1874 34209	(1)1.1916 37529	(1)1.1958 26074	(1)1.2000 00000
1/3	5.1924 94102	5.2048 27863	5.2171 03446	5.2293 21532	5.2414 82788
1/4	3.4397 90628	3.4459 16727	3.4520 10326	3.4580 71824	3.4641 01615
1/5	2.6867 39790	2.6905 67070	2.6943 72696	2.6981 56943	2.7019 20077

k	145	146	147	148	149
1	145	146	147	148	149
2	21025	21316	21609	21904	22201
3	30 48625	31 12136	31 76523	32 41792	33 07949
4	4420 50625	4543 71856	4669 48881	4797 85216	4928 84401
5	(10)6.4097 34063	(10)6.6338 29098	(10)6.8641 48551	(10)7.1008 21197	(10)7.3439 77575
6	(12)9.2941 14391	(12)9.6853 90482	(13)1.0090 29837	(13)1.0509 21537	(13)1.0942 52659
7	(15)1.3476 46587	(15)1.4140 67010	(15)1.4832 73860	(15)1.5553 63875	(15)1.6304 36461
8	(17)1.9540 87551	(17)2.0645 37835	(17)2.1804 12575	(17)2.3019 38535	(17)2.4293 50327
9	(19)2.8334 26948	(19)3.0142 25239	(19)3.2052 06485	(19)3.4068 69032	(19)3.6197 31988
10	(21)4.1084 69075	(21)4.4007 68850	(21)4.7116 53533	(21)5.0421 66167	(21)5.3934 00662
24	(51)7.4616 01544	(51)8.7997 13625	(52)1.0366 11527	(52)1.2197 79049	(52)1.4337 40132
1/2	(1)1.2041 59458	(1)1.2083 04597	(1)1.2124 35565	(1)1.2165 52506	(1)1.2206 55562
1/3	5.2535 87872	5.2656 37428	5.2776 32088	5.2895 72473	5.3014 59192
1/4	3.4701 00082	3.4760 67602	3.4820 04545	3.4879 11275	3.4937 88147
1/5	2.7056 62363	2.7093 84058	2.7130 85417	2.7167 66686	2.7204 28110

TABLE II.

POWERS AND ROOTS n^k

Table II

k	150	151	152	153	154
1	150	151	152	153	154
2	22500	22801	23104	23409	23716
3	33 75000	34 42951	35 11808	35 81577	36 52264
4	5062 50000	5198 85601	5337 94816	5479 81281	5624 48656
5	(10)7.5937 50000	(10)7.8502 72575	(10)8.1136 81203	(10)8.3841 13599	(10)8.6617 09302
6	(13)1.1390 62500	(13)1.1853 91159	(13)1.2332 79543	(13)1.2827 69381	(13)1.3339 03233
7	(15)1.7085 93750	(15)1.7899 40650	(15)1.8745 84905	(15)1.9626 37152	(15)2.0542 10978
8	(17)2.5628 90625	(17)2.7028 10381	(17)2.8493 69056	(17)3.0028 34843	(17)3.1634 84906
9	(19)3.8443 35938	(19)4.0812 43676	(19)4.3310 40965	(19)4.5943 37310	(19)4.8717 66756
10	(21)5.7665 03906	(21)6.1626 77950	(21)6.5831 82267	(21)7.0293 36085	(21)7.5025 20804
24	(52)1.6834 11220	(52)1.9744 52704	(52)2.3133 75387	(52)2.7076 61312	(52)3.1659 00782
1/2	(1)1.2247 44871	(1)1.2288 20573	(1)1.2328 82801	(1)1.2369 31688	(1)1.2409 67365
1/3	5.3132 92846	5.3250 74022	5.3368 03297	5.3484 81241	5.3601 08411
1/4	3.4996 35512	3.5054 53712	3.5112 43086	3.5170 03963	3.5227 36670
1/5	2.7240 69927	2.7276 92374	2.7312 95679	2.7348 80069	2.7384 45765

k	155	156	157	158	159
1	155	156	157	158	159
2	24025	24336	24649	24964	25281
3	37 23875	37 96416	38 69893	39 44312	40 19679
4	5772 00625	5922 40896	6075 73201	6232 01296	6391 28961
5	(10)8.9466 09688	(10)9.2389 57978	(10)9.5388 99256	(10)9.8465 80477	(11)1.0162 15048
6	(13)1.3867 24502	(13)1.4412 77445	(13)1.4976 07183	(13)1.5557 59715	(13)1.6157 81926
7	(15)2.1494 22977	(15)2.2483 92813	(15)2.3512 43278	(15)2.4581 00350	(15)2.5690 93263
8	(17)3.3316 05615	(17)3.5074 92789	(17)3.6914 51946	(17)3.8837 98553	(17)4.0848 58288
9	(19)5.1639 88703	(19)5.4716 88751	(19)5.7955 79555	(19)6.1364 01714	(19)6.4949 24678
10	(21)8.0041 82490	(21)8.5358 34451	(21)9.0990 59901	(21)9.6955 14709	(22)1.0326 93024
24	(52)3.6979 47627	(52)4.3150 94990	(52)5.0302 74186	(52)5.8582 79483	(52)6.8160 22003
1/2	(1)1.2449 89960	(1)1.2489 99600	(1)1.2529 96409	(1)1.2569 80509	(1)1.2609 52021
1/3	5.3716 85355	5.3832 12612	5.3946 90712	5.4061 20176	5.4175 01515
1/4	3.5284 41525	3.5341 18843	3.5397 68931	3.5453 92093	3.5509 88625
1/5	2.7419 92987	2.7455 21947	2.7490 32856	2.7525 25920	2.7560 01343

k	160	161	162	163	164
1	160	161	162	163	164
2	25600	25921	26244	26569	26896
3	40 96000	41 73281	42 51528	43 30747	44 10944
4	6553 60000	6718 98241	6887 47536	7059 11761	7233 94816
5	(11)1.0485 76000	(11)1.0817 56168	(11)1.1157 71008	(11)1.1506 36170	(11)1.1863 67498
6	(13)1.6777 21600	(13)1.7416 27430	(13)1.8075 49033	(13)1.8755 36958	(13)1.9456 42697
7	(15)2.6843 54560	(15)2.8040 20163	(15)2.9282 29434	(15)3.0571 25241	(15)3.1908 54023
8	(17)4.2949 67296	(17)4.5144 72463	(17)4.7437 31683	(17)4.9831 14143	(17)5.2330 00598
9	(19)6.8719 47674	(19)7.2683 00665	(19)7.6848 45327	(19)8.1224 76053	(19)8.5821 20981
10	(22)1.0995 11628	(22)1.1701 96407	(22)1.2449 44943	(22)1.3239 63597	(22)1.4074 67841
24	(52)7.9228 16251	(52)9.2007 03274	(53)1.0674 81480	(53)1.2373 78329	(53)1.4330 20335
1/2	(1)1.2649 11064	(1)1.2688 57754	(1)1.2727 92206	(1)1.2767 14533	(1)1.2806 24847
1/3	5.4288 35233	5.4401 21825	5.4513 61778	5.4625 55571	5.4737 03675
1/4	3.5565 58820	3.5621 02966	3.5676 21345	3.5731 14235	3.5785 81908
1/5	2.7594 59323	2.7629 00056	2.7663 23734	2.7697 30547	2.7731 20681

k	165	166	167	168	169
1	165	166	167	168	169
2	27225	27556	27889	28224	28561
3	44 92125	45 74296	46 57463	47 41632	48 26809
4	7412 00625	7593 33136	7777 96321	7965 94176	8157 30721
5	(11)1.2229 81031	(11)1.2604 93006	(11)1.2989 19856	(11)1.3382 78216	(11)1.3785 84918
6	(13)2.0179 18702	(13)2.0924 18390	(13)2.1691 96160	(13)2.2483 07402	(13)2.3298 08512
7	(15)3.3295 65858	(15)3.4734 14527	(15)3.6225 57587	(15)3.7771 56436	(15)3.9373 76386
8	(17)5.4937 83665	(17)5.7658 68114	(17)6.0496 71170	(17)6.3456 22812	(17)6.6541 66092
9	(19)9.0647 43047	(19)9.5713 41070	(20)1.0102 95085	(20)1.0660 64632	(20)1.1245 54070
10	(22)1.4956 82603	(22)1.5888 42618	(22)1.6871 92792	(22)1.7909 88583	(22)1.9004 96377
24	(53)1.6581 15050	(53)1.9168 76411	(53)2.2140 90189	(53)2.5551 87425	(53)2.9463 26763
1/2	(1)1.2845 23258	(1)1.2884 09873	(1)1.2922 84798	(1)1.2961 48140	(1)1.3000 00000
1/3	5.4848 06552	5.4958 64660	5.5068 78446	5.5178 48353	5.5287 74814
1/4	3.5840 24634	3.5894 42676	3.5948 36294	3.6002 05744	3.6055 51275
1/5	2.7764 94317	2.7798 51635	2.7831 92813	2.7865 18023	2.7898 27436

k	170	171	172	173	174
1	170	171	172	173	174
2	28900	29241	29584	29929	30276
3	49 13000	50 00211	50 88448	51 77717	52 68024
4	8352 10000	8550 36081	8752 13056	8957 45041	9166 36176
5	(11)1.4198 57000	(11)1.4621 11699	(11)1.5053 66456	(11)1.5496 38921	(11)1.5949 46946
6	(13)2.4137 56900	(13)2.5002 11004	(13)2.5892 30305	(13)2.6808 75333	(13)2.7752 07686
7	(15)4.1033 86730	(15)4.2753 60818	(15)4.4534 76124	(15)4.6379 14326	(15)4.8288 61374
8	(17)6.9757 57441	(17)7.3108 66998	(17)7.6599 78934	(17)8.0235 91785	(17)8.4022 18792
9	(20)1.1858 78765	(20)1.2501 58257	(20)1.3175 16377	(20)1.3880 81379	(20)1.4619 86070
10	(22)2.0159 93900	(22)2.1377 70619	(22)2.2661 28168	(22)2.4013 80785	(22)2.5438 55761
24	(53)3.3944 86713	(53)3.9075 68945	(53)4.4945 13878	(53)5.1654 29935	(53)5.9317 37979
1/2	(1)1.3038 40481	(1)1.3076 69683	(1)1.3114 87705	(1)1.3152 94644	(1)1.3190 90596
1/3	5.5396 58257	5.5504 99103	5.5612 97766	5.5720 54656	5.5827 70172
1/4	3.6108 73137	3.6161 71571	3.6214 46817	3.6266 99110	3.6319 28683
1/5	2.7931 21220	2.7963 99540	2.7996 62559	2.8029 10436	2.8061 43329

TABLE II.

Table II   **POWERS AND ROOTS** n^k *169*

k	175	176	177	178	179
1	175	176	177	178	179
2	30625	30976	31329	31684	32041
3	53 59375	54 51776	55 45233	56 39752	57 35339
4	9378 90625	9595 12576	9815 06241	(9)1.0038 75856	(9)1.0266 25681
5	(11)1.6413 08594	(11)1.6887 42134	(11)1.7372 66047	(11)1.7868 99024	(11)1.8376 59969
6	(13)2.8722 90039	(13)2.9721 86155	(13)3.0749 60902	(13)3.1806 80262	(13)3.2894 11344
7	(15)5.0265 07568	(15)5.2310 47634	(15)5.4426 80797	(15)5.6616 10867	(15)5.8880 46307
8	(17)8.7963 88245	(17)9.2066 43835	(17)9.6335 45011	(18)1.0077 66734	(18)1.0539 60289
9	(20)1.5393 67943	(20)1.6203 69315	(20)1.7051 37467	(20)1.7938 24787	(20)1.8865 88917
10	(22)2.6938 93900	(22)2.8518 49994	(22)3.0180 93317	(22)3.1930 08121	(22)3.3769 94162
24	(53)6.8063 32613	(53)7.8037 62212	(53)8.9404 29702	(54)1.0234 81638	(54)1.1707 73122
1/2	(1)1.3228 75656	(1)1.3266 49916	(1)1.3304 13470	(1)1.3341 66406	(1)1.3379 08816
1/3	5.5934 44710	5.6040 78661	5.6146 72408	5.6252 26328	5.6357 40794
1/4	3.6371 35763	3.6423 20574	3.6474 83337	3.6526 24271	3.6577 43589
1/5	2.8093 61392	2.8125 64777	2.8157 53634	2.8189 28111	2.8220 88352

k	180	181	182	183	184
1	180	181	182	183	184
2	32400	32761	33124	33489	33856
3	58 32000	59 29741	60 28568	61 28487	62 29504
4	(9)1.0497 60000	(9)1.0732 83121	(9)1.0971 99376	(9)1.1215 13121	(9)1.1462 28736
5	(11)1.8895 68000	(11)1.9426 42449	(11)1.9969 02864	(11)2.0523 69011	(11)2.1090 60874
6	(13)3.4012 22400	(13)3.5161 82833	(13)3.6343 63213	(13)3.7558 35291	(13)3.8806 72009
7	(15)6.1222 00320	(15)6.3642 90927	(15)6.6145 41048	(15)6.8731 78582	(15)7.1404 36496
8	(18)1.1019 96058	(18)1.1519 36658	(18)1.2038 46471	(18)1.2577 91681	(18)1.3138 40315
9	(20)1.9835 92904	(20)2.0850 05351	(20)2.1910 00577	(20)2.3017 58775	(20)2.4174 66180
10	(22)3.5704 67227	(22)3.7738 59685	(22)3.9876 21050	(22)4.2122 18559	(22)4.4481 37771
24	(54)1.3382 58845	(54)1.5285 71637	(54)1.7446 70074	(54)1.9898 76639	(54)2.2679 20111
1/2	(1)1.3416 40786	(1)1.3453 62405	(1)1.3490 73756	(1)1.3527 74926	(1)1.3564 65997
1/3	5.6462 16173	5.6566 52826	5.6670 51108	5.6774 11371	5.6877 33960
1/4	3.6628 41501	3.6679 18217	3.6729 73940	3.6780 08871	3.6830 23210
1/5	2.8252 34501	2.8283 66697	2.8314 85080	2.8345 89786	2.8376 80950

k	185	186	187	188	189
1	185	186	187	188	189
2	34225	34596	34969	35344	35721
3	63 31625	64 34856	65 39203	66 44672	67 51269
4	(9)1.1713 50625	(9)1.1968 83216	(9)1.2228 30961	(9)1.2491 98336	(9)1.2759 89841
5	(11)2.1669 98656	(11)2.2262 02782	(11)2.2866 93897	(11)2.3484 92872	(11)2.4116 20799
6	(13)4.0089 47514	(13)4.1407 37174	(13)4.2761 17588	(13)4.4151 66599	(13)4.5579 63311
7	(15)7.4165 52901	(15)7.7017 71144	(15)7.9963 39889	(15)8.3005 13206	(15)8.6145 50658
8	(18)1.3720 62287	(18)1.4325 29433	(18)1.4953 15559	(18)1.5604 96483	(18)1.6281 50074
9	(20)2.5383 15230	(20)2.6645 04745	(20)2.7962 40096	(20)2.9337 33387	(20)3.0772 03640
10	(22)4.6958 83176	(22)4.9559 78826	(22)5.2289 68979	(22)5.5154 18768	(22)5.8159 14881
24	(54)2.5829 82606	(54)2.9397 51775	(54)3.3434 78670	(54)3.8000 41874	(54)4.3160 18526
1/2	(1)1.3601 47051	(1)1.3638 18170	(1)1.3674 79433	(1)1.3711 30920	(1)1.3747 72708
1/3	5.6980 19215	5.7082 67473	5.7184 79065	5.7286 54316	5.7387 93548
1/4	3.6880 17151	3.6929 90888	3.6979 44609	3.7028 78502	3.7077 92751
1/5	2.8407 58702	2.8438 23174	2.8468 74493	2.8499 12786	2.8529 38178

k	190	191	192	193	194
1	190	191	192	193	194
2	36100	36481	36864	37249	37636
3	68 59000	69 67871	70 77888	71 89057	73 01384
4	(9)1.3032 10000	(9)1.3308 63361	(9)1.3589 54496	(9)1.3874 88001	(9)1.4164 68496
5	(11)2.4760 99000	(11)2.5419 49020	(11)2.6091 92632	(11)2.6778 51842	(11)2.7479 48882
6	(13)4.7045 88100	(13)4.8551 22627	(13)5.0096 49854	(13)5.1682 54055	(13)5.3310 20832
7	(15)8.9387 17390	(15)9.2732 84218	(15)9.6185 27720	(15)9.9747 30326	(16)1.0342 18041
8	(18)1.6983 56304	(18)1.7711 97286	(18)1.8467 57322	(18)1.9251 22953	(18)2.0063 83000
9	(20)3.2268 76978	(20)3.3829 86816	(20)3.5457 74059	(20)3.7154 87299	(20)3.8923 83020
10	(22)6.1310 66258	(22)6.4615 04818	(22)6.8078 86193	(22)7.1708 90487	(22)7.5512 23059
24	(54)4.8987 62931	(54)5.5564 93542	(54)6.2983 89130	(54)7.1346 95065	(54)8.0768 40718
1/2	(1)1.3784 04875	(1)1.3820 27496	(1)1.3856 40646	(1)1.3892 44399	(1)1.3928 38828
1/3	5.7488 97079	5.7589 65220	5.7689 98281	5.7789 96565	5.7889 60372
1/4	3.7126 87538	3.7175 63041	3.7224 19436	3.7272 56899	3.7320 75599
1/5	2.8559 50791	2.8589 50746	2.8619 38162	2.8649 13156	2.8678 75844

k	195	196	197	198	199
1	195	196	197	198	199
2	38025	38416	38809	39204	39601
3	74 14875	75 29536	76 45373	77 62392	78 80599
4	(9)1.4459 00625	(9)1.4757 89056	(9)1.5061 38481	(9)1.5369 53616	(9)1.5682 39201
5	(11)2.8195 06219	(11)2.8925 46550	(11)2.9670 92808	(11)3.0431 68160	(11)3.1207 96010
6	(13)5.4980 37127	(13)5.6693 91238	(13)5.8451 72831	(13)6.0254 72956	(13)6.2103 84060
7	(16)1.0721 17240	(16)1.1112 00683	(16)1.1514 99048	(16)1.1930 43645	(16)1.2358 66428
8	(18)2.0906 28617	(18)2.1779 53338	(18)2.2684 53124	(18)2.3622 26418	(18)2.4593 74192
9	(20)4.0767 25804	(20)4.2687 88542	(20)4.4688 52654	(20)4.6772 08307	(20)4.8941 54641
10	(22)7.9496 15318	(22)8.3668 25543	(22)8.8036 39729	(22)9.2608 72448	(22)9.7393 67736
24	(54)9.1375 69069	(55)1.0331 07971	(55)1.1673 18660	(55)1.3181 49187	(55)1.4875 57746
1/2	(1)1.3964 24004	(1)1.4000 00000	(1)1.4035 66885	(1)1.4071 24728	(1)1.4106 73598
1/3	5.7988 89998	5.8087 85734	5.8186 47867	5.8284 76683	5.8382 72461
1/4	3.7368 75706	3.7416 57387	3.7464 20805	3.7511 66123	3.7558 93499
1/5	2.8708 26340	2.8737 64756	2.8766 91203	2.8796 05790	2.8825 08624

TABLE II.

POWERS AND ROOTS n^k

Table II

k	200	201	202	203	204
1	200	201	202	203	204
2	40000	40401	40804	41209	41616
3	80 00000	81 20601	82 42408	83 65427	84 89664
4	(9)1.6000 00000	(9)1.6322 40801	(9)1.6649 66416	(9)1.6981 81681	(9)1.7318 91456
5	(11)3.2000 00000	(11)3.2808 04010	(11)3.3632 32160	(11)3.4473 08812	(11)3.5330 58570
6	(13)6.4000 00000	(13)6.5944 16060	(13)6.7937 28964	(13)6.9980 36889	(13)7.2074 39483
7	(16)1.2800 00000	(16)1.3254 77628	(16)1.3723 33251	(16)1.4206 01489	(16)1.4703 17655
8	(18)2.5600 00000	(18)2.6642 10032	(18)2.7721 13166	(18)2.8838 21022	(18)2.9994 48015
9	(20)5.1200 00000	(20)5.3550 62165	(20)5.5996 68596	(20)5.8541 56674	(20)6.1188 73951
10	(23)1.0240 00000	(23)1.0763 67495	(23)1.1311 33056	(23)1.1883 93805	(23)1.2482 50286
24	(55)1.6777 21600	(55)1.8910 60303	(55)2.1302 61246	(55)2.3983 07745	(55)2.6985 09916
1/2	(1)1.4142 13562	(1)1.4177 44688	(1)1.4212 67040	(1)1.4247 80685	(1)1.4282 85686
1/3	5.8480 35476	5.8577 66003	5.8674 64308	5.8771 30659	5.8867 65317
1/4	3.7606 03093	3.7652 95059	3.7699 69549	3.7746 26716	3.7792 66709
1/5	2.8853 99812	2.8882 79458	2.8911 47666	2.8940 04537	2.8968 50171

k	205	206	207	208	209
1	205	206	207	208	209
2	42025	42436	42849	43264	43681
3	86 15125	87 41816	88 69743	89 98912	91 29329
4	(9)1.7661 00625	(9)1.8008 14096	(9)1.8360 36801	(9)1.8717 73696	(9)1.9080 29761
5	(11)3.6205 06281	(11)3.7096 77038	(11)3.8005 96178	(11)3.8932 89288	(11)3.9877 82200
6	(13)7.4220 37877	(13)7.6419 34698	(13)7.8672 34089	(13)8.0980 41718	(13)8.3344 64799
7	(16)1.5215 17765	(16)1.5742 38548	(16)1.6285 17456	(16)1.6843 92677	(16)1.7419 03143
8	(18)3.1191 11418	(18)3.2429 31408	(18)3.3710 31135	(18)3.5035 36769	(18)3.6405 77569
9	(20)6.3941 78406	(20)6.6804 38701	(20)6.9780 34449	(20)7.2873 56480	(20)7.6088 07119
10	(23)1.3108 06573	(23)1.3761 70372	(23)1.4444 53131	(23)1.5157 70148	(23)1.5902 40688
24	(55)3.0345 38594	(55)3.4104 62581	(55)3.8307 89523	(55)4.3005 10765	(55)4.8251 50531
1/2	(1)1.4317 82106	(1)1.4352 70009	(1)1.4387 49457	(1)1.4422 20510	(1)1.4456 83229
1/3	5.8963 68540	5.9059 40584	5.9154 81700	5.9249 92137	5.9344 72140
1/4	3.7838 89674	3.7884 95756	3.7930 85099	3.7976 57844	3.8022 14131
1/5	2.8996 84668	2.9025 08125	2.9053 20638	2.9081 22302	2.9109 13212

k	210	211	212	213	214
1	210	211	212	213	214
2	44100	44521	44944	45369	45796
3	92 61000	93 93931	95 28128	96 63597	98 00344
4	(9)1.9448 10000	(9)1.9821 19441	(9)2.0199 63136	(9)2.0583 46161	(9)2.0972 73616
5	(11)4.0841 01000	(11)4.1822 72021	(11)4.2823 21848	(11)4.3842 77323	(11)4.4881 65538
6	(13)8.5766 12100	(13)8.8245 93963	(13)9.0785 22318	(13)9.3385 10698	(13)9.6046 74252
7	(16)1.8010 88541	(16)1.8619 89326	(16)1.9246 46732	(16)1.9891 02779	(16)2.0554 00290
8	(18)3.7822 85936	(18)3.9287 97478	(18)4.0802 51071	(18)4.2367 88919	(18)4.3985 56620
9	(20)7.9428 00466	(20)8.2897 62679	(20)8.6501 32270	(20)9.0243 60396	(20)9.4129 11168
10	(23)1.6679 88098	(23)1.7491 39925	(23)1.8338 28041	(23)1.9221 88764	(23)2.0143 62990
24	(55)5.4108 19838	(55)6.0642 75557	(55)6.7929 85105	(55)7.6051 97251	(55)8.5100 19601
1/2	(1)1.4491 37675	(1)1.4525 83905	(1)1.4560 21978	(1)1.4594 51952	(1)1.4628 73884
1/3	5.9439 21953	5.9533 41813	5.9627 31958	5.9720 92620	5.9814 24030
1/4	3.8067 54096	3.8112 77876	3.8157 85604	3.8202 77414	3.8247 53435
1/5	2.9136 93459	2.9164 63134	2.9192 22328	2.9219 71130	2.9247 09627

k	215	216	217	218	219
1	215	216	217	218	219
2	46225	46656	47089	47524	47961
3	99 38375	100 77696	102 18313	103 60232	105 03459
4	(9)2.1367 50625	(9)2.1767 82336	(9)2.2173 73921	(9)2.2585 30576	(9)2.3002 57521
5	(11)4.5940 13844	(11)4.7018 49846	(11)4.8117 01409	(11)4.9235 96656	(11)5.0375 63971
6	(13)9.8771 29764	(14)1.0155 99567	(14)1.0441 39206	(14)1.0733 44071	(14)1.1032 26510
7	(16)2.1235 82899	(16)2.1936 95064	(16)2.2657 82076	(16)2.3398 90075	(16)2.4160 66056
8	(18)4.5657 03233	(18)4.7383 81338	(18)4.9167 47106	(18)5.1009 60363	(18)5.2911 84663
9	(20)9.8162 61952	(21)1.0234 90369	(21)1.0669 34122	(21)1.1120 09359	(21)1.1587 69441
10	(23)2.1104 96320	(23)2.2107 39197	(23)2.3152 47045	(23)2.4241 80403	(23)2.5377 05076
24	(55)9.5175 03342	(56)1.0638 73589	(56)1.1885 94216	(56)1.3272 59512	(56)1.4813 53665
1/2	(1)1.4662 87830	(1)1.4696 93846	(1)1.4730 91986	(1)1.4764 82306	(1)1.4798 64859
1/3	5.9907 26415	6.0000 00000	6.0092 45007	6.0184 61655	6.0276 50160
1/4	3.8292 13796	3.8336 58625	3.8380 88048	3.8425 02187	3.8469 01167
1/5	2.9274 37906	2.9301 56052	2.9328 64149	2.9355 62280	2.9382 50529

k	220	221	222	223	224
1	220	221	222	223	224
2	48400	48841	49284	49729	50176
3	106 48000	107 93861	109 41048	110 89567	112 39424
4	(9)2.3425 60000	(9)2.3854 43281	(9)2.4289 12656	(9)2.4729 73441	(9)2.5176 30976
5	(11)5.1536 32000	(11)5.2718 29651	(11)5.3921 86096	(11)5.5147 30773	(11)5.6394 93386
6	(14)1.1337 99040	(14)1.1650 74353	(14)1.1970 65313	(14)1.2297 84962	(14)1.2632 46519
7	(16)2.4943 57888	(16)2.5748 14320	(16)2.6574 84996	(16)2.7424 20466	(16)2.8296 72201
8	(18)5.4875 87354	(18)5.6903 39647	(18)5.8996 16690	(18)6.1155 97640	(18)6.3384 65731
9	(21)1.2072 69218	(21)1.2575 65062	(21)1.3097 14905	(21)1.3637 78274	(21)1.4198 16324
10	(23)2.6559 92279	(23)2.7792 18787	(23)2.9075 67090	(23)3.0412 25550	(23)3.1803 88565
24	(56)1.6525 10926	(56)1.8425 30003	(56)2.0533 89736	(56)2.2872 66205	(56)2.5465 51362
1/2	(1)1.4832 39697	(1)1.4866 06875	(1)1.4899 66443	(1)1.4933 18452	(1)1.4966 62955
1/3	6.0368 10737	6.0459 43596	6.0550 48947	6.0641 26994	6.0731 77944
1/4	3.8512 85107	3.8556 54127	3.8600 08345	3.8643 47878	3.8686 72841
1/5	2.9409 28975	2.9435 97699	2.9462 56780	2.9489 06295	2.9515 46323

TABLE II.

k	225	226	227	228	229
1	225	226	227	228	229
2	50625	51076	51529	51984	52441
3	113 90625	115 43176	116 97083	118 52352	120 08989
4	(9)2.5628 90625	(9)2.6087 57776	(9)2.6552 37841	(9)2.7023 36256	(9)2.7500 58481
5	(11)5.7665 03906	(11)5.8957 92574	(11)6.0273 89899	(11)6.1613 26664	(11)6.2976 33921
6	(14)1.2974 63379	(14)1.3324 49122	(14)1.3682 17507	(14)1.4047 82479	(14)1.4421 58168
7	(16)2.9192 92603	(16)3.0113 35015	(16)3.1058 53741	(16)3.2029 04053	(16)3.3025 42205
8	(18)6.5684 08356	(18)6.8056 17134	(18)7.0502 87992	(18)7.3026 21240	(18)7.5628 21649
9	(21)1.4778 91880	(21)1.5380 69472	(21)1.6004 15374	(21)1.6649 97643	(21)1.7318 86158
10	(23)3.3252 56730	(23)3.4760 37007	(23)3.6329 42900	(23)3.7961 94626	(23)3.9660 19301
24	(56)2.8338 73334	(56)3.1521 18526	(56)3.5044 55686	(56)3.8943 62082	(56)4.3256 51988
1/2	(1)1.5000 00000	(1)1.5033 29638	(1)1.5066 51917	(1)1.5099 66887	(1)1.5132 74595
1/3	6.0822 01996	6.0911 99349	6.1001 70200	6.1091 14744	6.1180 33173
1/4	3.8729 83346	3.8772 79507	3.8815 61435	3.8858 29238	3.8900 83026
1/5	2.9541 76939	2.9567 98218	2.9594 10235	2.9620 13062	2.9646 06773

k	230	231	232	233	234
1	230	231	232	233	234
2	52900	53361	53824	54289	54756
3	121 67000	123 26391	124 87168	126 49337	128 12904
4	(9)2.7984 10000	(9)2.8473 96321	(9)2.8970 22976	(9)2.9472 95521	(9)2.9982 19536
5	(11)6.4363 43000	(11)6.5774 85502	(11)6.7210 93304	(11)6.8671 98564	(11)7.0158 33714
6	(14)1.4803 58890	(14)1.5193 99151	(14)1.5592 93647	(14)1.6000 57265	(14)1.6417 05089
7	(16)3.4048 25447	(16)3.5098 12038	(16)3.6175 61260	(16)3.7281 33428	(16)3.8415 89909
8	(18)7.8310 98528	(18)8.1076 65809	(18)8.3927 42123	(18)8.6865 50888	(18)8.9893 20386
9	(21)1.8011 52661	(21)1.8728 70802	(21)1.9471 16173	(21)2.0239 66357	(21)2.1035 00970
10	(23)4.1426 51121	(23)4.3263 31552	(23)4.5173 09521	(23)4.7158 41612	(23)4.9221 92271
24	(56)4.8025 07640	(56)5.3295 12896	(56)5.9116 89798	(56)6.5545 38287	(56)7.2640 79321
1/2	(1)1.5165 75089	(1)1.5198 68415	(1)1.5231 54621	(1)1.5264 33752	(1)1.5297 05854
1/3	6.1269 25675	6.1357 92440	6.1446 33651	6.1534 49494	6.1622 40148
1/4	3.8943 22905	3.8985 48980	3.9027 61357	3.9069 60138	3.9111 45426
1/5	2.9671 91438	2.9697 67129	2.9723 33915	2.9748 91866	2.9774 41049

k	235	236	237	238	239
1	235	236	237	238	239
2	55225	55696	56169	56644	57121
3	129 77875	131 44256	133 12053	134 81272	136 51919
4	(9)3.0498 00625	(9)3.1020 44416	(9)3.1549 56561	(9)3.2085 42736	(9)3.2628 08641
5	(11)7.1670 31469	(11)7.3208 24822	(11)7.4772 47050	(11)7.6363 31712	(11)7.7981 12652
6	(14)1.6842 52395	(14)1.7277 14658	(14)1.7721 07551	(14)1.8174 46947	(14)1.8637 48924
7	(16)3.9579 93129	(16)4.0774 06593	(16)4.1998 94895	(16)4.3255 23735	(16)4.4543 59928
8	(18)9.3012 83852	(18)9.6226 79559	(18)9.9537 50002	(19)1.0294 74649	(19)1.0645 92023
9	(21)2.1858 01705	(21)2.2709 52376	(21)2.3590 38964	(21)2.4501 49664	(21)2.5443 74934
10	(23)5.1366 34007	(23)5.3594 47607	(23)5.5909 22344	(23)5.8313 56201	(23)6.0810 56093
24	(56)8.0469 01671	(56)8.9102 12697	(56)9.8618 93410	(57)1.0910 55818	(57)1.2065 61943
1/2	(1)1.5329 70972	(1)1.5362 29150	(1)1.5394 80432	(1)1.5427 24862	(1)1.5459 62483
1/3	6.1710 05793	6.1797 46606	6.1884 62762	6.1971 54435	6.2058 21795
1/4	3.9153 17320	3.9194 75921	3.9236 21327	3.9277 53635	3.9318 72942
1/5	2.9799 81531	2.9825 13380	2.9850 36660	2.9875 51438	2.9900 57776

k	240	241	242	243	244
1	240	241	242	243	244
2	57600	58081	58564	59049	59536
3	138 24000	139 97521	141 72488	143 48907	145 26784
4	(9)3.3177 60000	(9)3.3734 02561	(9)3.4297 42096	(9)3.4867 84401	(9)3.5445 35296
5	(11)7.9626 24000	(11)8.1299 00172	(11)8.2999 75872	(11)8.4728 86094	(11)8.6486 66122
6	(14)1.9110 29760	(14)1.9593 05941	(14)2.0085 94161	(14)2.0589 11321	(14)2.1102 74534
7	(16)4.5864 71424	(16)4.7219 27319	(16)4.8607 97870	(16)5.0031 54510	(16)5.1490 69863
8	(19)1.1007 53142	(19)1.1379 84484	(19)1.1763 13085	(19)1.2157 66546	(19)1.2563 73046
9	(21)2.6418 07540	(21)2.7425 42606	(21)2.8466 77665	(21)2.9543 12707	(21)3.0655 50233
10	(23)6.3403 38097	(23)6.6095 27681	(23)6.8889 59948	(23)7.1789 79877	(23)7.4799 42569
24	(57)1.3337 35777	(57)1.4736 99791	(57)1.6276 79087	(57)1.7970 10300	(57)1.9831 51223
1/2	(1)1.5491 93338	(1)1.5524 17470	(1)1.5556 34919	(1)1.5588 45727	(1)1.5620 49935
1/3	6.2144 65012	6.2230 84253	6.2316 79684	6.2402 51469	6.2487 99770
1/4	3.9359 79343	3.9400 72930	3.9441 53798	3.9482 22039	3.9522 77742
1/5	2.9925 55740	2.9950 45390	2.9975 26790	3.0000 00000	3.0024 65081

k	245	246	247	248	249
1	245	246	247	248	249
2	60025	60516	61009	61504	62001
3	147 06125	148 86936	150 69223	152 52992	154 38249
4	(9)3.6030 00625	(9)3.6621 86256	(9)3.7220 98081	(9)3.7827 42016	(9)3.8441 24001
5	(11)8.8273 51531	(11)9.0089 78190	(11)9.1935 82260	(11)9.3812 00200	(11)9.5718 68762
6	(14)2.1627 01125	(14)2.2162 08635	(14)2.2708 14818	(14)2.3265 37650	(14)2.3833 95322
7	(16)5.2986 17757	(16)5.4518 73241	(16)5.6089 12601	(16)5.7698 13371	(16)5.9346 54351
8	(19)1.2981 61350	(19)1.3411 60817	(19)1.3854 01412	(19)1.4309 13716	(19)1.4777 28934
9	(21)3.1804 95308	(21)3.2992 55611	(21)3.4219 41489	(21)3.5486 66016	(21)3.6795 45044
10	(23)7.7922 13506	(23)8.1161 68802	(23)8.4521 95477	(23)8.8006 91719	(23)9.1620 67161
24	(57)2.1876 91225	(57)2.4123 62509	(57)2.6590 52293	(57)2.9298 15956	(57)3.2268 91257
1/2	(1)1.5652 47584	(1)1.5684 38714	(1)1.5716 23365	(1)1.5748 01575	(1)1.5779 73384
1/3	6.2573 24746	6.2658 26556	6.2743 05357	6.2827 61305	6.2911 94552
1/4	3.9563 20998	3.9603 51896	3.9643 70523	3.9683 76966	3.9723 71312
1/5	3.0049 22094	3.0073 71096	3.0098 12147	3.0122 45305	3.0146 70627

TABLE II.

POWERS AND ROOTS n^k

Table II

k	250	251	252	253	254
1	250	251	252	253	254
2	62500	63001	63504	64009	64516
3	156 25000	158 13251	160 03008	161 94277	163 87064
4	(9)3.9062 50000	(9)3.9691 26001	(9)4.0327 58016	(9)4.0971 52081	(9)4.1623 14256
5	(11)9.7656 25000	(11)9.9625 06263	(12)1.0162 55020	(12)1.0365 79476	(12)1.0572 27821
6	(14)2.4414 06250	(14)2.5005 89072	(14)2.5609 62650	(14)2.6225 46076	(14)2.6853 58665
7	(16)6.1035 15625	(16)6.2764 78570	(16)6.4536 25879	(16)6.6350 41571	(16)6.8208 11010
8	(19)1.5258 78906	(19)1.5753 96121	(19)1.6263 13722	(19)1.6786 65517	(19)1.7324 85997
9	(21)3.8146 97266	(21)3.9542 44264	(21)4.0983 10578	(21)4.2470 23759	(21)4.4005 14431
10	(23)9.5367 43164	(23)9.9251 53103	(24)1.0327 74266	(24)1.0744 97011	(24)1.1177 30666
24	(57)3.5527 13679	(57)3.9099 33001	(57)4.3014 31179	(57)4.7303 41643	(57)5.2000 70108
1/2	(1)1.5811 38830	(1)1.5842 97952	(1)1.5874 50787	(1)1.5905 97372	(1)1.5937 37745
1/3	6.2996 05249	6.3079 93549	6.3163 59598	6.3247 03543	6.3330 25531
1/4	3.9763 53644	3.9803 24047	3.9842 82604	3.9882 29397	3.9921 64507
1/5	3.0170 88168	3.0194 97986	3.0219 00136	3.0242 94671	3.0266 81647

k	255	256	257	258	259
1	255	256	257	258	259
2	65025	65536	66049	66564	67081
3	165 81375	167 77216	169 74593	171 73512	173 73979
4	(9)4.2282 50625	(9)4.2949 67296	(9)4.3624 70401	(9)4.4307 66096	(9)4.4998 60561
5	(12)1.0782 03909	(12)1.0995 11628	(12)1.1211 54893	(12)1.1431 37653	(12)1.1654 63886
6	(14)2.7494 19969	(14)2.8147 49767	(14)2.8813 68075	(14)2.9492 95144	(14)3.0185 51463
7	(16)7.0110 20921	(16)7.2057 59404	(16)7.4051 15953	(16)7.6091 81472	(16)7.8180 48289
8	(19)1.7878 10335	(19)1.8446 74407	(19)1.9031 14800	(19)1.9631 68820	(19)2.0248 74507
9	(21)4.5589 16354	(21)4.7223 66483	(21)4.8910 05036	(21)5.0649 75555	(21)5.2444 24973
10	(24)1.1625 23670	(24)1.2089 25820	(24)1.2569 88294	(24)1.3067 63693	(24)1.3583 06068
24	(57)5.7143 17018	(57)6.2771 01735	(57)6.8927 88615	(57)7.5661 15089	(57)8.3022 21920
1/2	(1)1.5968 71942	(1)1.6000 00000	(1)1.6031 21954	(1)1.6062 37840	(1)1.6093 47694
1/3	6.3413 25705	6.3496 04208	6.3578 61180	6.3660 96760	6.3743 11088
1/4	3.9960 88015	4.0000 00000	4.0039 00541	4.0077 89716	4.0116 67601
1/5	3.0290 61117	3.0314 33133	3.0337 97748	3.0361 55014	3.0385 04982

k	260	261	262	263	264
1	260	261	262	263	264
2	67600	68121	68644	69169	69696
3	175 76000	177 79581	179 84728	181 91447	183 99744
4	(9)4.5697 60000	(9)4.6404 70641	(9)4.7119 98736	(9)4.7843 50561	(9)4.8575 32416
5	(12)1.1881 37600	(12)1.2111 62837	(12)1.2345 43669	(12)1.2582 84198	(12)1.2823 88558
6	(14)3.0891 57760	(14)3.1611 35005	(14)3.2345 04412	(14)3.3092 87440	(14)3.3855 05793
7	(16)8.0318 10176	(16)8.2505 62364	(16)8.4744 01560	(16)8.7034 25966	(16)8.9377 35293
8	(19)2.0882 70646	(19)2.1533 96777	(19)2.2202 93209	(19)2.2890 01029	(19)2.3595 62117
9	(21)5.4295 03679	(21)5.6203 65588	(21)5.8171 68207	(21)6.0200 72706	(21)6.2292 43990
10	(24)1.4116 70957	(24)1.4669 15418	(24)1.5240 98070	(24)1.5832 79122	(24)1.6445 20413
24	(57)9.1066 85770	(57)9.9855 54265	(58)1.0945 38372	(58)1.1993 27974	(58)1.3136 94086
1/2	(1)1.6124 51550	(1)1.6155 49442	(1)1.6186 41406	(1)1.6217 27474	(1)1.6248 07681
1/3	6.3825 04299	6.3906 76528	6.3988 27910	6.4069 58577	6.4150 68660
1/4	4.0155 34273	4.0193 89807	4.0232 34278	4.0270 67760	4.0308 90325
1/5	3.0408 47703	3.0431 83226	3.0455 11602	3.0478 32879	3.0501 47105

k	265	266	267	268	269
1	265	266	267	268	269
2	70225	70756	71289	71824	72361
3	186 09625	188 21096	190 34163	192 48832	194 65109
4	(9)4.9315 50625	(9)5.0064 11536	(9)5.0821 21521	(9)5.1586 86976	(9)5.2361 14321
5	(12)1.3068 60916	(12)1.3317 05469	(12)1.3569 26446	(12)1.3825 28110	(12)1.4085 14752
6	(14)3.4631 81426	(14)3.5423 36546	(14)3.6229 93611	(14)3.7051 75334	(14)3.7889 04684
7	(16)9.1774 30780	(16)9.4226 15213	(16)9.6733 92942	(16)9.9298 69894	(17)1.0192 15360
8	(19)2.4320 19157	(19)2.5064 15647	(19)2.5827 95915	(19)2.6612 05132	(19)2.7416 89318
9	(21)6.4448 50765	(21)6.6670 65620	(21)6.8960 65094	(21)7.1320 29753	(21)7.3751 44266
10	(24)1.7078 85453	(24)1.7734 39455	(24)1.8412 49380	(24)1.9113 83974	(24)1.9839 13808
24	(58)1.4384 70548	(58)1.5745 60235	(58)1.7229 40472	(58)1.8846 68868	(58)2.0608 89564
1/2	(1)1.6278 82060	(1)1.6309 50643	(1)1.6340 13464	(1)1.6370 70554	(1)1.6401 21947
1/3	6.4231 58289	6.4312 27591	6.4392 76696	6.4473 05727	6.4553 14811
1/4	4.0347 02045	4.0385 02994	4.0422 93240	4.0460 72854	4.0498 41906
1/5	3.0524 54329	3.0547 54599	3.0570 47961	3.0593 34462	3.0616 14147

k	270	271	272	273	274
1	270	271	272	273	274
2	72900	73441	73984	74529	75076
3	196 83000	199 02511	201 23648	203 46417	205 70824
4	(9)5.3144 10000	(9)5.3935 80481	(9)5.4736 32256	(9)5.5545 71841	(9)5.6364 05776
5	(12)1.4348 90700	(12)1.4616 60310	(12)1.4888 27974	(12)1.5163 98113	(12)1.5443 75183
6	(14)3.8742 04890	(14)3.9610 99441	(14)4.0496 12088	(14)4.1397 66847	(14)4.2315 88000
7	(17)1.0460 35320	(17)1.0734 57949	(17)1.1014 94488	(17)1.1301 56349	(17)1.1594 55112
8	(19)2.8242 95365	(19)2.9090 71041	(19)2.9960 65007	(19)3.0853 26834	(19)3.1769 07007
9	(21)7.6255 97485	(21)7.8835 82520	(21)8.1492 96820	(21)8.4229 42256	(21)8.7047 25200
10	(24)2.0589 11321	(24)2.1364 50863	(24)2.2166 08735	(24)2.2994 63236	(24)2.3850 94705
24	(58)2.2528 39954	(58)2.4618 57897	(58)2.6893 89450	(58)2.9369 97176	(58)3.2063 69049
1/2	(1)1.6431 67673	(1)1.6462 07763	(1)1.6492 42250	(1)1.6522 71164	(1)1.6552 94536
1/3	6.4633 04070	6.4712 73627	6.4792 23603	6.4871 54117	6.4950 65288
1/4	4.0536 00464	4.0573 48596	4.0610 86370	4.0648 13851	4.0685 31106
1/5	3.0638 87063	3.0661 53254	3.0684 12765	3.0706 65640	3.0729 11923

TABLE II.

k	275	276	277	278	279
1	275	276	277	278	279
2	75625	76176	76729	77284	77841
3	207 96875	210 24576	212 53933	214 84952	217 17639
4	(9)5.7191 40625	(9)5.8027 82976	(9)5.8873 39441	(9)5.9728 16656	(9)6.0592 21281
5	(12)1.5727 63672	(12)1.6015 68101	(12)1.6307 93025	(12)1.6604 43030	(12)1.6905 22737
6	(14)4.3251 00098	(14)4.4203 27960	(14)4.5172 96680	(14)4.6160 31624	(14)4.7165 58437
7	(17)1.1894 02527	(17)1.2200 10517	(17)1.2512 91180	(17)1.2832 56792	(17)1.3159 19804
8	(19)3.2708 56949	(19)3.3672 29027	(19)3.4660 76569	(19)3.5674 53881	(19)3.6714 16253
9	(21)8.9948 56609	(21)9.2935 52114	(21)9.6010 32097	(21)9.9175 21788	(22)1.0243 25135
10	(24)2.4735 85568	(24)2.5650 20383	(24)2.6594 85891	(24)2.7570 71057	(24)2.8578 67126
24	(58)3.4993 28001	(58)3.8178 42160	(58)4.1640 35828	(58)4.5402 01230	(58)4.9488 11121
1/2	(1)1.6583 12395	(1)1.6613 24773	(1)1.6643 31698	(1)1.6673 33200	(1)1.6703 29309
1/3	6.5029 57234	6.5108 30071	6.5186 83915	6.5265 18879	6.5343 35077
1/4	4.0722 38199	4.0759 35196	4.0796 22161	4.0832 99156	4.0869 66245
1/5	3.0751 51657	3.0773 84885	3.0796 11650	3.0818 31992	3.0840 45954

k	280	281	282	283	284
1	280	281	282	283	284
2	78400	78961	79524	80089	80656
3	219 52000	221 88041	224 25768	226 65187	229 06304
4	(9)6.1465 60000	(9)6.2348 39521	(9)6.3240 66576	(9)6.4142 47921	(9)6.5053 90336
5	(12)1.7210 36800	(12)1.7519 89905	(12)1.7833 86774	(12)1.8152 32162	(12)1.8475 30855
6	(14)4.8189 03040	(14)4.9230 91634	(14)5.0291 50704	(14)5.1371 07017	(14)5.2469 87629
7	(17)1.3492 92851	(17)1.3833 88749	(17)1.4182 20498	(17)1.4538 01286	(17)1.4901 44487
8	(19)3.7780 19983	(19)3.8873 22385	(19)3.9993 81806	(19)4.1142 57639	(19)4.2320 10342
9	(22)1.0578 45595	(22)1.0923 37590	(22)1.1278 25669	(22)1.1643 34912	(22)1.2018 90937
10	(24)2.9619 67667	(24)3.0694 68629	(24)3.1804 68387	(24)3.2950 67801	(24)3.4133 70262
24	(58)5.3925 32264	(58)5.8742 39885	(58)6.3970 33126	(58)6.9642 51599	(58)7.5794 93086
1/2	(1)1.6733 20053	(1)1.6763 05461	(1)1.6792 85562	(1)1.6822 60384	(1)1.6852 29955
1/3	6.5421 32620	6.5499 11620	6.5576 72186	6.5654 14427	6.5731 38451
1/4	4.0906 23489	4.0942 70950	4.0979 08689	4.1015 36766	4.1051 55240
1/5	3.0862 53577	3.0884 54901	3.0906 49967	3.0928 38815	3.0950 21484

k	285	286	287	288	289
1	285	286	287	288	289
2	81225	81796	82369	82944	83521
3	231 49125	233 93656	236 39903	238 87872	241 37569
4	(9)6.5975 00625	(9)6.6905 85616	(9)6.7846 52161	(9)6.8797 07136	(9)6.9757 57441
5	(12)1.8802 87678	(12)1.9135 07486	(12)1.9471 95170	(12)1.9813 55655	(12)2.0159 93900
6	(14)5.3588 19883	(14)5.4726 31410	(14)5.5884 50138	(14)5.7063 04287	(14)5.8262 22372
7	(17)1.5272 63667	(17)1.5651 72583	(17)1.6038 85190	(17)1.6434 15635	(17)1.6837 78266
8	(19)4.3527 01450	(19)4.4763 93589	(19)4.6031 50495	(19)4.7330 37028	(19)4.8661 19188
9	(22)1.2405 19913	(22)1.2802 48566	(22)1.3211 04192	(22)1.3631 14664	(22)1.4063 08445
10	(24)3.5354 81753	(24)3.6615 10900	(24)3.7915 69031	(24)3.9257 70232	(24)4.0642 31407
24	(58)8.2466 32480	(58)8.9698 42039	(58)9.7536 13040	(59)1.0602 77893	(59)1.1522 54005
1/2	(1)1.6881 94302	(1)1.6911 53453	(1)1.6941 07435	(1)1.6970 56275	(1)1.7000 00000
1/3	6.5808 44365	6.5885 32275	6.5962 02284	6.6038 54498	6.6114 89018
1/4	4.1087 64171	4.1123 63618	4.1159 53637	4.1195 34288	4.1231 05626
1/5	3.0971 98013	3.0993 68441	3.1015 32807	3.1036 91148	3.1058 43502

k	290	291	292	293	294
1	290	291	292	293	294
2	84100	84681	85264	85849	86436
3	243 89000	246 42171	248 97088	251 53757	254 12184
4	(9)7.0728 10000	(9)7.1708 71761	(9)7.2699 49696	(9)7.3700 50801	(9)7.4711 82096
5	(12)2.0511 14900	(12)2.0867 23682	(12)2.1228 25311	(12)2.1594 24885	(12)2.1965 27536
6	(14)5.9482 33210	(14)6.0723 65916	(14)6.1986 49909	(14)6.3271 14912	(14)6.4577 90956
7	(17)1.7249 87631	(17)1.7670 58482	(17)1.8100 05773	(17)1.8538 44669	(17)1.8985 90541
8	(19)5.0024 64130	(19)5.1421 40181	(19)5.2852 16858	(19)5.4317 64881	(19)5.5818 56191
9	(22)1.4507 14598	(22)1.4963 62793	(22)1.5432 83323	(22)1.5915 07110	(22)1.6410 65720
10	(24)4.2070 72333	(24)4.3544 15727	(24)4.5063 87302	(24)4.6631 15833	(24)4.8247 33217
24	(59)1.2518 49008	(59)1.3596 64428	(59)1.4763 46962	(59)1.6025 91698	(59)1.7391 45550
1/2	(1)1.7029 38637	(1)1.7058 72211	(1)1.7088 00749	(1)1.7117 24277	(1)1.7146 42820
1/3	6.6191 05948	6.6267 05387	6.6342 87437	6.6418 52195	6.6493 99761
1/4	4.1266 67707	4.1302 20588	4.1337 64325	4.1372 98970	4.1408 24580
1/5	3.1079 89906	3.1101 30396	3.1122 65011	3.1143 93785	3.1165 16755

k	295	296	297	298	299
1	295	296	297	298	299
2	87025	87616	88209	88804	89401
3	256 72375	259 34336	261 98073	264 63592	267 30899
4	(9)7.5733 50625	(9)7.6765 63456	(9)7.7808 27681	(9)7.8861 50416	(9)7.9925 38801
5	(12)2.2341 38434	(12)2.2722 62783	(12)2.3109 05821	(12)2.3500 72824	(12)2.3897 69101
6	(14)6.5907 08381	(14)6.7258 97838	(14)6.8633 90289	(14)7.0032 17015	(14)7.1454 09613
7	(17)1.9442 58973	(17)1.9908 65760	(17)2.0384 26916	(17)2.0869 58671	(17)2.1364 77474
8	(19)5.7355 63969	(19)5.8929 62649	(19)6.0541 27940	(19)6.2191 36838	(19)6.3880 67649
9	(22)1.6919 91371	(22)1.7443 16944	(22)1.7980 75998	(22)1.8533 02778	(22)1.9100 32227
10	(24)4.9913 74544	(24)5.1631 78155	(24)5.3402 85715	(24)5.5228 42278	(24)5.7109 96358
24	(59)1.8868 10930	(59)2.0464 49657	(59)2.2189 87131	(59)2.4054 16789	(59)2.6068 04847
1/2	(1)1.7175 56404	(1)1.7204 65053	(1)1.7233 68794	(1)1.7262 67650	(1)1.7291 61647
1/3	6.6569 30232	6.6644 43703	6.6719 40272	6.6794 20032	6.6868 83077
1/4	4.1443 41207	4.1478 48904	4.1513 47726	4.1548 37723	4.1583 18947
1/5	3.1186 33956	3.1207 45423	3.1228 51191	3.1249 51295	3.1270 45768

TABLE II.

POWERS AND ROOTS n^k

Table II

k	300	301	302	303	304
1	300	301	302	303	304
2	90000	90601	91204	91809	92416
3	270 00000	272 70901	275 43608	278 18127	280 94464
4	(9)8.1000 00000	(9)8.2085 41201	(9)8.3181 69616	(9)8.4288 92481	(9)8.5407 17056
5	(12)2.4300 00000	(12)2.4707 70902	(12)2.5120 87224	(12)2.5539 54422	(12)2.5963 77985
6	(14)7.2900 00000	(14)7.4370 20414	(14)7.5865 03417	(14)7.7384 81898	(14)7.8929 89074
7	(17)2.1870 00000	(17)2.2385 43144	(17)2.2911 24032	(17)2.3447 60015	(17)2.3994 68679
8	(19)6.5610 00000	(19)6.7380 14865	(19)6.9191 94576	(19)7.1046 22846	(19)7.2943 84783
9	(22)1.9683 00000	(22)2.0281 42474	(22)2.0895 96762	(22)2.1527 00722	(22)2.2174 92974
10	(24)5.9049 00000	(24)6.1047 08848	(24)6.3105 82221	(24)6.5226 83188	(24)6.7411 78641
24	(59)2.8242 95365	(59)3.0591 15639	(59)3.3125 81949	(59)3.5861 05682	(59)3.8811 99856
1/2	(1)1.7320 50808	(1)1.7349 35157	(1)1.7378 14720	(1)1.7406 89519	(1)1.7435 59577
1/3	6.6943 29501	6.7017 59395	6.7091 72852	6.7165 69962	6.7239 50814
1/4	4.1617 91450	4.1652 55283	4.1687 10496	4.1721 57138	4.1755 95260
1/5	3.1291 34645	3.1312 17958	3.1332 95743	3.1353 68030	3.1374 34853

k	305	306	307	308	309
1	305	306	307	308	309
2	93025	93636	94249	94864	95481
3	283 72625	286 52616	289 34443	292 18112	295 03629
4	(9)8.6536 50625	(9)8.7677 00496	(9)8.8828 74001	(9)8.9991 78496	(9)9.1166 21361
5	(12)2.6393 63441	(12)2.6829 16352	(12)2.7270 42318	(12)2.7717 46977	(12)2.8170 36001
6	(14)8.0500 58494	(14)8.2097 24036	(14)8.3720 19917	(14)8.5369 80688	(14)8.7046 41242
7	(17)2.4552 67841	(17)2.5121 75555	(17)2.5702 10115	(17)2.6293 90052	(17)2.6897 34144
8	(19)7.4885 66914	(19)7.6872 57199	(19)7.8905 45052	(19)8.0985 21360	(19)8.3112 78504
9	(22)2.2840 12909	(22)2.3523 00703	(22)2.4223 97331	(22)2.4943 44579	(22)2.5681 85058
10	(24)6.9662 39372	(24)7.1980 40151	(24)7.4367 59806	(24)7.6825 81303	(24)7.9356 91828
24	(59)4.1994 86063	(59)4.5427 01868	(59)4.9127 08679	(59)5.3115 00125	(59)5.7412 10972
1/2	(1)1.7464 24920	(1)1.7492 85568	(1)1.7521 41547	(1)1.7549 92877	(1)1.7578 39583
1/3	6.7313 15497	6.7386 64101	6.7459 96712	6.7533 13417	6.7606 14302
1/4	4.1790 24910	4.1824 46136	4.1858 58988	4.1892 63512	4.1926 59756
1/5	3.1394 96244	3.1415 52236	3.1436 02859	3.1456 48146	3.1476 88127

k	310	311	312	313	314
1	310	311	312	313	314
2	96100	96721	97344	97969	98596
3	297 91000	300 80231	303 71328	306 64297	309 59144
4	(9)9.2352 10000	(9)9.3549 51841	(9)9.4758 54336	(9)9.5979 24961	(9)9.7211 71216
5	(12)2.8629 15100	(12)2.9093 90023	(12)2.9564 66553	(12)3.0041 50513	(12)3.0524 47762
6	(14)8.8750 36810	(14)9.0482 02970	(14)9.2241 75645	(14)9.4029 91105	(14)9.5846 85972
7	(17)2.7512 61411	(17)2.8139 91124	(17)2.8779 42801	(17)2.9431 36216	(17)3.0095 91395
8	(19)8.5289 10374	(19)8.7515 12395	(19)8.9791 81540	(19)9.2120 16356	(19)9.4501 16981
9	(22)2.6439 62216	(22)2.7217 20355	(22)2.8015 04640	(22)2.8833 61119	(22)2.9673 36732
10	(24)8.1962 82870	(24)8.4645 50303	(24)8.7406 94478	(24)9.0249 20304	(24)9.3174 37339
24	(59)6.2041 26610	(59)6.7026 93132	(59)7.2395 28072	(59)7.8174 31800	(59)8.4393 99655
1/2	(1)1.7606 81686	(1)1.7635 19209	(1)1.7663 52173	(1)1.7691 80601	(1)1.7720 04515
1/3	6.7678 99452	6.7751 68952	6.7824 22886	6.7896 61336	6.7968 84386
1/4	4.1960 47767	4.1994 27591	4.2027 99273	4.2061 62861	4.2095 18398
1/5	3.1497 22833	3.1517 52295	3.1537 76544	3.1557 95609	3.1578 09519

k	315	316	317	318	319
1	315	316	317	318	319
2	99225	99856	1 00489	1 01124	1 01761
3	312 55875	315 54496	318 55013	321 57432	324 61759
4	(9)9.8456 00625	(9)9.9712 20736	(10)1.0098 03912	(10)1.0226 06338	(10)1.0355 30112
5	(12)3.1013 64197	(12)3.1509 05753	(12)3.2010 78401	(12)3.2518 88154	(12)3.3033 41058
6	(14)9.7692 97220	(14)9.9568 62178	(15)1.0147 41853	(15)1.0341 00433	(15)1.0537 65797
7	(17)3.0773 28624	(17)3.1463 68448	(17)3.2167 31675	(17)3.2884 39376	(17)3.3615 12894
8	(19)9.6935 85167	(19)9.9425 24297	(20)1.0197 03941	(20)1.0457 23722	(20)1.0723 22613
9	(22)3.0534 79328	(22)3.1418 37678	(22)3.2324 61493	(22)3.3254 01435	(22)3.4207 09136
10	(24)9.6184 59882	(24)9.9282 07062	(25)1.0246 90293	(25)1.0574 77656	(25)1.0912 06214
24	(59)9.1086 34822	(59)9.8285 62028	(60)1.0602 84208	(60)1.1435 38734	(60)1.2330 37808
1/2	(1)1.7748 23935	(1)1.7776 38883	(1)1.7804 49381	(1)1.7832 55450	(1)1.7860 57110
1/3	6.8040 92116	6.8112 84608	6.8184 61941	6.8256 24197	6.8327 71452
1/4	4.2128 65931	4.2162 05502	4.2195 37156	4.2228 60938	4.2261 76889
1/5	3.1598 18306	3.1618 21997	3.1638 20622	3.1658 14209	3.1678 02787

k	320	321	322	323	324
1	320	321	322	323	324
2	1 02400	1 03041	1 03684	1 04329	1 04976
3	327 68000	330 76161	333 86248	336 98267	340 12224
4	(10)1.0485 76000	(10)1.0617 44768	(10)1.0750 37186	(10)1.0884 54024	(10)1.1019 96058
5	(12)3.3554 43200	(12)3.4082 00706	(12)3.4616 19738	(12)3.5157 06498	(12)3.5704 67227
6	(15)1.0737 41824	(15)1.0940 32426	(15)1.1146 41556	(15)1.1355 73199	(15)1.1568 31381
7	(17)3.4359 73837	(17)3.5118 44089	(17)3.5891 45809	(17)3.6679 01432	(17)3.7481 33676
8	(20)1.0995 11628	(20)1.1273 01953	(20)1.1557 04950	(20)1.1847 32163	(20)1.2143 95311
9	(22)3.5184 37209	(22)3.6186 39268	(22)3.7213 69940	(22)3.8266 84885	(22)3.9346 40808
10	(25)1.1258 99907	(25)1.1615 83205	(25)1.1982 81121	(25)1.2360 19218	(25)1.2748 23622
24	(60)1.3292 27996	(60)1.4325 86248	(60)1.5436 21862	(60)1.6628 78568	(60)1.7909 36736
1/2	(1)1.7888 54382	(1)1.7916 47287	(1)1.7944 35844	(1)1.7972 20076	(1)1.8000 00000
1/3	6.8399 03787	6.8470 21278	6.8541 24002	6.8612 12036	6.8682 85455
1/4	4.2294 85054	4.2327 85474	4.2360 78192	4.2393 63249	4.2426 40687
1/5	3.1697 86385	3.1717 65030	3.1737 38749	3.1757 07571	3.1776 71523

TABLE II.

k	325	326	327	328	329
1	325	326	327	328	329
2	1 05625	1 06276	1 06929	1 07584	1 08241
3	343 28125	346 45976	349 65783	352 87552	356 11289
4	(10)1.1156 64063	(10)1.1294 58818	(10)1.1433 81104	(10)1.1574 31706	(10)1.1716 11408
5	(12)3.6259 08203	(12)3.6820 35745	(12)3.7388 56210	(12)3.7963 75994	(12)3.8546 01533
6	(15)1.1784 20166	(15)1.2003 43653	(15)1.2226 05981	(15)1.2452 11326	(15)1.2681 63904
7	(17)3.8298 65540	(17)3.9131 20309	(17)3.9979 21557	(17)4.0842 93150	(17)4.1722 59245
8	(20)1.2447 06300	(20)1.2756 77221	(20)1.3073 20349	(20)1.3396 48153	(20)1.3726 73292
9	(22)4.0452 95476	(22)4.1587 07739	(22)4.2749 37542	(22)4.3940 45942	(22)4.5160 95129
10	(25)1.3147 21030	(25)1.3557 38723	(25)1.3979 04576	(25)1.4412 47069	(25)1.4857 95298
24	(60)1.9284 15722	(60)2.0759 76350	(60)2.2343 23554	(60)2.4042 09169	(60)2.5864 34894
1/2	(1)1.8027 75638	(1)1.8055 47009	(1)1.8083 14132	(1)1.8110 77028	(1)1.8138 35715
1/3	6.8753 44335	6.8823 88750	6.8894 18774	6.8964 34481	6.9034 35942
1/4	4.2459 10547	4.2491 72871	4.2524 27697	4.2556 75067	4.2589 15020
1/5	3.1796 30632	3.1815 84924	3.1835 34426	3.1854 79164	3.1874 19165

k	330	331	332	333	334
1	330	331	332	333	334
2	1 08900	1 09561	1 10224	1 10889	1 11556
3	359 37000	362 64691	365 94368	369 26037	372 59704
4	(10)1.1859 21000	(10)1.2003 61272	(10)1.2149 33018	(10)1.2296 37032	(10)1.2444 74114
5	(12)3.9135 39300	(12)3.9731 95811	(12)4.0335 77618	(12)4.0946 91317	(12)4.1565 43539
6	(15)1.2914 67969	(15)1.3151 27813	(15)1.3391 47769	(15)1.3635 32209	(15)1.3882 85542
7	(17)4.2618 44298	(17)4.3530 73062	(17)4.4459 70594	(17)4.5405 62254	(17)4.6368 73711
8	(20)1.4064 08618	(20)1.4408 67184	(20)1.4760 62237	(20)1.5120 07231	(20)1.5487 15819
9	(22)4.6411 48440	(22)4.7692 70378	(22)4.9005 26628	(22)5.0349 84078	(22)5.1727 10837
10	(25)1.5315 78985	(25)1.5786 28495	(25)1.6269 74840	(25)1.6766 49698	(25)1.7276 85420
24	(60)2.7818 55434	(60)2.9913 81825	(60)3.2159 84959	(60)3.4566 99320	(60)3.7146 26935
1/2	(1)1.8165 90212	(1)1.8193 40540	(1)1.8220 86716	(1)1.8248 28759	(1)1.8275 66688
1/3	6.9104 23230	6.9173 96417	6.9243 55573	6.9313 00768	6.9382 32074
1/4	4.2621 47595	4.2653 72832	4.2685 90770	4.2718 01446	4.2750 04899
1/5	3.1893 54454	3.1912 85058	3.1932 11001	3.1951 32308	3.1970 49006

k	335	336	337	338	339
1	335	336	337	338	339
2	1 12225	1 12896	1 13569	1 14244	1 14921
3	375 95375	379 33056	382 72753	386 14472	389 58219
4	(10)1.2594 45063	(10)1.2745 50682	(10)1.2897 91776	(10)1.3051 69154	(10)1.3206 83624
5	(12)4.2191 40959	(12)4.2824 90290	(12)4.3465 98285	(12)4.4114 71739	(12)4.4771 17486
6	(15)1.4134 12221	(15)1.4389 16737	(15)1.4648 03622	(15)1.4910 77448	(15)1.5177 42828
7	(17)4.7349 30942	(17)4.8347 60238	(17)4.9363 88207	(17)5.0398 41774	(17)5.1451 48186
8	(20)1.5862 01865	(20)1.6244 79440	(20)1.6635 62826	(20)1.7034 66520	(20)1.7442 05235
9	(22)5.3137 76249	(22)5.4582 50918	(22)5.6062 06723	(22)5.7577 16836	(22)5.9128 55747
10	(25)1.7801 15044	(25)1.8339 72309	(25)1.8892 91666	(25)1.9461 08291	(25)2.0044 58098
24	(60)3.9909 41565	(60)4.2868 93134	(60)4.6038 12427	(60)4.9431 16051	(60)5.3063 11693
1/2	(1)1.8303 00522	(1)1.8330 30278	(1)1.8357 55975	(1)1.8384 77643	(1)1.8411 95264
1/3	6.9451 49558	6.9520 53290	6.9589 43337	6.9658 19768	6.9726 82649
1/4	4.2782 01166	4.2813 90286	4.2845 72295	4.2877 47230	4.2909 15128
1/5	3.1989 61118	3.2008 68669	3.2027 71684	3.2046 70186	3.2065 64201

k	340	341	342	343	344
1	340	341	342	343	344
2	1 15600	1 16281	1 16964	1 17649	1 18336
3	393 04000	396 51821	400 01688	403 53607	407 07584
4	(10)1.3363 36000	(10)1.3521 27096	(10)1.3680 57730	(10)1.3841 28720	(10)1.4003 40890
5	(12)4.5435 42400	(12)4.6107 53398	(12)4.6787 57435	(12)4.7475 61510	(12)4.8171 72660
6	(15)1.5448 04416	(15)1.5722 66909	(15)1.6001 35043	(15)1.6284 13598	(15)1.6571 07395
7	(17)5.2523 35014	(17)5.3614 30158	(17)5.4724 61847	(17)5.5854 58641	(17)5.7004 49439
8	(20)1.7857 93905	(20)1.8282 47684	(20)1.8715 81952	(20)1.9158 12314	(20)1.9609 54607
9	(22)6.0716 99277	(22)6.2343 24602	(22)6.4008 10274	(22)6.5712 36236	(22)6.7456 83848
10	(25)2.0643 77754	(25)2.1259 04689	(25)2.1890 77114	(25)2.2539 34029	(25)2.3205 15244
24	(60)5.6950 03680	(60)6.1108 98859	(60)6.5558 12822	(60)7.0316 76479	(60)7.5405 43015
1/2	(1)1.8439 08891	(1)1.8466 18531	(1)1.8493 24201	(1)1.8520 25918	(1)1.8547 23699
1/3	6.9795 32047	6.9863 68028	6.9931 90657	7.0000 00000	7.0067 96121
1/4	4.2940 76026	4.2972 29958	4.3003 76961	4.3035 17071	4.3066 50321
1/5	3.2084 53751	3.2103 38860	3.2122 19552	3.2140 95850	3.2159 67776

k	345	346	347	348	349
1	345	346	347	348	349
2	1 19025	1 19716	1 20409	1 21104	1 21801
3	410 63625	414 21736	417 81923	421 44192	425 08549
4	(10)1.4166 95063	(10)1.4331 92066	(10)1.4498 32728	(10)1.4666 17882	(10)1.4835 48360
5	(12)4.8875 97966	(12)4.9588 44547	(12)5.0309 19567	(12)5.1038 30228	(12)5.1775 83777
6	(15)1.6862 21298	(15)1.7157 60213	(15)1.7457 29090	(15)1.7761 32919	(15)1.8069 76738
7	(17)5.8174 63479	(17)5.9365 30338	(17)6.0576 79941	(17)6.1809 42559	(17)6.3063 48816
8	(20)2.0070 24900	(20)2.0540 39497	(20)2.1020 14939	(20)2.1509 68011	(20)2.2009 15737
9	(22)6.9242 35905	(22)7.1069 76659	(22)7.2939 91840	(22)7.4853 68677	(22)7.6811 95921
10	(25)2.3888 61387	(25)2.4590 13924	(25)2.5310 15168	(25)2.6049 08300	(25)2.6807 37377
24	(60)8.0845 95243	(60)8.6661 53376	(60)9.2876 83235	(60)9.9518 04932	(61)1.0661 30203
1/2	(1)1.8574 17562	(1)1.8601 07524	(1)1.8627 93601	(1)1.8654 75811	(1)1.8681 54169
1/3	7.0135 79083	7.0203 48952	7.0271 05788	7.0338 49656	7.0405 80617
1/4	4.3097 76748	4.3128 96386	4.3160 09269	4.3191 15431	4.3222 14906
1/5	3.2178 35355	3.2196 98608	3.2215 57557	3.2234 12226	3.2252 62636

TABLE II.

Table II

POWERS AND ROOTS n^k

k	350	351	352	353	354
1	350	351	352	353	354
2	1 22500	1 23201	1 23904	1 24609	1 25316
3	428 75000	432 43551	436 14208	439 86977	443 61864
4	(10)1.5006 25000	(10)1.5178 48640	(10)1.5352 20122	(10)1.5527 40288	(10)1.5704 09986
5	(12)5.2521 87500	(12)5.3276 48727	(12)5.4039 74828	(12)5.4811 73217	(12)5.5592 51349
6	(15)1.8382 65625	(15)1.8700 04703	(15)1.9021 99139	(15)1.9348 54146	(15)1.9679 74978
7	(17)6.4339 29688	(17)6.5637 16508	(17)6.6957 40971	(17)6.8300 35134	(17)6.9666 31421
8	(20)2.2518 75391	(20)2.3038 64494	(20)2.3569 00822	(20)2.4110 02402	(20)2.4661 87523
9	(22)7.8815 63867	(22)8.0865 64375	(22)8.2962 90893	(22)8.5108 38480	(22)8.7303 03831
10	(25)2.7585 47354	(25)2.8383 84096	(25)2.9202 94394	(25)3.0043 25983	(25)3.0905 27556
24	(61)1.1419 13124	(61)1.2228 43263	(61)1.3092 54042	(61)1.4014 99442	(61)1.4999 55202
1/2	(1)1.8708 28693	(1)1.8734 99400	(1)1.8761 66304	(1)1.8788 29423	(1)1.8814 88772
1/3	7.0472 98732	7.0540 04063	7.0606 96671	7.0673 76615	7.0740 43955
1/4	4.3253 07727	4.3283 93928	4.3314 73541	4.3345 46600	4.3376 13137
1/5	3.2271 08809	3.2289 50768	3.2307 88532	3.2326 22125	3.2344 51567

k	355	356	357	358	359
1	355	356	357	358	359
2	1 26025	1 26736	1 27449	1 28164	1 28881
3	447 38875	451 18016	454 99293	458 82712	462 68279
4	(10)1.5882 30063	(10)1.6062 01370	(10)1.6243 24760	(10)1.6426 01090	(10)1.6610 31216
5	(12)5.6382 16722	(12)5.7180 76876	(12)5.7988 39394	(12)5.8805 11901	(12)5.9631 02066
6	(15)2.0015 66936	(15)2.0356 35368	(15)2.0701 85663	(15)2.1052 23260	(15)2.1407 53642
7	(17)7.1055 62624	(17)7.2468 61909	(17)7.3905 62819	(17)7.5366 99273	(17)7.6853 05573
8	(20)2.5224 74731	(20)2.5798 82840	(20)2.6384 30926	(20)2.6981 38340	(20)2.7590 24701
9	(22)8.9547 85297	(22)9.1843 82909	(22)9.4191 98407	(22)9.6593 35256	(22)9.9048 98676
10	(25)3.1789 48780	(25)3.2696 40316	(25)3.3626 53831	(25)3.4580 42022	(25)3.5558 58625
24	(61)1.6050 20092	(61)1.7171 17251	(61)1.8366 95605	(61)1.9642 31355	(61)2.1002 29556
1/2	(1)1.8841 44368	(1)1.8867 96226	(1)1.8894 44363	(1)1.8920 88793	(1)1.8947 29532
1/3	7.0806 98751	7.0873 41061	7.0939 70945	7.1005 88459	7.1071 93661
1/4	4.3406 73183	4.3437 26771	4.3467 73933	4.3498 14700	4.3528 49104
1/5	3.2362 76880	3.2380 98084	3.2399 15199	3.2417 28247	3.2435 37249

k	360	361	362	363	364
1	360	361	362	363	364
2	1 29600	1 30321	1 31044	1 31769	1 32496
3	466 56000	470 45881	474 37928	478 32147	482 28544
4	(10)1.6796 16000	(10)1.6983 56304	(10)1.7172 52994	(10)1.7363 06936	(10)1.7555 19002
5	(12)6.0466 17600	(12)6.1310 66258	(12)6.2164 55837	(12)6.3027 94178	(12)6.3900 89166
6	(15)2.1767 82336	(15)2.2133 14919	(15)2.2503 57013	(15)2.2879 14287	(15)2.3259 92456
7	(17)7.8364 16410	(17)7.9900 66858	(17)8.1462 92387	(17)8.3051 28860	(17)8.4666 12541
8	(20)2.8211 09907	(20)2.8844 14136	(20)2.9489 57844	(20)3.0147 61776	(20)3.0818 46965
9	(23)1.0155 99567	(23)1.0412 73503	(23)1.0675 22740	(23)1.0943 58525	(23)1.1217 92295
10	(25)3.6561 58440	(25)3.7589 97346	(25)3.8644 32317	(25)3.9725 21445	(25)4.0833 23955
24	(61)2.2452 25771	(61)2.3997 87825	(61)2.5645 17652	(61)2.7400 53237	(61)2.9270 70667
1/2	(1)1.8973 66596	(1)1.9000 00000	(1)1.9026 29759	(1)1.9052 55888	(1)1.9078 78403
1/3	7.1137 86609	7.1203 67359	7.1269 35967	7.1334 92490	7.1400 36982
1/4	4.3558 77175	4.3588 98944	4.3619 14441	4.3649 23697	4.3679 26743
1/5	3.2453 42223	3.2471 43191	3.2489 40172	3.2507 33187	3.2525 22254

k	365	366	367	368	369
1	365	366	367	368	369
2	1 33225	1 33956	1 34689	1 35424	1 36161
3	486 27125	490 27896	494 30863	498 36032	502 43409
4	(10)1.7748 90063	(10)1.7944 20994	(10)1.8141 12672	(10)1.8339 65978	(10)1.8539 81792
5	(12)6.4783 48728	(12)6.5675 80837	(12)6.6577 93507	(12)6.7489 94798	(12)6.8411 92813
6	(15)2.3645 97286	(15)2.4037 34586	(15)2.4434 10217	(15)2.4836 30086	(15)2.5244 00148
7	(17)8.6307 80093	(17)8.7976 68585	(17)8.9673 15496	(17)9.1397 58715	(17)9.3150 36546
8	(20)3.1502 34734	(20)3.2199 46702	(20)3.2910 04787	(20)3.3634 31207	(20)3.4372 48485
9	(23)1.1498 35678	(23)1.1785 00493	(23)1.2077 98757	(23)1.2377 42684	(23)1.2683 44691
10	(25)4.1969 00224	(25)4.3133 11804	(25)4.4326 21438	(25)4.5548 93078	(25)4.6801 91910
24	(61)3.1262 86296	(61)3.3384 59019	(61)3.5643 92671	(61)3.8049 38558	(61)4.0609 98114
1/2	(1)1.9104 97317	(1)1.9131 12647	(1)1.9157 24406	(1)1.9183 32609	(1)1.9209 37271
1/3	7.1465 69499	7.1530 90095	7.1595 98825	7.1660 95742	7.1725 80900
1/4	4.3709 23607	4.3739 14319	4.3768 98909	4.3798 77406	4.3828 49839
1/5	3.2543 07394	3.2560 88625	3.2578 65967	3.2596 39439	3.2614 09059

k	370	371	372	373	374
1	370	371	372	373	374
2	1 36900	1 37641	1 38384	1 39129	1 39876
3	506 53000	510 64811	514 78848	518 95117	523 13624
4	(10)1.8741 61000	(10)1.8945 04488	(10)1.9150 13146	(10)1.9356 87864	(10)1.9565 29538
5	(12)6.9343 95700	(12)7.0286 11651	(12)7.1238 48902	(12)7.2201 15733	(12)7.3174 20471
6	(15)2.5657 26409	(15)2.6076 14922	(15)2.6500 71791	(15)2.6931 03168	(15)2.7367 15256
7	(17)9.4931 87713	(17)9.6742 51362	(17)9.8582 67064	(18)1.0045 27482	(18)1.0235 31506
8	(20)3.5124 79454	(20)3.5891 47255	(20)3.6672 75348	(20)3.7468 87507	(20)3.8280 07832
9	(23)1.2996 17398	(23)1.3315 73632	(23)1.3642 26429	(23)1.3975 89040	(23)1.4316 74929
10	(25)4.8085 84372	(25)4.9401 38174	(25)5.0749 22317	(25)5.2130 07120	(25)5.3544 64234
24	(61)4.3335 25711	(61)4.6235 31606	(61)4.9320 85051	(61)5.2603 17567	(61)5.6094 26383
1/2	(1)1.9235 38406	(1)1.9261 36028	(1)1.9287 30152	(1)1.9313 20792	(1)1.9339 07961
1/3	7.1790 54352	7.1855 16151	7.1919 66348	7.1984 04996	7.2048 32147
1/4	4.3858 16237	4.3887 76627	4.3917 31039	4.3946 79501	4.3976 22040
1/5	3.2631 74848	3.2649 36822	3.2666 95001	3.2684 49404	3.2702 00047

TABLE II.

k	375	376	377	378	379
1					
2	1 40625	1 41376	1 42129	1 42884	1 43641
3	527 34375	531 57376	535 82633	540 10152	544 39939
4	(10)1.9775 39063	(10)1.9987 17338	(10)2.0200 65264	(10)2.0415 83746	(10)2.0632 73688
5	(12)7.4157 71484	(12)7.5151 77189	(12)7.6156 46046	(12)7.7171 86558	(12)7.8198 07278
6	(15)2.7809 14307	(15)2.8257 06623	(15)2.8710 98559	(15)2.9170 96519	(15)2.9637 06958
7	(18)1.0428 42865	(18)1.0624 65690	(18)1.0824 04157	(18)1.1026 62484	(18)1.1232 44937
8	(20)3.9106 60744	(20)3.9948 70996	(20)4.0806 63671	(20)4.1680 64190	(20)4.2570 98312
9	(23)1.4664 97779	(23)1.5020 71494	(23)1.5384 10204	(23)1.5755 28264	(23)1.6134 40260
10	(25)5.4993 66671	(25)5.6477 88819	(25)5.7998 06469	(25)5.9554 96838	(25)6.1149 38586
24	(61)5.9806 78067	(61)6.3754 12334	(61)6.7950 46060	(61)7.2410 77507	(61)7.7150 90756
1/2	(1)1.9364 91673	(1)1.9390 71943	(1)1.9416 48784	(1)1.9442 22210	(1)1.9467 92233
1/3	7.2112 47852	7.2176 52160	7.2240 45124	7.2304 26792	7.2367 97216
1/4	4.4005 58684	4.4034 89461	4.4064 14397	4.4093 33520	4.4122 46858
1/5	3.2719 46950	3.2736 90130	3.2754 29605	3.2771 65392	3.2788 97510

k	380	381	382	383	384
1					
2	1 44400	1 45161	1 45924	1 46689	1 47456
3	548 72000	553 06341	557 42968	561 81887	566 23104
4	(10)2.0851 36000	(10)2.1071 71592	(10)2.1293 81378	(10)2.1517 66272	(10)2.1743 27194
5	(12)7.9235 16800	(12)8.0283 23766	(12)8.1342 36862	(12)8.2412 64822	(12)8.3494 16423
6	(15)3.0109 36384	(15)3.0587 91355	(15)3.1072 78481	(15)3.1564 04427	(15)3.2061 75907
7	(18)1.1441 55826	(18)1.1653 99506	(18)1.1869 80380	(18)1.2089 02895	(18)1.2311 71548
8	(20)4.3477 92138	(20)4.4401 72119	(20)4.5342 65051	(20)4.6300 98090	(20)4.7276 98745
9	(23)1.6521 61013	(23)1.6917 05577	(23)1.7320 89250	(23)1.7733 27568	(23)1.8154 36318
10	(25)6.2782 11848	(25)6.4453 98249	(25)6.6165 80933	(25)6.7918 44587	(25)6.9712 75461
24	(61)8.2187 60383	(61)8.7538 56362	(61)9.3222 49236	(61)9.9259 15535	(62)1.0566 94349
1/2	(1)1.9493 58869	(1)1.9519 22130	(1)1.9544 82029	(1)1.9570 38579	(1)1.9595 91794
1/3	7.2431 56443	7.2495 04524	7.2558 41507	7.2621 67440	7.2684 82371
1/4	4.4151 54436	4.4180 56280	4.4209 52418	4.4238 42876	4.4267 27679
1/5	3.2806 25976	3.2823 50807	3.2840 72019	3.2857 89631	3.2875 03659

k	385	386	387	388	389
1					
2	1 48225	1 48996	1 49769	1 50544	1 51321
3	570 66625	575 12456	579 60603	584 11072	588 63869
4	(10)2.1970 65063	(10)2.2199 80802	(10)2.2430 75336	(10)2.2663 49594	(10)2.2898 04504
5	(12)8.4587 00491	(12)8.5691 25894	(12)8.6807 01551	(12)8.7934 36423	(12)8.9073 39521
6	(15)3.2565 99689	(15)3.3076 82595	(15)3.3594 31500	(15)3.4118 53332	(15)3.4649 55074
7	(18)1.2537 90880	(18)1.2767 65482	(18)1.3000 99991	(18)1.3237 99093	(18)1.3478 67524
8	(20)4.8270 94889	(20)4.9283 14759	(20)5.0313 86963	(20)5.1363 40480	(20)5.2432 04667
9	(23)1.8584 31532	(23)1.9023 29497	(23)1.9471 46755	(23)1.9929 00106	(23)2.0396 06615
10	(25)7.1549 61399	(25)7.3429 91859	(25)7.5354 57941	(25)7.7324 52413	(25)7.9340 69734
24	(62)1.1247 53901	(62)1.1970 03202	(62)1.2736 88303	(62)1.3550 69013	(62)1.4414 19629
1/2	(1)1.9621 41687	(1)1.9646 88270	(1)1.9672 31557	(1)1.9697 71560	(1)1.9723 08292
1/3	7.2747 86349	7.2810 79420	7.2873 61631	7.2936 33030	7.2998 93662
1/4	4.4296 06853	4.4324 80423	4.4353 48416	4.4382 10856	4.4410 67768
1/5	3.2892 14120	3.2909 21030	3.2926 24406	3.2943 24265	3.2960 20622

k	390	391	392	393	394
1					
2	1 52100	1 52881	1 53664	1 54449	1 55236
3	593 19000	597 76471	602 36288	606 98457	611 62984
4	(10)2.3134 41000	(10)2.3372 60016	(10)2.3612 62490	(10)2.3854 49360	(10)2.4098 21570
5	(12)9.0224 19900	(12)9.1386 86663	(12)9.2561 48959	(12)9.3748 15985	(12)9.4946 96984
6	(15)3.5187 43761	(15)3.5732 26485	(15)3.6284 10392	(15)3.6843 02682	(15)3.7409 10612
7	(18)1.3723 10067	(18)1.3971 31556	(18)1.4223 36874	(18)1.4479 30954	(18)1.4739 18781
8	(20)5.3520 09260	(20)5.4627 84383	(20)5.5755 60545	(20)5.6903 68650	(20)5.8072 39997
9	(23)2.0872 83612	(23)2.1359 48694	(23)2.1856 19734	(23)2.2363 14879	(23)2.2880 52559
10	(25)8.1404 06085	(25)8.3515 59392	(25)8.5676 29356	(25)8.7887 17476	(25)9.0149 27082
24	(62)1.5330 29700	(62)1.6302 04837	(62)1.7332 67559	(62)1.8425 58176	(62)1.9584 35730
1/2	(1)1.9748 41766	(1)1.9773 71993	(1)1.9798 98987	(1)1.9824 22760	(1)1.9849 43324
1/3	7.3061 43574	7.3123 82812	7.3186 11340	7.3248 29445	7.3310 36930
1/4	4.4439 19178	4.4467 65109	4.4496 05586	4.4524 40634	4.4552 70277
1/5	3.2977 13494	3.2994 02898	3.3010 88848	3.3027 71361	3.3044 50453

k	395	396	397	398	399
1					
2	1 56025	1 56816	1 57609	1 58404	1 59201
3	616 29875	620 99136	625 70773	630 44792	635 21199
4	(10)2.4343 80063	(10)2.4591 25786	(10)2.4840 59688	(10)2.5091 82722	(10)2.5344 95840
5	(12)9.6158 01247	(12)9.7381 38111	(12)9.8617 16962	(12)9.9865 47232	(13)1.0112 63840
6	(15)3.7982 41493	(15)3.8563 02692	(15)3.9151 01634	(15)3.9746 45798	(15)4.0349 42722
7	(18)1.5003 05390	(18)1.5270 95866	(18)1.5542 95349	(18)1.5819 09028	(18)1.6099 42146
8	(20)5.9262 06289	(20)6.0472 99629	(20)6.1705 52534	(20)6.2959 97930	(20)6.4236 69163
9	(23)2.3408 51484	(23)2.3947 30653	(23)2.4497 09356	(23)2.5058 07176	(23)2.5630 43996
10	(25)9.2463 63362	(25)9.4831 33387	(25)9.7253 46143	(25)9.9731 12562	(26)1.0226 54554
24	(62)2.0812 78965	(62)2.2114 87364	(62)2.3494 82217	(62)2.4957 07762	(62)2.6506 32365
1/2	(1)1.9874 60691	(1)1.9899 74874	(1)1.9924 85885	(1)1.9949 93734	(1)1.9974 98436
1/3	7.3372 33921	7.3434 20462	7.3495 96597	7.3557 62368	7.3619 17821
1/4	4.4580 94538	4.4609 13443	4.4637 27013	4.4665 35273	4.4693 38246
1/5	3.3061 26138	3.3077 98433	3.3094 67354	3.3111 32914	3.3127 95131

TABLE II.

POWERS AND ROOTS n^k

k	400	401	402	403	404
1	400	401	402	403	404
2	1 60000	1 60801	1 61604	1 62409	1 63216
3	640 00000	644 81201	649 64808	654 50827	659 39264
4	(10)2.5600 00000	(10)2.5856 96160	(10)2.6115 85282	(10)2.6376 68328	(10)2.6639 46266
5	(13)1.0240 00000	(13)1.0368 64160	(13)1.0498 57283	(13)1.0629 80336	(13)1.0762 34291
6	(15)4.0960 00000	(15)4.1578 25282	(15)4.2204 26278	(15)4.2838 10755	(15)4.3479 86537
7	(18)1.6384 00000	(18)1.6672 87938	(18)1.6966 11364	(18)1.7263 75734	(18)1.7565 86561
8	(20)6.5536 00000	(20)6.6858 24632	(20)6.8203 77683	(20)6.9572 94209	(20)7.0966 09706
9	(23)2.6214 40000	(23)2.6810 15678	(23)2.7417 91829	(23)2.8037 89566	(23)2.8670 30321
10	(26)1.0485 76000	(26)1.0750 87287	(26)1.1022 00315	(26)1.1299 27195	(26)1.1582 80250
24	(62)2.8147 49767	(62)2.9885 80393	(62)3.1726 72718	(62)3.3676 04703	(62)3.5739 85306
1/2	(1)2.0000 00000	(1)2.0024 98439	(1)2.0049 93766	(1)2.0074 85990	(1)2.0099 75124
1/3	7.3680 62997	7.3741 97940	7.3803 22692	7.3864 37295	7.3925 41792
1/4	4.4721 35955	4.4749 28423	4.4777 15674	4.4804 97729	4.4832 74611
1/5	3.3144 54017	3.3161 09590	3.3177 61862	3.3194 10850	3.3210 56568

k	405	406	407	408	409
1	405	406	407	408	409
2	1 64025	1 64836	1 65649	1 66464	1 67281
3	664 30125	669 23416	674 19143	679 17312	684 17929
4	(10)2.6904 20063	(10)2.7170 90690	(10)2.7439 59120	(10)2.7710 26330	(10)2.7982 93296
5	(13)1.0896 20125	(13)1.1031 38820	(13)1.1167 91362	(13)1.1305 78742	(13)1.1445 01958
6	(15)4.4129 61508	(15)4.4787 43609	(15)4.5453 40843	(15)4.6127 61269	(15)4.6810 13009
7	(18)1.7872 49411	(18)1.8183 69905	(18)1.8499 53723	(18)1.8820 06598	(18)1.9145 34321
8	(20)7.2383 60113	(20)7.3825 81816	(20)7.5293 11653	(20)7.6785 86919	(20)7.8304 45371
9	(23)2.9315 35846	(23)2.9973 28217	(23)3.0644 29843	(23)3.1328 63463	(23)3.2026 52157
10	(26)1.1872 72017	(26)1.2169 15256	(26)1.2472 22946	(26)1.2782 08293	(26)1.3098 84732
24	(62)3.7924 56055	(62)4.0236 92707	(62)4.2684 06980	(62)4.5273 48373	(62)4.8013 06073
1/2	(1)2.0124 61180	(1)2.0149 44168	(1)2.0174 24100	(1)2.0199 00988	(1)2.0223 74842
1/3	7.3986 36223	7.4047 20630	7.4107 95055	7.4168 59539	7.4229 14120
1/4	4.4860 46344	4.4888 12948	4.4915 74446	4.4943 30860	4.4970 82211
1/5	3.3226 99030	3.3243 38251	3.3259 74245	3.3276 07026	3.3292 36609

k	410	411	412	413	414
1	410	411	412	413	414
2	1 68100	1 68921	1 69744	1 70569	1 71396
3	689 21000	694 26531	699 34528	704 44997	709 57944
4	(10)2.8257 61000	(10)2.8534 30424	(10)2.8813 02554	(10)2.9093 78376	(10)2.9376 58882
5	(13)1.1585 62010	(13)1.1727 59904	(13)1.1870 96652	(13)1.2015 73269	(13)1.2161 90777
6	(15)4.7501 04241	(15)4.8200 43207	(15)4.8908 38207	(15)4.9624 97602	(15)5.0350 29817
7	(18)1.9475 42739	(18)1.9810 37758	(18)2.0150 25341	(18)2.0495 11510	(18)2.0845 02344
8	(20)7.9849 25229	(20)8.1420 65185	(20)8.3019 04405	(20)8.4644 82535	(20)8.6298 39705
9	(23)3.2738 19344	(23)3.3463 88791	(23)3.4203 84615	(23)3.4958 31287	(23)3.5727 53638
10	(26)1.3422 65931	(26)1.3753 65793	(26)1.4091 98461	(26)1.4437 78322	(26)1.4791 20006
24	(62)5.0911 10945	(62)5.3976 37632	(62)5.7218 06738	(62)6.0645 87127	(62)6.4269 98328
1/2	(1)2.0248 45673	(1)2.0273 13493	(1)2.0297 78313	(1)2.0322 40143	(1)2.0346 98995
1/3	7.4289 58841	7.4349 93742	7.4410 18861	7.4470 34238	7.4530 39914
1/4	4.4998 28522	4.5025 69814	4.5053 06108	4.5080 37426	4.5107 63788
1/5	3.3308 63008	3.3324 86236	3.3341 06308	3.3357 23237	3.3373 37037

k	415	416	417	418	419
1	415	416	417	418	419
2	1 72225	1 73056	1 73889	1 74724	1 75561
3	714 73375	719 91296	725 11713	730 34632	735 60059
4	(10)2.9661 45063	(10)2.9948 37914	(10)3.0237 38432	(10)3.0528 47618	(10)3.0821 66472
5	(13)1.2309 50201	(13)1.2458 52572	(13)1.2608 98926	(13)1.2760 90304	(13)1.2914 27752
6	(15)5.1084 43334	(15)5.1827 46700	(15)5.2579 48522	(15)5.3340 57471	(15)5.4110 82280
7	(18)2.1200 03984	(18)2.1560 22627	(18)2.1925 64534	(18)2.2296 36023	(18)2.2672 43475
8	(20)8.7980 16532	(20)8.9690 54129	(20)9.1429 94106	(20)9.3198 78576	(20)9.4997 50162
9	(23)3.6511 76861	(23)3.7311 26518	(23)3.8126 28542	(23)3.8957 09245	(23)3.9803 95318
10	(26)1.5152 38397	(26)1.5521 48631	(26)1.5898 66102	(26)1.6284 06464	(26)1.6677 85638
24	(62)6.8101 13045	(62)7.2150 59801	(62)7.6430 25069	(62)8.0952 59269	(62)8.5730 73581
1/2	(1)2.0371 54879	(1)2.0396 07805	(1)2.0420 57786	(1)2.0445 04830	(1)2.0469 48949
1/3	7.4590 35926	7.4650 22314	7.4709 99115	7.4769 66370	7.4829 24114
1/4	4.5134 85215	4.5162 01729	4.5189 13349	4.5216 20097	4.5243 21992
1/5	3.3389 47722	3.3405 55305	3.3421 59799	3.3437 61218	3.3453 59575

k	420	421	422	423	424
1	420	421	422	423	424
2	1 76400	1 77241	1 78084	1 78929	1 79776
3	740 88000	746 18461	751 51448	756 86967	762 25024
4	(10)3.1116 96000	(10)3.1414 37208	(10)3.1713 91106	(10)3.2015 58704	(10)3.2319 41018
5	(13)1.3069 12320	(13)1.3225 45065	(13)1.3383 27047	(13)1.3542 59332	(13)1.3703 42991
6	(15)5.4890 31744	(15)5.5679 14722	(15)5.6477 40136	(15)5.7285 16974	(15)5.8102 54284
7	(18)2.3053 93332	(18)2.3440 92098	(18)2.3833 46338	(18)2.4231 62680	(18)2.4635 47816
8	(20)9.6826 51996	(20)9.8686 27732	(21)1.0057 72154	(21)1.0249 97814	(21)1.0445 44274
9	(23)4.0667 13838	(23)4.1546 92275	(23)4.2443 58492	(23)4.3357 40751	(23)4.4288 67722
10	(26)1.7080 19812	(26)1.7491 25448	(26)1.7911 19284	(26)1.8340 18338	(26)1.8778 39914
24	(62)9.0778 49315	(62)9.6110 38126	(63)1.0174 16609	(63)1.0768 83734	(63)1.1396 73784
1/2	(1)2.0493 90153	(1)2.0518 28453	(1)2.0542 63858	(1)2.0566 96380	(1)2.0591 26028
1/3	7.4888 72387	7.4948 11226	7.5007 40668	7.5066 60749	7.5125 71508
1/4	4.5270 19056	4.5297 11307	4.5323 98767	4.5350 81455	4.5377 59390
1/5	3.3469 54883	3.3485 47155	3.3501 36405	3.3517 22644	3.3533 05887

TABLE II.

k	425	426	427	428	429
1	425	426	427	428	429
2	1 80625	1 81476	1 82329	1 83184	1 84041
3	767 65625	773 08776	778 54483	784 02752	789 53589
4	(10)3.2625 39063	(10)3.2933 53858	(10)3.3243 86424	(10)3.3556 37786	(10)3.3871 08968
5	(13)1.3865 79102	(13)1.4029 68743	(13)1.4195 13003	(13)1.4362 12972	(13)1.4530 69747
6	(15)5.8929 61182	(15)5.9766 46847	(15)6.0613 20523	(15)6.1469 91521	(15)6.2336 69216
7	(18)2.5045 08502	(18)2.5460 51557	(18)2.5881 83863	(18)2.6309 12371	(18)2.6742 44094
8	(21)1.0644 16113	(21)1.0846 17963	(21)1.1051 54510	(21)1.1260 30495	(21)1.1472 50716
9	(23)4.5237 68482	(23)4.6204 72523	(23)4.7190 09756	(23)4.8194 10518	(23)4.9217 05572
10	(26)1.9226 01605	(26)1.9683 21295	(26)2.0150 17166	(26)2.0627 07702	(26)2.1114 11691
24	(63)1.2059 63938	(63)1.2759 40370	(63)1.3497 98685	(63)1.4277 44370	(63)1.5099 93273
1/2	(1)2.0615 52813	(1)2.0639 76744	(1)2.0663 97832	(1)2.0688 16087	(1)2.0712 31518
1/3	7.5184 72981	7.5243 65204	7.5302 48212	7.5361 22043	7.5419 86732
1/4	4.5404 32593	4.5431 01082	4.5457 64877	4.5484 23998	4.5510 78463
1/5	3.3548 86145	3.3564 63431	3.3580 37758	3.3596 09138	3.3611 77583

k	430	431	432	433	434
1	430	431	432	433	434
2	1 84900	1 85761	1 86624	1 87489	1 88356
3	795 07000	800 62991	806 21568	811 82737	817 46504
4	(10)3.4188 01000	(10)3.4507 14912	(10)3.4828 51738	(10)3.5152 12512	(10)3.5477 98274
5	(13)1.4700 84430	(13)1.4872 58127	(13)1.5045 91951	(13)1.5220 87018	(13)1.5397 44451
6	(15)6.3213 63049	(15)6.4100 82528	(15)6.4998 37227	(15)6.5906 36787	(15)6.6824 90916
7	(18)2.7181 86111	(18)2.7627 45570	(18)2.8079 29682	(18)2.8537 45729	(18)2.9002 01058
8	(21)1.1688 20028	(21)1.1907 43340	(21)1.2130 25623	(21)1.2356 71901	(21)1.2586 87259
9	(23)5.0259 26119	(23)5.1321 03797	(23)5.2402 70690	(23)5.3504 59329	(23)5.4627 02704
10	(26)2.1611 48231	(26)2.2119 36737	(26)2.2637 96938	(26)2.3167 48890	(26)2.3708 12974
24	(63)1.5967 72093	(63)1.6883 18906	(63)1.7848 83700	(63)1.8867 28946	(63)1.9941 30189
1/2	(1)2.0736 44135	(1)2.0760 53949	(1)2.0784 60969	(1)2.0808 65205	(1)2.0832 66666
1/3	7.5478 42314	7.5536 88825	7.5595 26299	7.5653 54772	7.5711 74278
1/4	4.5537 28292	4.5563 73502	4.5590 14114	4.5616 50145	4.5642 81614
1/5	3.3627 43107	3.3643 05720	3.3658 65436	3.3674 22267	3.3689 76223

k	435	436	437	438	439
1	435	436	437	438	439
2	1 89225	1 90096	1 90969	1 91844	1 92721
3	823 12875	828 81856	834 53453	840 27672	846 04519
4	(10)3.5806 10063	(10)3.6136 48922	(10)3.6469 15896	(10)3.6804 12034	(10)3.7141 38384
5	(13)1.5575 65377	(13)1.5755 50930	(13)1.5937 02247	(13)1.6120 20471	(13)1.6305 06751
6	(15)6.7754 09391	(15)6.8694 02054	(15)6.9644 78818	(15)7.0606 49662	(15)7.1579 24635
7	(18)2.9473 03085	(18)2.9950 59296	(18)3.0434 77243	(18)3.0925 64552	(18)3.1423 28915
8	(21)1.2820 76842	(21)1.3058 45853	(21)1.3299 99555	(21)1.3545 43274	(21)1.3794 82394
9	(23)5.5770 34263	(23)5.6934 87918	(23)5.8120 98057	(23)5.9328 99539	(23)6.0559 27708
10	(26)2.4260 09904	(26)2.4823 60732	(26)2.5398 86851	(26)2.5986 09998	(26)2.6585 52264
24	(63)2.1073 76666	(63)2.2267 71952	(63)2.3526 34640	(63)2.4852 99040	(63)2.6251 15920
1/2	(1)2.0856 65361	(1)2.0880 61302	(1)2.0904 54496	(1)2.0928 44954	(1)2.0952 32684
1/3	7.5769 84852	7.5827 86527	7.5885 79338	7.5943 63318	7.6001 38502
1/4	4.5669 08540	4.5695 30941	4.5721 48834	4.5747 62238	4.5773 71171
1/5	3.3705 27318	3.3720 75562	3.3736 20969	3.3751 63549	3.3767 03314

k	440	441	442	443	444
1	440	441	442	443	444
2	1 93600	1 94481	1 95364	1 96249	1 97136
3	851 84000	857 66121	863 50888	869 38307	875 28384
4	(10)3.7480 96000	(10)3.7822 85936	(10)3.8167 09250	(10)3.8513 67000	(10)3.8862 60250
5	(13)1.6491 62240	(13)1.6679 88098	(13)1.6869 85488	(13)1.7061 55581	(13)1.7254 99551
6	(15)7.2563 13856	(15)7.3558 27511	(15)7.4564 75858	(15)7.5582 69224	(15)7.6612 18006
7	(18)3.1927 78097	(18)3.2439 19933	(18)3.2957 62329	(18)3.3483 13266	(18)3.4015 80795
8	(21)1.4048 22363	(21)1.4305 68690	(21)1.4567 26950	(21)1.4833 02777	(21)1.5103 01873
9	(23)6.1812 18395	(23)6.3088 07924	(23)6.4387 33117	(23)6.5710 31302	(23)6.7057 40315
10	(26)2.7197 36094	(26)2.7821 84294	(26)2.8459 20038	(26)2.9109 66867	(26)2.9773 48700
24	(63)2.7724 53276	(63)2.9276 97132	(63)3.0912 52385	(63)3.2635 43677	(63)3.4450 16313
1/2	(1)2.0976 17696	(1)2.1000 00000	(1)2.1023 79604	(1)2.1047 56518	(1)2.1071 30751
1/3	7.6059 04492	7.6116 62611	7.6174 11603	7.6231 51930	7.6288 83626
1/4	4.5799 75651	4.5825 75695	4.5851 71321	4.5877 62546	4.5903 49388
1/5	3.3782 40276	3.3797 74445	3.3813 05834	3.3828 34454	3.3843 60316

k	445	446	447	448	449
1	445	446	447	448	449
2	1 98025	1 98916	1 99809	2 00704	2 01601
3	881 21125	887 16536	893 14623	899 15392	905 18849
4	(10)3.9213 90063	(10)3.9567 57506	(10)3.9923 63648	(10)4.0282 09562	(10)4.0642 96320
5	(13)1.7450 18578	(13)1.7647 13847	(13)1.7845 86551	(13)1.8046 37884	(13)1.8248 69048
6	(15)7.7653 32671	(15)7.8706 23760	(15)7.9771 01882	(15)8.0847 77719	(15)8.1936 62024
7	(18)3.4555 73039	(18)3.5102 98197	(18)3.5657 64541	(18)3.6219 80418	(18)3.6789 54249
8	(21)1.5377 30002	(21)1.5655 92996	(21)1.5938 06750	(21)1.6226 47227	(21)1.6518 50458
9	(23)6.8428 98510	(23)6.9825 44761	(23)7.1247 18472	(23)7.2694 59578	(23)7.4168 08555
10	(26)3.0450 89837	(26)3.1142 14964	(26)3.1847 49157	(26)3.2567 17891	(26)3.3301 47041
24	(63)3.6361 37215	(63)3.8373 95917	(63)4.0493 05610	(63)4.2724 04226	(63)4.5072 55570
1/2	(1)2.1095 02311	(1)2.1118 71208	(1)2.1142 37451	(1)2.1166 01049	(1)2.1189 62010
1/3	7.6346 06721	7.6403 21250	7.6460 27242	7.6517 24731	7.6574 13748
1/4	4.5929 31864	4.5955 09991	4.5980 83787	4.6006 53268	4.6032 18450
1/5	3.3858 83431	3.3874 03811	3.3889 21465	3.3904 36406	3.3919 48644

TABLE II.

POWERS AND ROOTS n^k

Table II

k	450	451	452	453	454
1	450	451	452	453	454
2	2 02500	2 03401	2 04304	2 05209	2 06116
3	911 25000	917 33851	923 45408	929 59677	935 76664
4	(10)4.1006 25000	(10)4.1371 96680	(10)4.1740 12442	(10)4.2110 73368	(10)4.2483 80546
5	(13)1.8452 81250	(13)1.8658 75703	(13)1.8866 53624	(13)1.9076 16236	(13)1.9287 64768
6	(15)8.3037 65625	(15)8.4150 99419	(15)8.5276 74379	(15)8.6415 01548	(15)8.7565 92045
7	(18)3.7366 94531	(18)3.7952 09838	(18)3.8545 08819	(18)3.9146 00201	(18)3.9754 92789
8	(21)1.6815 12539	(21)1.7116 39637	(21)1.7422 37986	(21)1.7733 13891	(21)1.8048 73726
9	(23)7.5668 06426	(23)7.7194 94763	(23)7.8749 15698	(23)8.0331 11927	(23)8.1941 26716
10	(26)3.4050 62892	(26)3.4814 92138	(26)3.5594 61895	(26)3.6389 99703	(26)3.7201 33529
24	(63)4.7544 50505	(63)5.0146 08183	(63)5.2883 77338	(63)5.5764 37619	(63)5.8795 01000
1/2	(1)2.1213 20344	(1)2.1236 76058	(1)2.1260 29163	(1)2.1283 79665	(1)2.1307 27575
1/3	7.6630 94324	7.6687 66491	7.6744 30279	7.6800 85719	7.6857 32843
1/4	4.6057 79352	4.6083 35988	4.6108 88377	4.6134 36534	4.6159 80476
1/5	3.3934 58190	3.3949 65055	3.3964 69249	3.3979 70784	3.3994 69669

k	455	456	457	458	459
1	455	456	457	458	459
2	2 07025	2 07936	2 08849	2 09764	2 10681
3	941 96375	948 18816	954 43993	960 71912	967 02579
4	(10)4.2859 35063	(10)4.3237 38010	(10)4.3617 90480	(10)4.4000 93570	(10)4.4386 48376
5	(13)1.9501 00453	(13)1.9716 24532	(13)1.9933 38249	(13)2.0152 42855	(13)2.0373 39605
6	(15)8.8729 57063	(15)8.9906 07868	(15)9.1095 55800	(15)9.2298 12275	(15)9.3513 88785
7	(18)4.0371 95464	(18)4.0997 17188	(18)4.1630 67001	(18)4.2272 54022	(18)4.2922 87452
8	(21)1.8369 23936	(21)1.8694 71038	(21)1.9025 21619	(21)1.9360 82342	(21)1.9701 59941
9	(23)8.3580 03909	(23)8.5247 87931	(23)8.6945 23800	(23)8.8672 57127	(23)9.0430 34128
10	(26)3.8028 91778	(26)3.8873 03297	(26)3.9733 97377	(26)4.0612 03764	(26)4.1507 52665
24	(63)6.1983 13235	(63)6.5336 55383	(63)6.8863 45396	(63)7.2572 39774	(63)7.6472 35292
1/2	(1)2.1330 72901	(1)2.1354 15650	(1)2.1377 55833	(1)2.1400 93456	(1)2.1424 28529
1/3	7.6913 71681	7.6970 02263	7.7026 24618	7.7082 38778	7.7138 44772
1/4	4.6185 20218	4.6210 55778	4.6235 87171	4.6261 14413	4.6286 37519
1/5	3.4009 65915	3.4024 59532	3.4039 50532	3.4054 38923	3.4069 24718

k	460	461	462	463	464
1	460	461	462	463	464
2	2 11600	2 12521	2 13444	2 14369	2 15296
3	973 36000	979 72181	986 11128	992 52847	998 97344
4	(10)4.4774 56000	(10)4.5165 17544	(10)4.5558 34114	(10)4.5954 06816	(10)4.6352 36762
5	(13)2.0596 29760	(13)2.0821 14588	(13)2.1047 95360	(13)2.1276 73356	(13)2.1507 49857
6	(15)9.4742 96896	(15)9.5985 48250	(15)9.7241 54565	(15)9.8511 27638	(15)9.9794 79338
7	(18)4.3581 76572	(18)4.4249 30743	(18)4.4925 59409	(18)4.5610 72096	(18)4.6304 78413
8	(21)2.0047 61223	(21)2.0398 93073	(21)2.0755 62447	(21)2.1117 76381	(21)2.1485 41984
9	(23)9.2219 01627	(23)9.4039 07065	(23)9.5890 98505	(23)9.7775 24642	(23)9.9692 34804
10	(26)4.2420 74748	(26)4.3352 01157	(26)4.4301 63510	(26)4.5269 93909	(26)4.6257 24949
24	(63)8.0572 70802	(63)8.4883 29103	(63)8.9414 38903	(63)9.4176 76852	(63)9.9181 69666
1/2	(1)2.1447 61059	(1)2.1470 91055	(1)2.1494 18526	(1)2.1517 43479	(1)2.1540 65923
1/3	7.7194 42629	7.7250 32380	7.7306 14052	7.7361 87677	7.7417 53281
1/4	4.6311 56507	4.6336 71390	4.6361 82186	4.6386 88909	4.6411 91574
1/5	3.4084 07924	3.4098 88554	3.4113 66616	3.4128 42121	3.4143 15079

k	465	466	467	468	469
1	465	466	467	468	469
2	2 16225	2 17156	2 18089	2 19024	2 19961
3	1005 44625	1011 94696	1018 47563	1025 03232	1031 61709
4	(10)4.6753 25063	(10)4.7156 72834	(10)4.7562 81192	(10)4.7971 51258	(10)4.8382 84152
5	(13)2.1740 26154	(13)2.1975 03540	(13)2.2211 83317	(13)2.2450 66789	(13)2.2691 55267
6	(16)1.0109 22162	(16)1.0240 36650	(16)1.0372 92609	(16)1.0506 91257	(16)1.0642 33820
7	(18)4.7007 88052	(18)4.7720 10788	(18)4.8441 56484	(18)4.9172 35083	(18)4.9912 56618
8	(21)2.1858 66444	(21)2.2237 57027	(21)2.2622 21078	(21)2.3012 66019	(21)2.3408 99354
9	(24)1.0164 27896	(24)1.0362 70775	(24)1.0564 57243	(24)1.0769 92497	(24)1.0978 81797
10	(26)4.7263 89719	(26)4.8290 21810	(26)4.9336 55326	(26)5.0403 24885	(26)5.1490 65627
24	(64)1.0444 09634	(64)1.0996 69046	(64)1.1577 24259	(64)1.2187 10278	(64)1.2827 68318
1/2	(1)2.1563 85865	(1)2.1587 03314	(1)2.1610 18278	(1)2.1633 30765	(1)2.1656 40783
1/3	7.7473 10895	7.7528 60547	7.7584 02264	7.7639 36077	7.7694 62012
1/4	4.6436 90198	4.6461 84795	4.6486 75380	4.6511 61968	4.6536 44575
1/5	3.4157 85500	3.4172 53393	3.4187 18768	3.4201 81635	3.4216 42003

k	470	471	472	473	474
1	470	471	472	473	474
2	2 20900	2 21841	2 22784	2 23729	2 24676
3	1038 23000	1044 87111	1051 54048	1058 23817	1064 96424
4	(10)4.8796 81000	(10)4.9213 42928	(10)4.9632 71066	(10)5.0054 66544	(10)5.0479 30498
5	(13)2.2934 50070	(13)2.3179 52519	(13)2.3426 63943	(13)2.3675 85675	(13)2.3927 19056
6	(16)1.0779 21533	(16)1.0917 55637	(16)1.1057 37381	(16)1.1198 68024	(16)1.1341 48832
7	(18)5.0662 31205	(18)5.1421 69048	(18)5.2190 80439	(18)5.2969 75756	(18)5.3758 65466
8	(21)2.3811 28666	(21)2.4219 61622	(21)2.4634 05967	(21)2.5054 69532	(21)2.5481 60231
9	(24)1.1191 30473	(24)1.1407 43924	(24)1.1627 27616	(24)1.1850 87089	(24)1.2078 27949
10	(26)5.2599 13224	(26)5.3729 03881	(26)5.4880 74350	(26)5.6054 61930	(26)5.7251 04480
24	(64)1.3500 46075	(64)1.4206 98007	(64)1.4948 85630	(64)1.5727 77826	(64)1.6545 51159
1/2	(1)2.1679 48339	(1)2.1702 53441	(1)2.1725 56098	(1)2.1748 56317	(1)2.1771 54106
1/3	7.7749 80097	7.7804 90361	7.7859 92832	7.7914 87536	7.7969 74500
1/4	4.6561 23215	4.6585 97902	4.6610 68652	4.6635 35480	4.6659 98399
1/5	3.4230 99883	3.4245 55283	3.4260 08213	3.4274 58683	3.4289 06701

TABLE II.

Table II
181

POWERS AND ROOTS n^k

k	475	476	477	478	479
1	475	476	477	478	479
2	2 25625	2 26576	2 27529	2 28484	2 29441
3	1071 71875	1078 50176	1085 31333	1092 15352	1099 02239
4	(10)5.0906 64063	(10)5.1336 68378	(10)5.1769 44584	(10)5.2204 93826	(10)5.2643 17248
5	(13)2.4180 65430	(13)2.4436 26148	(13)2.4694 02567	(13)2.4953 96049	(13)2.5216 07962
6	(16)1.1485 81079	(16)1.1631 66046	(16)1.1779 05024	(16)1.1927 99311	(16)1.2078 50214
7	(18)5.4557 60126	(18)5.5366 70380	(18)5.6186 06966	(18)5.7015 80708	(18)5.7856 02524
8	(21)2.5914 86060	(21)2.6354 55101	(21)2.6800 75523	(21)2.7253 55578	(21)2.7713 03609
9	(24)1.2309 55878	(24)1.2544 76628	(24)1.2783 96024	(24)1.3027 19966	(24)1.3274 54429
10	(26)5.8470 40422	(26)5.9713 08750	(26)6.0979 49036	(26)6.2270 01440	(26)6.3585 06713
24	(64)1.7403 90207	(64)1.8304 87912	(64)1.9250 45935	(64)2.0242 75033	(64)2.1283 95451
1/2	(1)2.1794 49472	(1)2.1817 42423	(1)2.1840 32967	(1)2.1863 21111	(1)2.1886 06863
1/3	7.8024 53753	7.8079 25322	7.8133 89232	7.8188 45511	7.8242 94186
1/4	4.6684 57424	4.6709 12569	4.6733 63849	4.6758 11278	4.6782 54870
1/5	3.4303 52278	3.4317 95422	3.4332 36143	3.4346 74449	3.4361 10350

k	480	481	482	483	484
1	480	481	482	483	484
2	2 30400	2 31361	2 32324	2 33289	2 34256
3	1105 92000	1112 84641	1119 80168	1126 78587	1133 79904
4	(10)5.3084 16000	(10)5.3527 91232	(10)5.3974 44098	(10)5.4423 75752	(10)5.4875 87354
5	(13)2.5480 39680	(13)2.5746 92583	(13)2.6015 68055	(13)2.6286 67488	(13)2.6559 92279
6	(16)1.2230 59046	(16)1.2384 27132	(16)1.2539 55803	(16)1.2696 46397	(16)1.2855 00263
7	(18)5.8706 83423	(18)5.9568 34506	(18)6.0440 66968	(18)6.1323 92097	(18)6.2218 21273
8	(21)2.8179 28043	(21)2.8652 37397	(21)2.9132 40279	(21)2.9619 45383	(21)3.0113 61496
9	(24)1.3526 05461	(24)1.3781 79188	(24)1.4041 81814	(24)1.4306 19620	(24)1.4574 98964
10	(26)6.4925 06211	(26)6.6290 41895	(26)6.7681 56345	(26)6.9098 92764	(26)7.0542 94987
24	(64)2.2376 37322	(64)2.3522 41094	(64)2.4724 57971	(64)2.5985 50361	(64)2.7307 92362
1/2	(1)2.1908 90230	(1)2.1931 71220	(1)2.1954 49840	(1)2.1977 26098	(1)2.2000 00000
1/3	7.8297 35282	7.8351 68827	7.8405 94846	7.8460 13365	7.8514 24411
1/4	4.6806 94639	4.6831 30598	4.6855 62762	4.6879 91145	4.6904 15760
1/5	3.4375 43855	3.4389 74973	3.4404 03713	3.4418 30083	3.4432 54092

k	485	486	487	488	489
1	485	486	487	488	489
2	2 35225	2 36196	2 37169	2 38144	2 39121
3	1140 84125	1147 91256	1155 01303	1162 14272	1169 30169
4	(10)5.5330 80063	(10)5.5788 55042	(10)5.6249 13456	(10)5.6712 56474	(10)5.7178 85264
5	(13)2.6835 43830	(13)2.7113 23550	(13)2.7393 32853	(13)2.7675 73159	(13)2.7960 45894
6	(16)1.3015 18758	(16)1.3177 03245	(16)1.3340 55099	(16)1.3505 75702	(16)1.3672 66442
7	(18)6.3123 65975	(18)6.4040 37773	(18)6.4968 48334	(18)6.5908 09424	(18)6.6859 32903
8	(21)3.0614 97498	(21)3.1123 62358	(21)3.1639 65139	(21)3.2163 14999	(21)3.2694 21189
9	(24)1.4848 26286	(24)1.5126 08106	(24)1.5408 51023	(24)1.5695 61719	(24)1.5987 46962
10	(26)7.2014 07489	(26)7.3512 75394	(26)7.5039 44480	(26)7.6594 61191	(26)7.8178 72642
24	(64)2.8694 70250	(64)3.0148 82996	(64)3.1673 42798	(64)3.3271 75643	(64)3.4947 21879
1/2	(1)2.2022 71555	(1)2.2045 40769	(1)2.2068 07649	(1)2.2090 72203	(1)2.2113 34439
1/3	7.8568 28008	7.8622 24183	7.8676 12960	7.8729 94366	7.8783 68425
1/4	4.6928 36620	4.6952 53740	4.6976 67133	4.7000 76812	4.7024 82790
1/5	3.4446 75750	3.4460 95065	3.4475 12045	3.4489 26700	3.4503 39037

k	490	491	492	493	494
1	490	491	492	493	494
2	2 40100	2 41081	2 42064	2 43049	2 44036
3	1176 49000	1183 70771	1190 95488	1198 23157	1205 53784
4	(10)5.7648 01000	(10)5.8120 04856	(10)5.8594 98010	(10)5.9072 81640	(10)5.9553 56930
5	(13)2.8247 52490	(13)2.8536 94384	(13)2.8828 73021	(13)2.9122 89849	(13)2.9419 46323
6	(16)1.3841 28720	(16)1.4011 63943	(16)1.4183 73526	(16)1.4357 58895	(16)1.4533 21484
7	(18)6.7822 30728	(18)6.8797 14959	(18)6.9783 97749	(18)7.0782 91354	(18)7.1794 08129
8	(21)3.3232 93057	(21)3.3779 40045	(21)3.4333 71692	(21)3.4895 97638	(21)3.5466 27616
9	(24)1.6284 13598	(24)1.6585 68562	(24)1.6892 18873	(24)1.7203 71635	(24)1.7520 34042
10	(26)7.9792 26630	(26)8.1435 71639	(26)8.3109 56854	(26)8.4814 32162	(26)8.6550 48169
24	(64)3.6703 36822	(64)3.8543 91376	(64)4.0472 72689	(64)4.2493 84825	(64)4.4611 49467
1/2	(1)2.2135 94362	(1)2.2158 51981	(1)2.2181 07301	(1)2.2203 60331	(1)2.2226 11077
1/3	7.8837 35163	7.8890 94604	7.8944 46773	7.8997 91695	7.9051 29393
1/4	4.7048 85081	4.7072 83697	4.7096 78653	4.7120 69960	4.7144 57633
1/5	3.4517 49066	3.4531 56794	3.4545 62231	3.4559 65384	3.4573 66263

k	495	496	497	498	499
1	495	496	497	498	499
2	2 45025	2 46016	2 47009	2 48004	2 49001
3	1212 87375	1220 23936	1227 63473	1235 05992	1242 51499
4	(10)6.0037 25063	(10)6.0523 87226	(10)6.1013 44608	(10)6.1505 98402	(10)6.2001 49800
5	(13)2.9718 43906	(13)3.0019 84064	(13)3.0323 68270	(13)3.0629 98004	(13)3.0938 74750
6	(16)1.4710 62733	(16)1.4889 84096	(16)1.5070 87030	(16)1.5253 73006	(16)1.5438 43500
7	(18)7.2817 60531	(18)7.3853 61115	(18)7.4902 22541	(18)7.5963 57570	(18)7.7037 79067
8	(21)3.6044 71463	(21)3.6631 39113	(21)3.7226 40603	(21)3.7829 86070	(21)3.8441 85754
9	(24)1.7842 13374	(24)1.8169 17000	(24)1.8501 52380	(24)1.8839 27063	(24)1.9182 48691
10	(26)8.8318 56201	(26)9.0119 08320	(26)9.1952 57326	(26)9.3819 56772	(26)9.5720 60970
24	(64)4.6830 06649	(64)4.9154 15513	(64)5.1588 55098	(64)5.4138 25162	(64)5.6808 47029
1/2	(1)2.2248 59546	(1)2.2271 05745	(1)2.2293 49681	(1)2.2315 91360	(1)2.2338 30790
1/3	7.9104 59893	7.9157 83219	7.9210 99395	7.9264 08444	7.9317 10391
1/4	4.7168 41683	4.7192 22124	4.7215 98967	4.7239 72227	4.7263 41916
1/5	3.4587 64874	3.4601 61227	3.4615 55329	3.4629 47190	3.4643 36816

TABLE II.

POWERS AND ROOTS n^k

Table II

k					
1	500	501	502	503	504
2	2 50000	2 51001	2 52004	2 53009	2 54016
3	1250 00000	1257 51501	1265 06008	1272 63527	1280 24064
4	(10)6.2500 00000	(10)6.3001 50200	(10)6.3506 01602	(10)6.4013 55408	(10)6.4524 12826
5	(13)3.1250 00000	(13)3.1563 75250	(13)3.1880 02004	(13)3.2198 81770	(13)3.2520 16064
6	(16)1.5625 00000	(16)1.5813 44000	(16)1.6003 77006	(16)1.6196 00530	(16)1.6390 16096
7	(18)7.8125 00000	(18)7.9225 33442	(18)8.0338 92570	(18)8.1465 90668	(18)8.2606 41125
8	(21)3.9062 50000	(21)3.9691 89254	(21)4.0330 14070	(21)4.0977 35106	(21)4.1633 63127
9	(24)1.9531 25000	(24)1.9885 63816	(24)2.0245 73063	(24)2.0611 60758	(24)2.0983 35016
10	(26)9.7656 25000	(26)9.9627 04720	(27)1.0163 35678	(27)1.0367 63861	(27)1.0575 60848
24	(64)5.9604 64478	(64)6.2532 44659	(64)6.5597 79050	(64)6.8806 84448	(64)7.2166 04000
1/2	(1)2.2360 67977	(1)2.2383 02929	(1)2.2405 35650	(1)2.2427 66149	(1)2.2449 94432
1/3	7.9370 05260	7.9422 93073	7.9475 73855	7.9528 47628	7.9581 14416
1/4	4.7287 08045	4.7310 70628	4.7334 29676	4.7357 85203	4.7381 37221
1/5	3.4657 24216	3.4671 09398	3.4684 92370	3.4698 73139	3.4712 51715

k					
1	505	506	507	508	509
2	2 55025	2 56036	2 57049	2 58064	2 59081
3	1287 87625	1295 54216	1303 23843	1310 96512	1318 72229
4	(10)6.5037 75063	(10)6.5554 43330	(10)6.6074 18840	(10)6.6597 02810	(10)6.7122 96456
5	(13)3.2844 06407	(13)3.3170 54325	(13)3.3499 61352	(13)3.3831 29027	(13)3.4165 58896
6	(16)1.6586 25235	(16)1.6784 29488	(16)1.6984 30405	(16)1.7186 29546	(16)1.7390 28478
7	(18)8.3760 57438	(18)8.4928 53211	(18)8.6110 42156	(18)8.7306 38093	(18)8.8516 54954
8	(21)4.2299 09006	(21)4.2973 83725	(21)4.3657 98373	(21)4.4351 64151	(21)4.5054 92371
9	(24)2.1361 04048	(24)2.1744 76165	(24)2.2134 59775	(24)2.2530 63389	(24)2.2932 95617
10	(27)1.0787 32544	(27)1.1002 84939	(27)1.1222 24106	(27)1.1445 56202	(27)1.1672 87469
24	(64)7.5682 08268	(64)7.9361 96349	(64)8.3212 97020	(64)8.7242 69942	(64)9.1459 06897
1/2	(1)2.2472 20505	(1)2.2494 44376	(1)2.2516 66050	(1)2.2538 85534	(1)2.2561 02835
1/3	7.9633 74242	7.9686 27129	7.9738 73099	7.9791 12176	7.9843 44383
1/4	4.7404 85740	4.7428 30775	4.7451 72336	4.7475 10436	4.7498 45086
1/5	3.4726 28104	3.4740 02314	3.4753 74353	3.4767 44229	3.4781 11950

k					
1	510	511	512	513	514
2	2 60100	2 61121	2 62144	2 63169	2 64196
3	1326 51000	1334 32831	1342 17728	1350 05697	1357 96744
4	(10)6.7652 01000	(10)6.8184 17664	(10)6.8719 47674	(10)6.9257 92256	(10)6.9799 52642
5	(13)3.4502 52510	(13)3.4842 11426	(13)3.5184 37209	(13)3.5529 31427	(13)3.5876 95658
6	(16)1.7596 28780	(16)1.7804 32039	(16)1.8014 39851	(16)1.8226 53822	(16)1.8440 75568
7	(18)8.9741 06779	(18)9.0980 07719	(18)9.2233 72037	(18)9.3502 14108	(18)9.4785 48420
8	(21)4.5767 94457	(21)4.6490 81944	(21)4.7223 66483	(21)4.7966 59837	(21)4.8719 73888
9	(24)2.3341 65173	(24)2.3756 80873	(24)2.4178 51639	(24)2.4606 86497	(24)2.5041 94578
10	(27)1.1904 24238	(27)1.2139 72926	(27)1.2379 40039	(27)1.2623 32173	(27)1.2871 56013
24	(64)9.5870 33090	(65)1.0048 50848	(65)1.0531 22917	(65)1.1036 12886	(65)1.1564 18034
1/2	(1)2.2583 17958	(1)2.2605 30911	(1)2.2627 41700	(1)2.2649 50331	(1)2.2671 56810
1/3	7.9895 69740	7.9947 88272	8.0000 00000	8.0052 04946	8.0104 03133
1/4	4.7521 76299	4.7545 04087	4.7568 28460	4.7591 49431	4.7614 67011
1/5	3.4794 77522	3.4808 40954	3.4822 02253	3.4835 61427	3.4849 18483

k					
1	515	516	517	518	519
2	2 65225	2 66256	2 67289	2 68324	2 69361
3	1365 90875	1373 88096	1381 88413	1389 91832	1397 98359
4	(10)7.0344 30063	(10)7.0892 25754	(10)7.1443 40952	(10)7.1997 76898	(10)7.2555 34832
5	(13)3.6227 31482	(13)3.6580 40489	(13)3.6936 24272	(13)3.7294 84433	(13)3.7656 22578
6	(16)1.8657 06713	(16)1.8875 48892	(16)1.9096 03749	(16)1.9318 72936	(16)1.9543 58118
7	(18)9.6083 89574	(18)9.7397 52284	(18)9.8726 51381	(19)1.0007 10981	(19)1.0143 11863
8	(21)4.9483 20630	(21)5.0257 12179	(21)5.1041 60764	(21)5.1836 78738	(21)5.2642 78570
9	(24)2.5483 85125	(24)2.5932 67484	(24)2.6388 51115	(24)2.6851 45586	(24)2.7321 60578
10	(27)1.3124 18339	(27)1.3381 26022	(27)1.3642 86026	(27)1.3909 05414	(27)1.4179 91340
24	(65)1.2116 39706	(65)1.2693 83471	(65)1.3297 59294	(65)1.3928 81704	(65)1.4588 69982
1/2	(1)2.2693 61144	(1)2.2715 63338	(1)2.2737 63400	(1)2.2759 61335	(1)2.2781 57150
1/3	8.0155 94581	8.0207 79314	8.0259 57353	8.0311 28718	8.0362 93433
1/4	4.7637 81212	4.7660 92045	4.7683 99522	4.7707 03654	4.7730 04452
1/5	3.4862 73428	3.4876 26271	3.4889 77017	3.4903 25675	3.4916 72252

k					
1	520	521	522	523	524
2	2 70400	2 71441	2 72484	2 73529	2 74576
3	1406 08000	1414 20761	1422 36648	1430 55667	1438 77824
4	(10)7.3116 16000	(10)7.3680 21648	(10)7.4247 53026	(10)7.4818 11384	(10)7.5391 97978
5	(13)3.8020 40320	(13)3.8387 39279	(13)3.8757 21079	(13)3.9129 87354	(13)3.9505 39740
6	(16)1.9770 60966	(16)1.9999 83164	(16)2.0231 26403	(16)2.0464 92386	(16)2.0700 82824
7	(19)1.0280 71703	(19)1.0419 91229	(19)1.0560 71983	(19)1.0703 15518	(19)1.0847 23400
8	(21)5.3459 72853	(21)5.4287 74301	(21)5.5126 95749	(21)5.5977 50159	(21)5.6839 50615
9	(24)2.7799 05884	(24)2.8283 91411	(24)2.8776 27181	(24)2.9276 23333	(24)2.9783 90122
10	(27)1.4455 51059	(27)1.4735 91925	(27)1.5021 21389	(27)1.5311 47003	(27)1.5606 76424
24	(65)1.5278 48342	(65)1.5999 46126	(65)1.6752 98008	(65)1.7540 44200	(65)1.8363 30669
1/2	(1)2.2803 50850	(1)2.2825 42442	(1)2.2847 31932	(1)2.2869 19325	(1)2.2891 04628
1/3	8.0414 51517	8.0466 02993	8.0517 47881	8.0568 86203	8.0620 17979
1/4	4.7753 01928	4.7775 96092	4.7798 86957	4.7821 74532	4.7844 58829
1/5	3.4930 16754	3.4943 59190	3.4956 99566	3.4970 37889	3.4983 74167

TABLE II.

Table II — POWERS AND ROOTS n^k — 183

k					
1	525	526	527	528	529
2	2 75625	2 76676	2 77729	2 78784	2 79841
3	1447 03125	1455 31576	1463 63183	1471 97952	1480 35889
4	(10)7.5969 14063	(10)7.6549 60898	(10)7.7133 39744	(10)7.7720 51866	(10)7.8310 98528
5	(13)3.9883 79883	(13)4.0265 09432	(13)4.0649 30045	(13)4.1036 43385	(13)4.1426 51121
6	(16)2.0938 99438	(16)2.1179 43961	(16)2.1422 18134	(16)2.1667 23707	(16)2.1914 62443
7	(19)1.0992 97205	(19)1.1140 38524	(19)1.1289 48957	(19)1.1440 30117	(19)1.1592 83632
8	(21)5.7713 10327	(21)5.8598 42634	(21)5.9495 61001	(21)6.0404 79020	(21)6.1326 10416
9	(24)3.0299 37922	(24)3.0822 77226	(24)3.1354 18647	(24)3.1893 72923	(24)3.2441 50910
10	(27)1.5907 17409	(27)1.6212 77821	(27)1.6523 65627	(27)1.6839 88903	(27)1.7161 55831
24	(65)1.9223 09365	(65)2.0121 38448	(65)2.1059 82534	(65)2.2040 12944	(65)2.3064 07963
1/2	(1)2.2912 87847	(1)2.2934 68988	(1)2.2956 48057	(1)2.2978 25059	(1)2.3000 00000
1/3	8.0671 43230	8.0722 61977	8.0773 74241	8.0824 80041	8.0875 79399
1/4	4.7867 39859	4.7890 17632	4.7912 92160	4.7935 63454	4.7958 31523
1/5	3.4997 08406	3.5010 40614	3.5023 70797	3.5036 98962	3.5050 25117

k					
1	530	531	532	533	534
2	2 80900	2 81961	2 83024	2 84089	2 85156
3	1488 77000	1497 21291	1505 68768	1514 19437	1522 73304
4	(10)7.8904 81000	(10)7.9502 00552	(10)8.0102 58458	(10)8.0706 55992	(10)8.1313 94434
5	(13)4.1819 54930	(13)4.2215 56493	(13)4.2614 57499	(13)4.3016 59644	(13)4.3421 64628
6	(16)2.2164 36113	(16)2.2416 46498	(16)2.2670 95390	(16)2.2927 84590	(16)2.3187 15911
7	(19)1.1747 11140	(19)1.1903 14290	(19)1.2060 94747	(19)1.2220 54187	(19)1.2381 94297
8	(21)6.2259 69041	(21)6.3205 68882	(21)6.4164 24056	(21)6.5135 48814	(21)6.6119 57543
9	(24)3.2997 63592	(24)3.3562 22076	(24)3.4135 37598	(24)3.4717 21518	(24)3.5307 85328
10	(27)1.7488 74704	(27)1.7821 53922	(27)1.8160 02002	(27)1.8504 27569	(27)1.8854 39365
24	(65)2.4133 53110	(65)2.5250 41417	(65)2.6416 73716	(65)2.7634 58943	(65)2.8906 14446
1/2	(1)2.3021 72887	(1)2.3043 43724	(1)2.3065 12519	(1)2.3086 79276	(1)2.3108 44002
1/3	8.0926 72335	8.0977 58868	8.1028 39019	8.1079 12808	8.1129 80255
1/4	4.7980 96379	4.8003 58033	4.8026 16494	4.8048 71774	4.8071 23882
1/5	3.5063 49267	3.5076 71420	3.5089 91583	3.5103 09762	3.5116 25964

k					
1	535	536	537	538	539
2	2 86225	2 87296	2 88369	2 89444	2 90521
3	1531 30375	1539 90656	1548 54153	1557 20872	1565 90819
4	(10)8.1924 75063	(10)8.2538 99162	(10)8.3156 68016	(10)8.3777 82914	(10)8.4402 45144
5	(13)4.3829 74158	(13)4.4240 89951	(13)4.4655 13725	(13)4.5072 47208	(13)4.5492 92133
6	(16)2.3448 91175	(16)2.3713 12214	(16)2.3979 80870	(16)2.4248 98998	(16)2.4520 68460
7	(19)1.2545 16778	(19)1.2710 23346	(19)1.2877 15727	(19)1.3045 95661	(19)1.3216 64900
8	(21)6.7116 64765	(21)6.8126 85137	(21)6.9150 33455	(21)7.0187 24655	(21)7.1237 73809
9	(24)3.5907 40649	(24)3.6515 99233	(24)3.7133 72966	(24)3.7770 73864	(24)3.8397 14083
10	(27)1.9210 46247	(27)1.9572 57189	(27)1.9940 81282	(27)2.0315 27739	(27)2.0696 05891
24	(65)3.0233 66304	(65)3.1619 49669	(65)3.3066 09101	(65)3.4575 98937	(65)3.6151 83652
1/2	(1)2.3130 06701	(1)2.3151 67381	(1)2.3173 26045	(1)2.3194 82701	(1)2.3216 37353
1/3	8.1180 41379	8.1230 96201	8.1281 44739	8.1331 87014	8.1382 23044
1/4	4.8093 72829	4.8116 18626	4.8138 61283	4.8161 00810	4.8183 37217
1/5	3.5129 40196	3.5142 52463	3.5155 62774	3.5168 71134	3.5181 77550

k					
1	540	541	542	543	544
2	2 91600	2 92681	2 93764	2 94849	2 95936
3	1574 64000	1583 40421	1592 20088	1601 03007	1609 89184
4	(10)8.5030 56000	(10)8.5662 16776	(10)8.6297 28770	(10)8.6935 93280	(10)8.7578 11610
5	(13)4.5916 50240	(13)4.6343 23276	(13)4.6773 12993	(13)4.7206 21151	(13)4.7642 49516
6	(16)2.4794 91130	(16)2.5071 68892	(16)2.5351 03642	(16)2.5632 97285	(16)2.5917 51736
7	(19)1.3389 25210	(19)1.3563 78371	(19)1.3740 26114	(19)1.3918 70426	(19)1.4099 12945
8	(21)7.2301 96134	(21)7.3380 06986	(21)7.4472 21864	(21)7.5578 56412	(21)7.6699 26419
9	(24)3.9043 05912	(24)3.9698 61779	(24)4.0363 94250	(24)4.1039 16032	(24)4.1724 39972
10	(27)2.1083 25193	(27)2.1476 95223	(27)2.1877 25684	(27)2.2284 26405	(27)2.2698 07345
24	(65)3.7796 38253	(65)3.9512 48669	(65)4.1303 12169	(65)4.3171 37789	(65)4.5120 46770
1/2	(1)2.3237 90008	(1)2.3259 40670	(1)2.3280 89345	(1)2.3302 36040	(1)2.3323 80758
1/3	8.1432 52850	8.1482 76449	8.1532 93862	8.1583 05107	8.1633 10204
1/4	4.8205 70514	4.8228 00711	4.8250 27819	4.8272 51847	4.8294 72806
1/5	3.5194 82029	3.5207 84576	3.5220 85199	3.5233 83903	3.5246 80696

k					
1	545	546	547	548	549
2	2 97025	2 98116	2 99209	3 00304	3 01401
3	1618 78625	1627 71336	1636 67323	1645 66592	1654 69149
4	(10)8.8223 85063	(10)8.8873 14946	(10)8.9526 02568	(10)9.0182 49242	(10)9.0842 56280
5	(13)4.8081 99859	(13)4.8524 73960	(13)4.8970 73605	(13)4.9420 00584	(13)4.9872 56698
6	(16)2.6204 68923	(16)2.6494 50782	(16)2.6786 99262	(16)2.7082 16320	(16)2.7380 03927
7	(19)1.4281 55563	(19)1.4466 00127	(19)1.4652 48496	(19)1.4841 02543	(19)1.5031 64156
8	(21)7.7834 47819	(21)7.8984 36694	(21)8.0149 09274	(21)8.1328 81938	(21)8.2523 71216
9	(24)4.2419 79061	(24)4.3125 46435	(24)4.3841 55373	(24)4.4568 19302	(24)4.5305 51798
10	(27)2.3118 78588	(27)2.3546 50354	(27)2.3981 32989	(27)2.4423 36978	(27)2.4872 72937
24	(65)4.7153 73024	(65)4.9274 63602	(65)5.1486 79188	(65)5.3793 94612	(65)5.6199 99369
1/2	(1)2.3345 23506	(1)2.3366 64289	(1)2.3388 03113	(1)2.3409 39982	(1)2.3430 74903
1/3	8.1683 09170	8.1733 02026	8.1782 88788	8.1832 69477	8.1882 44110
1/4	4.8316 90704	4.8339 05553	4.8361 17361	4.8383 26138	4.8405 31895
1/5	3.5259 75582	3.5272 68570	3.5285 59664	3.5298 48871	3.5311 36198

TABLE II.

Table **II**

POWERS AND ROOTS n^k

k					
1	550	551	552	553	554
2	3 02500	3 03601	3 04704	3 05809	3 06916
3	1663 75000	1672 84151	1681 96608	1691 12377	1700 31464
4	(10)9.1506 25000	(10)9.2173 56720	(10)9.2844 52762	(10)9.3519 14448	(10)9.4197 43106
5	(13)5.0328 43750	(13)5.0787 63553	(13)5.1250 17924	(13)5.1716 08690	(13)5.2185 37681
6	(16)2.7680 64063	(16)2.7983 98718	(16)2.8290 09894	(16)2.8598 99605	(16)2.8910 69875
7	(19)1.5224 35234	(19)1.5419 17693	(19)1.5616 13462	(19)1.5815 24482	(19)1.6016 52711
8	(21)8.3733 93789	(21)8.4959 66491	(21)8.6201 06308	(21)8.7458 30384	(21)8.8731 56018
9	(24)4.6053 66584	(24)4.6812 77536	(24)4.7582 98682	(24)4.8364 44203	(24)4.9157 28434
10	(27)2.5329 51621	(27)2.5793 83922	(27)2.6265 80873	(27)2.6745 53644	(27)2.7233 13552
24	(65)5.8708 98173	(65)6.1325 11516	(65)6.4052 76258	(65)6.6896 46227	(65)6.9860 92851
1/2	(1)2.3452 07880	(1)2.3473 38919	(1)2.3494 68025	(1)2.3515 95203	(1)2.3537 20459
1/3	8.1932 12706	8.1981 75283	8.2031 31859	8.2080 82453	8.2130 27082
1/4	4.8427 34641	4.8449 34384	4.8471 31136	4.8493 24905	4.8515 15700
1/5	3.5324 21650	3.5337 05234	3.5349 86956	3.5362 66821	3.5375 44836

k					
1	555	556	557	558	559
2	3 08025	3 09136	3 10249	3 11364	3 12481
3	1709 53875	1718 79616	1728 08693	1737 41112	1746 76879
4	(10)9.4879 40063	(10)9.5565 06650	(10)9.6254 44200	(10)9.6947 54050	(10)9.7644 37536
5	(13)5.2658 06735	(13)5.3134 17697	(13)5.3613 72419	(13)5.4096 72760	(13)5.4583 20583
6	(16)2.9225 22738	(16)2.9542 60240	(16)2.9862 84438	(16)3.0185 97400	(16)3.0512 01206
7	(19)1.6220 00119	(19)1.6425 68693	(19)1.6633 60432	(19)1.6843 77349	(19)1.7056 21474
8	(21)9.0021 00663	(21)9.1326 81934	(21)9.2649 17605	(21)9.3988 25608	(21)9.5344 24040
9	(24)4.9961 65868	(24)5.0777 71156	(24)5.1605 59106	(24)5.2445 44689	(24)5.3297 43038
10	(27)2.7728 72057	(27)2.8232 40762	(27)2.8744 31422	(27)2.9264 55937	(27)2.9793 26358
24	(65)7.2951 05803	(65)7.6171 93672	(65)7.9528 84664	(65)8.3027 27311	(65)8.6672 91224
1/2	(1)2.3558 43798	(1)2.3579 65225	(1)2.3600 84744	(1)2.3622 02362	(1)2.3643 18084
1/3	8.2179 65765	8.2228 98519	8.2278 25361	8.2327 46311	8.2376 61384
1/4	4.8537 03532	4.8558 88409	4.8580 70341	4.8602 49337	4.8624 25407
1/5	3.5388 21007	3.5400 95340	3.5413 67840	3.5426 38514	3.5439 07368

k					
1	560	561	562	563	564
2	3 13600	3 14721	3 15844	3 16969	3 18096
3	1756 16000	1765 58481	1775 04328	1784 53547	1794 06144
4	(10)9.8344 96000	(10)9.9049 30784	(10)9.9757 43234	(11)1.0046 93470	(11)1.0118 50652
5	(13)5.5073 17760	(13)5.5566 66170	(13)5.6063 67697	(13)5.6564 24234	(13)5.7068 37678
6	(16)3.0840 97946	(16)3.1172 89721	(16)3.1507 78646	(16)3.1845 66844	(16)3.2186 56450
7	(19)1.7270 94850	(19)1.7487 99534	(19)1.7707 37599	(19)1.7929 11133	(19)1.8153 22238
8	(21)9.6717 31157	(21)9.8107 65384	(21)9.9515 45306	(22)1.0094 08968	(22)1.0238 41742
9	(24)5.4161 69448	(24)5.5038 39380	(24)5.5927 48462	(24)5.6829 72489	(24)5.7744 67426
10	(27)3.0330 54891	(27)3.0876 53892	(27)3.1431 35876	(27)3.1995 13511	(27)3.2567 99629
24	(65)9.0471 67858	(65)9.4429 71309	(65)9.8553 39138	(66)1.0284 93323	(66)1.0732 44065
1/2	(1)2.3664 31913	(1)2.3685 43856	(1)2.3706 53918	(1)2.3727 62104	(1)2.3748 68417
1/3	8.2425 70600	8.2474 73974	8.2523 71525	8.2572 63270	8.2621 49226
1/4	4.8645 98558	4.8667 68801	4.8689 36145	4.8711 00598	4.8732 62170
1/5	3.5451 74407	3.5464 39637	3.5477 03064	3.5489 64695	3.5502 24533

k					
1	565	566	567	568	569
2	3 19225	3 20356	3 21489	3 22624	3 23761
3	1803 62125	1813 21496	1822 84263	1832 50432	1842 20009
4	(11)1.0190 46006	(11)1.0262 79667	(11)1.0335 51771	(11)1.0408 62454	(11)1.0482 11851
5	(13)5.7576 09935	(13)5.8087 42917	(13)5.8602 38543	(13)5.9120 98737	(13)5.9643 25433
6	(16)3.2530 49613	(16)3.2877 48491	(16)3.3227 55254	(16)3.3580 72083	(16)3.3937 01172
7	(19)1.8379 73032	(19)1.8608 65646	(19)1.8840 02229	(19)1.9073 84943	(19)1.9310 15967
8	(22)1.0384 54763	(22)1.0532 49956	(22)1.0682 29264	(22)1.0833 94648	(22)1.0987 48085
9	(24)5.8672 69410	(24)5.9613 94749	(24)6.0568 59926	(24)6.1536 81599	(24)6.2518 76604
10	(27)3.3150 07217	(27)3.3741 49428	(27)3.4342 39578	(27)3.4952 91148	(27)3.5573 17788
24	(66)1.1198 57461	(66)1.1684 07534	(66)1.2189 71112	(66)1.2716 27927	(66)1.3264 60719
1/2	(1)2.3769 72865	(1)2.3790 75451	(1)2.3811 76180	(1)2.3832 75058	(1)2.3853 72088
1/3	8.2670 29409	8.2719 03838	8.2767 72529	8.2816 35499	8.2864 92764
1/4	4.8754 20869	4.8775 76704	4.8797 29685	4.8818 79820	4.8840 27117
1/5	3.5514 82586	3.5527 38859	3.5539 93358	3.5552 46087	3.5564 97054

k					
1	570	571	572	573	574
2	3 24900	3 26041	3 27184	3 28329	3 29476
3	1851 93000	1861 69411	1871 49248	1881 32517	1891 19224
4	(11)1.0556 00100	(11)1.0630 27337	(11)1.0704 93699	(11)1.0779 99322	(11)1.0855 44384
5	(13)6.0169 20570	(13)6.0698 86093	(13)6.1232 23956	(13)6.1769 36117	(13)6.2310 24545
6	(16)3.4296 44725	(16)3.4659 04959	(16)3.5024 84103	(16)3.5393 84395	(16)3.5766 08089
7	(19)1.9548 97493	(19)1.9790 31732	(19)2.0034 20907	(19)2.0280 67258	(19)2.0529 73043
8	(22)1.1142 91571	(22)1.1300 27119	(22)1.1459 56759	(22)1.1620 82539	(22)1.1784 06527
9	(24)6.3514 61955	(24)6.4524 54848	(24)6.5548 72660	(24)6.6587 32949	(24)6.7640 53463
10	(27)3.6203 33315	(27)3.6843 51718	(27)3.7493 87161	(27)3.8154 53980	(27)3.8825 66688
24	(66)1.3835 55344	(66)1.4430 00887	(66)1.5048 89774	(66)1.5693 17896	(66)1.6363 84728
1/2	(1)2.3874 67277	(1)2.3895 60629	(1)2.3916 52149	(1)2.3937 41841	(1)2.3958 29710
1/3	8.2913 44342	8.2961 90248	8.3010 30501	8.3058 65115	8.3106 94107
1/4	4.8861 71586	4.8883 13236	4.8904 52074	4.8925 88109	4.8947 21351
1/5	3.5577 46263	3.5589 93720	3.5602 39430	3.5614 83400	3.5627 25633

TABLE II.

Table II POWERS AND ROOTS n^k *185*

k	575	576	577	578	579
1	575	576	577	578	579
2	3 30625	3 31776	3 32929	3 34084	3 35241
3	1901 09375	1911 02976	1921 00033	1931 00552	1941 04539
4	(11)1.0931 28906	(11)1.1007 53142	(11)1.1084 17190	(11)1.1161 21191	(11)1.1238 65281
5	(13)6.2854 91211	(13)6.3403 38097	(13)6.3955 67189	(13)6.4511 80481	(13)6.5071 79976
6	(16)3.6141 57446	(16)3.6520 34744	(16)3.6902 42268	(16)3.7287 82318	(16)3.7676 57206
7	(19)2.0781 40532	(19)2.1035 72012	(19)2.1292 69789	(19)2.1552 36180	(19)2.1814 73522
8	(22)1.1949 30806	(22)1.2116 57479	(22)1.2285 88668	(22)1.2457 26512	(22)1.2630 73169
9	(24)6.8708 52133	(24)6.9791 47080	(24)7.0889 56614	(24)7.2002 99239	(24)7.3131 93651
10	(27)3.9507 39976	(27)4.0199 88718	(27)4.0903 27966	(27)4.1617 72960	(27)4.2343 39124
24	(66)1.7061 93459	(66)1.7788 51122	(66)1.8544 68735	(66)1.9331 61432	(66)2.0150 48620
1/2	(1)2.3979 15762	(1)2.4000 00000	(1)2.4020 82430	(1)2.4041 63056	(1)2.4062 41883
1/3	8.3155 17494	8.3203 35292	8.3251 47517	8.3299 54185	8.3347 55313
1/4	4.8968 51807	4.8989 79486	4.9011 04396	4.9032 26546	4.9053 45944
1/5	3.5639 66137	3.5652 04916	3.5664 41976	3.5676 77321	3.5689 10958

k	580	581	582	583	584
1	580	581	582	583	584
2	3 36400	3 37561	3 38724	3 39889	3 41056
3	1951 12000	1961 22941	1971 37368	1981 55287	1991 76704
4	(11)1.1316 49600	(11)1.1394 74287	(11)1.1473 39482	(11)1.1552 45323	(11)1.1631 91951
5	(13)6.5635 67680	(13)6.6203 45609	(13)6.6775 15784	(13)6.7350 80234	(13)6.7930 40996
6	(16)3.8068 69254	(16)3.8464 20799	(16)3.8863 14186	(16)3.9265 51777	(16)3.9671 35942
7	(19)2.2079 84168	(19)2.2347 70484	(19)2.2618 34856	(19)2.2891 79686	(19)2.3168 07390
8	(22)1.2806 30817	(22)1.2984 01651	(22)1.3163 87886	(22)1.3345 91757	(22)1.3530 15516
9	(24)7.4276 58740	(24)7.5437 13594	(24)7.6613 77499	(24)7.7806 69942	(24)7.9016 10612
10	(27)4.3080 42069	(27)4.3828 97598	(27)4.4589 21704	(27)4.5361 30576	(27)4.6145 40597
24	(66)2.1002 54121	(66)2.1889 06331	(66)2.2811 38380	(66)2.3770 88299	(66)2.4768 99188
1/2	(1)2.4083 18916	(1)2.4103 94159	(1)2.4124 67616	(1)2.4145 39294	(1)2.4166 09195
1/3	8.3395 50915	8.3443 41009	8.3491 25609	8.3539 04732	8.3586 78393
1/4	4.9074 62599	4.9095 76518	4.9116 87710	4.9137 96184	4.9159 01946
1/5	3.5701 42892	3.5713 73127	3.5726 01670	3.5738 28526	3.5750 53698

k	585	586	587	588	589
1	585	586	587	588	589
2	3 42225	3 43396	3 44569	3 45744	3 46921
3	2002 01625	2012 30056	2022 62003	2032 97472	2043 36469
4	(11)1.1711 79506	(11)1.1792 08128	(11)1.1872 77958	(11)1.1953 89135	(11)1.2035 41802
5	(13)6.8514 00112	(13)6.9101 59631	(13)6.9693 21611	(13)7.0288 88116	(13)7.0888 61216
6	(16)4.0080 69065	(16)4.0493 53544	(16)4.0909 91786	(16)4.1329 86212	(16)4.1753 39256
7	(19)2.3447 20403	(19)2.3729 21177	(19)2.4014 12178	(19)2.4301 95893	(19)2.4592 74822
8	(22)1.3716 61436	(22)1.3905 31810	(22)1.4096 28949	(22)1.4289 55185	(22)1.4485 12870
9	(24)8.0242 19400	(24)8.1485 16404	(24)8.2745 21928	(24)8.4022 56487	(24)8.5317 40805
10	(27)4.6941 68349	(27)4.7750 30613	(27)4.8571 44372	(27)4.9405 26815	(27)5.0251 95334
24	(66)2.5807 19397	(66)2.6887 02707	(66)2.8010 08521	(66)2.9178 02055	(66)3.0392 54545
1/2	(1)2.4186 77324	(1)2.4207 43687	(1)2.4228 08288	(1)2.4248 71131	(1)2.4269 32220
1/3	8.3634 46607	8.3682 09391	8.3729 66760	8.3777 18728	8.3824 65312
1/4	4.9180 05007	4.9201 05372	4.9222 03051	4.9242 98052	4.9263 90382
1/5	3.5762 77194	3.5774 99018	3.5787 19175	3.5799 37670	3.5811 54508

k	590	591	592	593	594
1	590	591	592	593	594
2	3 48100	3 49281	3 50464	3 51649	3 52836
3	2053 79000	2064 25071	2074 74688	2085 27857	2095 84584
4	(11)1.2117 36100	(11)1.2199 72170	(11)1.2282 50153	(11)1.2365 70192	(11)1.2449 32429
5	(13)7.1492 42990	(13)7.2100 35522	(13)7.2712 40906	(13)7.3328 62239	(13)7.3948 98628
6	(16)4.2180 53364	(16)4.2611 30994	(16)4.3045 74616	(16)4.3483 86715	(16)4.3925 69785
7	(19)2.4886 51485	(19)2.5183 28417	(19)2.5483 08173	(19)2.5785 93322	(19)2.6091 86452
8	(22)1.4683 04376	(22)1.4883 32095	(22)1.5085 98438	(22)1.5291 05840	(22)1.5498 56753
9	(24)8.6629 95819	(24)8.7960 42679	(24)8.9309 02754	(24)9.0675 97630	(24)9.2061 49111
10	(27)5.1111 67533	(27)5.1984 61223	(27)5.2870 94431	(27)5.3770 85394	(27)5.4684 52572
24	(66)3.1655 43453	(66)3.2968 52680	(66)3.4333 72793	(66)3.5753 01250	(66)3.7228 42640
1/2	(1)2.4289 91560	(1)2.4310 49156	(1)2.4331 05012	(1)2.4351 59132	(1)2.4372 11521
1/3	8.3872 06527	8.3919 42387	8.3966 72908	8.4013 98104	8.4061 17992
1/4	4.9284 80050	4.9305 67063	4.9326 51429	4.9347 33156	4.9368 12252
1/5	3.5823 69695	3.5835 83235	3.5847 95134	3.5860 05396	3.5872 14026

k	595	596	597	598	599
1	595	596	597	598	599
2	3 54025	3 55216	3 56409	3 57604	3 58801
3	2106 44875	2117 08736	2127 76173	2138 47192	2149 21799
4	(11)1.2533 37006	(11)1.2617 84067	(11)1.2702 73753	(11)1.2788 06208	(11)1.2873 81576
5	(13)7.4573 55187	(13)7.5202 33037	(13)7.5835 34304	(13)7.6472 61125	(13)7.7114 15640
6	(16)4.4371 26336	(16)4.4820 58890	(16)4.5273 69980	(16)4.5730 62153	(16)4.6191 37969
7	(19)2.6400 90170	(19)2.6713 07098	(19)2.7028 39878	(19)2.7346 91167	(19)2.7668 63643
8	(22)1.5708 53651	(22)1.5920 99031	(22)1.6135 95407	(22)1.6353 45318	(22)1.6573 51322
9	(24)9.3465 79225	(24)9.4889 10223	(24)9.6331 64589	(24)9.7793 65002	(24)9.9275 34420
10	(27)5.5612 14639	(27)5.6553 90493	(27)5.7509 99254	(27)5.8480 60271	(27)5.9465 93118
24	(66)3.8762 08928	(66)4.0356 19703	(66)4.2013 02448	(66)4.3734 92798	(66)4.5524 34829
1/2	(1)2.4392 62184	(1)2.4413 11123	(1)2.4433 58345	(1)2.4454 03852	(1)2.4474 47650
1/3	8.4108 32585	8.4155 41899	8.4202 45948	8.4249 44747	8.4296 38310
1/4	4.9388 88725	4.9409 62581	4.9430 33830	4.9451 02478	4.9471 68534
1/5	3.5884 21030	3.5896 26411	3.5908 30176	3.5920 32329	3.5932 32875

TABLE II.

POWERS AND ROOTS n^k

Table II

k					
1	600	601	602	603	604
2	3 60000	3 61201	3 62404	3 63609	3 64816
3	2160 00000	2170 81801	2181 67208	2192 56227	2203 48864
4	(11)1.2960 00000	(11)1.3046 61624	(11)1.3133 66592	(11)1.3221 15049	(11)1.3309 07139
5	(13)7.7760 00000	(13)7.8410 16360	(13)7.9064 66885	(13)7.9723 53744	(13)8.0386 79117
6	(16)4.6656 00000	(16)4.7124 50833	(16)4.7596 93065	(16)4.8073 29308	(16)4.8553 62187
7	(19)2.7993 60000	(19)2.8321 82950	(19)2.8653 35225	(19)2.8988 19573	(19)2.9326 38761
8	(22)1.6796 16000	(22)1.7021 41953	(22)1.7249 31805	(22)1.7479 88202	(22)1.7713 13811
9	(25)1.0077 69600	(25)1.0229 87314	(25)1.0384 08947	(25)1.0540 36886	(25)1.0698 73542
10	(27)6.0466 17600	(27)6.1481 53756	(27)6.2512 21860	(27)6.3558 42422	(27)6.4620 36194
24	(66)4.7383 81338	(66)4.9315 94142	(66)5.1323 44384	(66)5.3409 12849	(66)5.5575 90288
1/2	(1)2.4494 89743	(1)2.4515 30134	(1)2.4535 68829	(1)2.4556 05832	(1)2.4576 41145
1/3	8.4343 26653	8.4390 09789	8.4436 87734	8.4483 60500	8.4530 28104
1/4	4.9492 32004	4.9512 92896	4.9533 51218	4.9554 06978	4.9574 60182
1/5	3.5944 31819	3.5956 29165	3.5968 24918	3.5980 19083	3.5992 11665

k					
1	605	606	607	608	609
2	3 66025	3 67236	3 68449	3 69664	3 70881
3	2214 45125	2225 45016	2236 48543	2247 55712	2258 66529
4	(11)1.3397 43006	(11)1.3486 22797	(11)1.3575 46656	(11)1.3665 14729	(11)1.3755 27162
5	(13)8.1054 45188	(13)8.1726 54150	(13)8.2403 08202	(13)8.3084 09552	(13)8.3769 60414
6	(16)4.9037 94339	(16)4.9526 28415	(16)5.0018 67079	(16)5.0515 13008	(16)5.1015 68892
7	(19)2.9667 95575	(19)3.0012 92819	(19)3.0361 33317	(19)3.0713 19909	(19)3.1068 55455
8	(22)1.7949 11323	(22)1.8187 83448	(22)1.8429 32923	(22)1.8673 62504	(22)1.8920 74972
9	(25)1.0859 21350	(25)1.1021 82770	(25)1.1186 60284	(25)1.1353 56403	(25)1.1522 73658
10	(27)6.5698 24169	(27)6.6792 27585	(27)6.7902 67926	(27)6.9029 66929	(27)7.0173 46578
24	(66)5.7826 77757	(66)6.0164 86963	(66)6.2593 40623	(66)6.5115 72833	(66)6.7735 29447
1/2	(1)2.4596 74775	(1)2.4617 06725	(1)2.4637 36999	(1)2.4657 65601	(1)2.4677 92536
1/3	8.4576 90558	8.4623 47878	8.4670 00076	8.4716 47168	8.4762 89168
1/4	4.9595 10838	4.9615 58954	4.9636 04536	4.9656 47592	4.9676 88130
1/5	3.6004 02669	3.6015 92098	3.6027 79959	3.6039 66255	3.6051 50991

k					
1	610	611	612	613	614
2	3 72100	3 73321	3 74544	3 75769	3 76996
3	2269 81000	2280 99131	2292 20928	2303 46397	2314 75544
4	(11)1.3845 84100	(11)1.3936 85690	(11)1.4028 32079	(11)1.4120 23414	(11)1.4212 59840
5	(13)8.4459 63010	(13)8.5154 19568	(13)8.5853 32326	(13)8.6557 03525	(13)8.7265 35419
6	(16)5.1520 37436	(16)5.2029 21356	(16)5.2542 23383	(16)5.3059 46261	(16)5.3580 92747
7	(19)3.1427 42836	(19)3.1789 84949	(19)3.2155 84711	(19)3.2525 45058	(19)3.2898 68947
8	(22)1.9170 73130	(22)1.9423 59804	(22)1.9679 37843	(22)1.9938 10121	(22)2.0199 79533
9	(25)1.1694 14609	(25)1.1867 81840	(25)1.2043 77960	(25)1.2222 05604	(25)1.2402 67433
10	(27)7.1334 29117	(27)7.2512 37043	(27)7.3707 93114	(27)7.4921 20352	(27)7.6152 42041
24	(66)7.0455 68477	(66)7.3280 60494	(66)7.6213 89047	(66)7.9259 51097	(66)8.2421 57465
1/2	(1)2.4698 17807	(1)2.4718 41419	(1)2.4738 63375	(1)2.4758 83681	(1)2.4779 02339
1/3	8.4809 26088	8.4855 57944	8.4901 84749	8.4948 06516	8.4994 23260
1/4	4.9697 26156	4.9717 61679	4.9737 94704	4.9758 25239	4.9778 53291
1/5	3.6063 34171	3.6075 15802	3.6086 95885	3.6098 74428	3.6110 51433

k					
1	615	616	617	618	619
2	3 78225	3 79456	3 80689	3 81924	3 83161
3	2326 08375	2337 44896	2348 85113	2360 29032	2371 76659
4	(11)1.4305 41506	(11)1.4398 68559	(11)1.4492 41147	(11)1.4586 59418	(11)1.4681 23519
5	(13)8.7978 30263	(13)8.8695 90326	(13)8.9418 17878	(13)9.0145 15202	(13)9.0876 84584
6	(16)5.4106 65612	(16)5.4636 67641	(16)5.5171 01631	(16)5.5709 70395	(16)5.6252 76757
7	(19)3.3275 59351	(19)3.3656 19267	(19)3.4040 51706	(19)3.4428 59704	(19)3.4820 46313
8	(22)2.0464 49001	(22)2.0732 21468	(22)2.1002 99903	(22)2.1276 87297	(22)2.1553 86668
9	(25)1.2585 66136	(25)1.2771 04424	(25)1.2958 85040	(25)1.3149 10750	(25)1.3341 84347
10	(27)7.7401 81734	(27)7.8669 63254	(27)7.9956 10697	(27)8.1261 48432	(27)8.2586 01110
24	(66)8.5704 33286	(66)8.9112 18488	(66)9.2649 68280	(66)9.6321 53659	(67)1.0013 26192
1/2	(1)2.4799 19354	(1)2.4819 34729	(1)2.4839 48470	(1)2.4859 60579	(1)2.4879 71061
1/3	8.5040 34993	8.5086 41730	8.5132 43484	8.5178 40269	8.5224 32097
1/4	4.9798 78868	4.9819 01975	4.9839 22621	4.9859 40813	4.9879 56556
1/5	3.6122 26906	3.6134 00850	3.6145 73271	3.6157 44173	3.6169 13560

k					
1	620	621	622	623	624
2	3 84400	3 85641	3 86884	3 88129	3 89376
3	2383 28000	2394 83061	2406 41848	2418 04367	2429 70624
4	(11)1.4776 33600	(11)1.4871 89809	(11)1.4967 92295	(11)1.5064 41206	(11)1.5161 36694
5	(13)9.1613 28320	(13)9.2354 48713	(13)9.3100 48072	(13)9.3851 28716	(13)9.4606 92969
6	(16)5.6800 23558	(16)5.7352 13651	(16)5.7908 49901	(16)5.8469 35116	(16)5.9034 72413
7	(19)3.5216 14606	(19)3.5615 67677	(19)3.6019 08638	(19)3.6426 40623	(19)3.6837 66786
8	(22)2.1834 01056	(22)2.2117 33527	(22)2.2403 87173	(22)2.2693 65108	(22)2.2986 70474
9	(25)1.3537 08655	(25)1.3734 86521	(25)1.3935 20822	(25)1.4138 14463	(25)1.4343 70376
10	(27)8.3929 93659	(27)8.5293 51293	(27)8.6676 99511	(27)8.8080 64101	(27)8.9504 71145
24	(67)1.0408 79722	(67)1.0819 28109	(67)1.1245 25305	(67)1.1687 27115	(67)1.2145 91262
1/2	(1)2.4899 79920	(1)2.4919 87159	(1)2.4939 92783	(1)2.4959 96795	(1)2.4979 99199
1/3	8.5270 18983	8.5316 00940	8.5361 77980	8.5407 50116	8.5453 17363
1/4	4.9899 69859	4.9919 80728	4.9939 89170	4.9959 95191	4.9979 98799
1/5	3.6180 81437	3.6192 47808	3.6204 12677	3.6215 76049	3.6227 37928

TABLE II.

Table II   187

POWERS AND ROOTS n^k

k	625	626	627	628	629
1	625	626	627	628	629
2	3 90625	3 91876	3 93129	3 94384	3 95641
3	2441 40625	2453 14376	2464 91883	2476 73152	2488 58189
4	(11)1.5258 78906	(11)1.5356 67994	(11)1.5455 04106	(11)1.5553 87395	(11)1.5653 18009
5	(13)9.5367 43164	(13)9.6132 81641	(13)9.6903 10747	(13)9.7678 32838	(13)9.8458 50275
6	(16)5.9604 64478	(16)6.0179 14307	(16)6.0758 24838	(16)6.1341 99022	(16)6.1930 39823
7	(19)3.7252 90298	(19)3.7672 14356	(19)3.8095 42174	(19)3.8522 76986	(19)3.8954 22049
8	(22)2.3283 06437	(22)2.3582 76187	(22)2.3885 82943	(22)2.4192 29947	(22)2.4502 20469
9	(25)1.4551 91523	(25)1.4762 80893	(25)1.4976 41505	(25)1.5192 76407	(25)1.5411 88675
10	(27)9.0949 47018	(27)9.2415 18391	(27)9.3902 12238	(27)9.5410 55835	(27)9.6940 76765
24	(67)1.2621 77448	(67)1.3115 47419	(67)1.3627 65028	(67)1.4158 96309	(67)1.4710 09545
1/2	(1)2.5000 00000	(1)2.5019 99201	(1)2.5039 96805	(1)2.5059 92817	(1)2.5079 87241
1/3	8.5498 79733	8.5544 37239	8.5589 89894	8.5635 37711	8.5680 80703
1/4	5.0000 00000	5.0019 98801	5.0039 95209	5.0059 89230	5.0079 80871
1/5	3.6238 98318	3.6250 57224	3.6262 14650	3.6273 70600	3.6285 25079

k	630	631	632	633	634
1	630	631	632	633	634
2	3 96900	3 98161	3 99424	4 00689	4 01956
3	2500 47000	2512 39591	2524 35968	2536 36137	2548 40104
4	(11)1.5752 96100	(11)1.5853 21819	(11)1.5953 95318	(11)1.6055 16747	(11)1.6156 86259
5	(13)9.9243 65430	(14)1.0003 38068	(14)1.0082 89841	(14)1.0162 92101	(14)1.0243 45088
6	(16)6.2523 50221	(16)6.3121 33209	(16)6.3723 91794	(16)6.4331 28999	(16)6.4943 47861
7	(19)3.9389 80639	(19)3.9829 56055	(19)4.0273 51614	(19)4.0721 70657	(19)4.1174 16544
8	(22)2.4815 57803	(22)2.5132 45270	(22)2.5452 86220	(22)2.5776 84026	(22)2.6104 42089
9	(25)1.5633 81416	(25)1.5858 57766	(25)1.6086 20891	(25)1.6316 73988	(25)1.6550 20284
10	(27)9.8493 02919	(28)1.0006 76250	(28)1.0166 48643	(28)1.0328 49635	(28)1.0492 82860
24	(67)1.5281 75339	(67)1.5874 66692	(67)1.6489 59081	(67)1.7127 30535	(67)1.7788 61719
1/2	(1)2.5099 80080	(1)2.5119 71337	(1)2.5139 61018	(1)2.5159 49125	(1)2.5179 35662
1/3	8.5726 18882	8.5771 52262	8.5816 80854	8.5862 04672	8.5907 23728
1/4	5.0099 70139	5.0119 57040	5.0139 41581	5.0159 23768	5.0179 03608
1/5	3.6296 78090	3.6308 29638	3.6319 79727	3.6331 28361	3.6342 75544

k	635	636	637	638	639
1	635	636	637	638	639
2	4 03225	4 04496	4 05769	4 07044	4 08321
3	2560 47875	2572 59456	2584 74853	2596 94072	2609 17119
4	(11)1.6259 04006	(11)1.6361 70140	(11)1.6464 84814	(11)1.6568 48179	(11)1.6672 60390
5	(14)1.0324 49044	(14)1.0406 04209	(14)1.0488 10826	(14)1.0570 69138	(14)1.0653 79389
6	(16)6.5560 51429	(16)6.6182 42770	(16)6.6809 24963	(16)6.7441 01103	(16)6.8077 74299
7	(19)4.1630 92658	(19)4.2092 02402	(19)4.2557 49202	(19)4.3027 36504	(19)4.3501 67777
8	(22)2.6435 63838	(22)2.6770 52728	(22)2.7109 12241	(22)2.7451 45889	(22)2.7797 57209
9	(25)1.6786 63037	(25)1.7026 05535	(25)1.7268 51098	(25)1.7514 03077	(25)1.7762 64857
10	(28)1.0659 51028	(28)1.0828 57120	(28)1.1000 04149	(28)1.1173 95163	(28)1.1350 33244
24	(67)1.8474 36020	(67)1.9185 39634	(67)1.9922 61654	(67)2.0686 94164	(67)2.1479 32334
1/2	(1)2.5199 20634	(1)2.5219 04043	(1)2.5238 85893	(1)2.5258 66188	(1)2.5278 44932
1/3	8.5952 38034	8.5997 47604	8.6042 52449	8.6087 52582	8.6132 48015
1/4	5.0198 81108	5.0218 56273	5.0238 29110	5.0257 99626	5.0277 67827
1/5	3.6354 21280	3.6365 65574	3.6377 08430	3.6388 49851	3.6399 89842

k	640	641	642	643	644
1	640	641	642	643	644
2	4 09600	4 10881	4 12164	4 13449	4 14736
3	2621 44000	2633 74721	2646 09288	2658 47707	2670 89984
4	(11)1.6777 21600	(11)1.6882 31962	(11)1.6987 91629	(11)1.7094 00756	(11)1.7200 59497
5	(14)1.0737 41824	(14)1.0821 56687	(14)1.0906 24226	(14)1.0991 44686	(14)1.1077 18316
6	(16)6.8719 47674	(16)6.9366 24366	(16)7.0018 07530	(16)7.0675 00332	(16)7.1337 05955
7	(19)4.3980 46511	(19)4.4463 76219	(19)4.4951 60434	(19)4.5444 02713	(19)4.5941 06635
8	(22)2.8147 49767	(22)2.8501 27156	(22)2.8858 92999	(22)2.9220 50945	(22)2.9586 04673
9	(25)1.8014 39851	(25)1.8269 31507	(25)1.8527 43305	(25)1.8788 78757	(25)1.9053 41409
10	(28)1.1529 21505	(28)1.1710 63096	(28)1.1894 61202	(28)1.2081 19041	(28)1.2270 39868
24	(67)2.2300 74520	(67)2.3152 22362	(67)2.4034 80891	(67)2.4949 58638	(67)2.5897 67740
1/2	(1)2.5298 22128	(1)2.5317 97780	(1)2.5337 71892	(1)2.5357 44467	(1)2.5377 15508
1/3	8.6177 38760	8.6222 24830	8.6267 06237	8.6311 82992	8.6356 55108
1/4	5.0297 33719	5.0316 97308	5.0336 58602	5.0356 17605	5.0375 74325
1/5	3.6411 28406	3.6422 65548	3.6434 01272	3.6445 35581	3.6456 68481

k	645	646	647	648	649
1	645	646	647	648	649
2	4 16025	4 17316	4 18609	4 19904	4 21201
3	2683 36125	2695 86136	2708 40023	2720 97792	2733 59449
4	(11)1.7307 68006	(11)1.7415 26439	(11)1.7523 34949	(11)1.7631 93692	(11)1.7741 02824
5	(14)1.1163 45364	(14)1.1250 26079	(14)1.1337 60712	(14)1.1425 49513	(14)1.1513 92733
6	(16)7.2004 27598	(16)7.2676 68472	(16)7.3354 31806	(16)7.4037 20841	(16)7.4725 38836
7	(19)4.6442 75801	(19)4.6949 13833	(19)4.7460 24378	(19)4.7976 11105	(19)4.8496 77704
8	(22)2.9955 57891	(22)3.0329 14336	(22)3.0706 77773	(22)3.1088 51996	(22)3.1474 40830
9	(25)1.9321 34840	(25)1.9592 62661	(25)1.9867 28519	(25)2.0145 36093	(25)2.0426 89099
10	(28)1.2462 26972	(28)1.2656 83679	(28)1.2854 13352	(28)1.3054 19389	(28)1.3257 05225
24	(67)2.6880 24057	(67)2.7898 47292	(67)2.8953 61105	(67)3.0046 93247	(67)3.1179 75679
1/2	(1)2.5396 85020	(1)2.5416 53005	(1)2.5436 19468	(1)2.5455 84412	(1)2.5475 47841
1/3	8.6401 22598	8.6445 85472	8.6490 43742	8.6534 97422	8.6579 46522
1/4	5.0395 28767	5.0414 80939	5.0434 30845	5.0453 78492	5.0473 23886
1/5	3.6467 99973	3.6479 30063	3.6490 58755	3.6501 86051	3.6513 11957

TABLE II.

POWERS AND ROOTS n^k

Table II

k	650	651	652	653	654
1	650	651	652	653	654
2	4 22500	4 23801	4 25104	4 26409	4 27716
3	2746 25000	2758 94451	2771 67808	2784 45077	2797 26264
4	(11)1.7850 62500	(11)1.7960 72876	(11)1.8071 34108	(11)1.8182 46353	(11)1.8294 09767
5	(14)1.1602 90625	(14)1.1692 43442	(14)1.1782 51439	(14)1.1873 14868	(14)1.1964 33987
6	(16)7.5418 89063	(16)7.6117 74809	(16)7.6821 99379	(16)7.7531 66091	(16)7.8246 78277
7	(19)4.9022 27891	(19)4.9552 65401	(19)5.0087 93995	(19)5.0628 17457	(19)5.1173 39593
8	(22)3.1864 48129	(22)3.2258 77776	(22)3.2657 33685	(22)3.3060 19800	(22)3.3467 40094
9	(25)2.0711 91284	(25)2.1000 46432	(25)2.1292 58363	(25)2.1588 30929	(25)2.1887 68021
10	(28)1.3462 74334	(28)1.3671 30227	(28)1.3882 76452	(28)1.4097 16597	(28)1.4314 54286
24	(67)3.2353 44710	(67)3.3569 41134	(67)3.4829 10364	(67)3.6134 02582	(67)3.7485 72888
1/2	(1)2.5495 09757	(1)2.5514 70164	(1)2.5534 29067	(1)2.5553 86468	(1)2.5573 42371
1/3	8.6623 91053	8.6668 31029	8.6712 66440	8.6756 97359	8.6801 23736
1/4	5.0492 67033	5.0512 07939	5.0531 46611	5.0550 83054	5.0570 17274
1/5	3.6524 36476	3.6535 59612	3.6546 81368	3.6558 01749	3.6569 20758

k	655	656	657	658	659
1	655	656	657	658	659
2	4 29025	4 30336	4 31649	4 32964	4 34281
3	2810 11375	2823 00416	2835 93393	2848 90312	2861 91179
4	(11)1.8406 24506	(11)1.8518 90729	(11)1.8632 08592	(11)1.8745 78253	(11)1.8859 99870
5	(14)1.2056 09052	(14)1.2148 40318	(14)1.2241 28045	(14)1.2334 72490	(14)1.2428 73914
6	(16)7.8967 39288	(16)7.9693 52487	(16)8.0425 21255	(16)8.1162 48987	(16)8.1905 39094
7	(19)5.1723 64234	(19)5.2278 95232	(19)5.2839 36465	(19)5.3404 91834	(19)5.3975 65263
8	(22)3.3878 98573	(22)3.4294 99272	(22)3.4715 46257	(22)3.5140 43626	(22)3.5569 95508
9	(25)2.2190 73565	(25)2.2497 51522	(25)2.2808 05891	(25)2.3122 40706	(25)2.3440 60040
10	(28)1.4534 93185	(28)1.4758 36999	(28)1.4984 89470	(28)1.5214 54385	(28)1.5447 35566
24	(67)3.8885 81447	(67)4.0335 93654	(67)4.1837 80288	(67)4.3393 17689	(67)4.5003 87920
1/2	(1)2.5592 96778	(1)2.5612 49695	(1)2.5632 01124	(1)2.5651 51068	(1)2.5670 99531
1/3	8.6845 45603	8.6889 62971	8.6933 75853	8.6977 84260	8.7021 88202
1/4	5.0589 49277	5.0608 79069	5.0628 06656	5.0647 32044	5.0666 55239
1/5	3.6580 38399	3.6591 54676	3.6602 69592	3.6613 83152	3.6624 95358

k	660	661	662	663	664
1	660	661	662	663	664
2	4 35600	4 36921	4 38244	4 39569	4 40896
3	2874 96000	2888 04781	2901 17528	2914 34247	2927 54944
4	(11)1.8974 73600	(11)1.9089 99602	(11)1.9205 78035	(11)1.9322 09058	(11)1.9438 92828
5	(14)1.2523 32576	(14)1.2618 48737	(14)1.2714 22659	(14)1.2810 54605	(14)1.2907 44838
6	(16)8.2653 95002	(16)8.3408 20153	(16)8.4168 18005	(16)8.4933 92032	(16)8.5705 45724
7	(19)5.4551 60701	(19)5.5132 82121	(19)5.5719 33519	(19)5.6311 18918	(19)5.6908 42360
8	(22)3.6004 06063	(22)3.6442 79482	(22)3.6886 19990	(22)3.7334 31842	(22)3.7787 19327
9	(25)2.3762 68001	(25)2.4088 68738	(25)2.4418 66433	(25)2.4752 65311	(25)2.5090 69633
10	(28)1.5683 36881	(28)1.5922 62236	(28)1.6165 15579	(28)1.6411 00901	(28)1.6660 22237
24	(67)4.6671 78950	(67)4.8398 84834	(67)5.0187 05901	(67)5.2038 48947	(67)5.3955 27431
1/2	(1)2.5690 46516	(1)2.5709 92026	(1)2.5729 36066	(1)2.5748 78638	(1)2.5768 19745
1/3	8.7065 87691	8.7109 82739	8.7153 73356	8.7197 59553	8.7241 41343
1/4	5.0685 76246	5.0704 95071	5.0724 11720	5.0743 26200	5.0762 38514
1/5	3.6636 06215	3.6647 15727	3.6658 23896	3.6669 30727	3.6680 36224

k	665	666	667	668	669
1	665	666	667	668	669
2	4 42225	4 43556	4 44889	4 46224	4 47561
3	2940 79625	2954 08296	2967 40963	2980 77632	2994 18309
4	(11)1.9556 29506	(11)1.9674 19251	(11)1.9792 62223	(11)1.9911 58582	(11)2.0031 08487
5	(14)1.3004 93622	(14)1.3103 01221	(14)1.3201 67903	(14)1.3300 93933	(14)1.3400 79578
6	(16)8.6482 82584	(16)8.7266 06135	(16)8.8055 19912	(16)8.8850 27470	(16)8.9651 32376
7	(19)5.7511 07918	(19)5.8119 19686	(19)5.8732 81781	(19)5.9351 98350	(19)5.9976 73560
8	(22)3.8244 86766	(22)3.8707 38511	(22)3.9174 78948	(22)3.9647 12498	(22)4.0124 43612
9	(25)2.5432 83699	(25)2.5779 11848	(25)2.6129 58458	(25)2.6484 27948	(25)2.6843 24776
10	(28)1.6912 83660	(28)1.7168 89291	(28)1.7428 43292	(28)1.7691 49870	(28)1.7958 13275
24	(67)5.5939 61683	(67)5.7993 79113	(67)6.0120 14426	(67)6.2321 09844	(67)6.4599 15340
1/2	(1)2.5787 59392	(1)2.5806 97580	(1)2.5826 34314	(1)2.5845 69597	(1)2.5865 03431
1/3	8.7285 18735	8.7328 91741	8.7372 60372	8.7416 24639	8.7459 84552
1/4	5.0781 48670	5.0800 56673	5.0819 62528	5.0838 66242	5.0857 67819
1/5	3.6691 40389	3.6702 43226	3.6713 44740	3.6724 44934	3.6735 43810

k	670	671	672	673	674
1	670	671	672	673	674
2	4 48900	4 50241	4 51584	4 52929	4 54276
3	3007 63000	3021 11711	3034 64448	3048 21217	3061 82024
4	(11)2.0151 12100	(11)2.0271 69581	(11)2.0392 81091	(11)2.0514 46790	(11)2.0636 66842
5	(14)1.3501 25107	(14)1.3602 30789	(14)1.3703 96893	(14)1.3806 23690	(14)1.3909 11451
6	(16)9.0458 38217	(16)9.1271 48592	(16)9.2090 67120	(16)9.2915 97433	(16)9.3747 43182
7	(19)6.0607 11605	(19)6.1243 16705	(19)6.1884 93105	(19)6.2532 45073	(19)6.3185 76905
8	(22)4.0606 76776	(22)4.1094 16509	(22)4.1586 57366	(22)4.2084 33934	(22)4.2587 20834
9	(25)2.7206 53440	(25)2.7574 18478	(25)2.7946 24470	(25)2.8322 76038	(25)2.8703 77842
10	(28)1.8228 37805	(28)1.8502 27799	(28)1.8779 87644	(28)1.9061 21773	(28)1.9346 34665
24	(67)6.6956 88867	(67)6.9396 96605	(67)7.1922 13208	(67)7.4535 22063	(67)7.7239 15552
1/2	(1)2.5884 35821	(1)2.5903 66769	(1)2.5922 96279	(1)2.5942 24354	(1)2.5961 50997
1/3	8.7503 40123	8.7546 91362	8.7590 38280	8.7633 80887	8.7677 19196
1/4	5.0876 67266	5.0895 64588	5.0914 59790	5.0933 52878	5.0952 43858
1/5	3.6746 41374	3.6757 37627	3.6768 32575	3.6779 26219	3.6790 18565

TABLE II.

Table II

189

POWERS AND ROOTS n^k

k	675	676	677	678	679
1	675	676	677	678	679
2	4 55625	4 56976	4 58329	4 59684	4 61041
3	3075 46875	3089 15776	3102 88733	3116 65752	3130 46839
4	(11)2.0759 41406	(11)2.0882 70646	(11)2.1006 54722	(11)2.1130 93799	(11)2.1255 88037
5	(14)1.4012 60449	(14)1.4116 70957	(14)1.4221 43247	(14)1.4326 77595	(14)1.4432 74277
6	(16)9.4585 08032	(16)9.5428 95666	(16)9.6279 09783	(16)9.7135 54097	(16)9.7998 32341
7	(19)6.3844 92922	(19)6.4509 97470	(19)6.5180 94923	(19)6.5857 89678	(19)6.6540 86159
8	(22)4.3095 32722	(22)4.3608 74290	(22)4.4127 50263	(22)4.4651 65402	(22)4.5181 24502
9	(25)2.9089 34587	(25)2.9479 51020	(25)2.9874 31928	(25)3.0273 82142	(25)3.0678 06537
10	(28)1.9635 30847	(28)1.9928 14890	(28)2.0224 91415	(28)2.0525 65092	(28)2.0830 40639
24	(67)8.0036 95322	(67)8.2931 72571	(67)8.5926 68325	(67)8.9025 13744	(67)9.2230 50418
1/2	(1)2.5980 76211	(1)2.6000 00000	(1)2.6019 22366	(1)2.6038 43313	(1)2.6057 62844
1/3	8.7720 53215	8.7763 82955	8.7807 08428	8.7850 29644	8.7893 46612
1/4	5.0971 32735	5.0990 19514	5.1009 04200	5.1027 86801	5.1046 67319
1/5	3.6801 09614	3.6811 99371	3.6822 87840	3.6833 75023	3.6844 60923

k	680	681	682	683	684
1	680	681	682	683	684
2	4 62400	4 63761	4 65124	4 66489	4 67856
3	3144 32000	3158 21241	3172 14568	3186 11987	3200 13504
4	(11)2.1381 37600	(11)2.1507 42651	(11)2.1634 03354	(11)2.1761 19871	(11)2.1888 92367
5	(14)1.4539 33568	(14)1.4646 55745	(14)1.4754 41087	(14)1.4862 89872	(14)1.4972 02379
6	(16)9.8867 48262	(16)9.9743 05627	(17)1.0062 50822	(17)1.0151 35983	(17)1.0240 86427
7	(19)6.7229 88818	(19)6.7925 02132	(19)6.8626 30603	(19)6.9333 78761	(19)7.0047 51164
8	(22)4.5716 32397	(22)4.6256 93952	(22)4.6803 14071	(22)4.7354 97694	(22)4.7912 49796
9	(25)3.1087 10030	(25)3.1500 97581	(25)3.1919 74196	(25)3.2343 44925	(25)3.2772 14860
10	(28)2.1139 22820	(28)2.1452 16453	(28)2.1769 26402	(28)2.2090 57584	(28)2.2416 14965
24	(67)9.5546 30685	(67)9.8976 17949	(68)1.0252 38701	(68)1.0619 32441	(68)1.0998 82878
1/2	(1)2.6076 80962	(1)2.6095 97670	(1)2.6115 12971	(1)2.6134 26869	(1)2.6153 39366
1/3	8.7936 59344	8.7979 67850	8.8022 72141	8.8065 72225	8.8108 68115
1/4	5.1065 45762	5.1084 22134	5.1102 96441	5.1121 68688	5.1140 38880
1/5	3.6855 45546	3.6866 28893	3.6877 10968	3.6887 91774	3.6898 71315

k	685	686	687	688	689
1	685	686	687	688	689
2	4 69225	4 70596	4 71969	4 73344	4 74721
3	3214 19125	3228 28856	3242 42703	3256 60672	3270 82769
4	(11)2.2017 21006	(11)2.2146 05952	(11)2.2275 47370	(11)2.2405 45423	(11)2.2536 00278
5	(14)1.5081 78889	(14)1.5192 19683	(14)1.5303 25043	(14)1.5414 95251	(14)1.5527 30592
6	(17)1.0331 02539	(17)1.0421 84703	(17)1.0513 33304	(17)1.0605 48733	(17)1.0698 31378
7	(19)7.0767 52393	(19)7.1493 87060	(19)7.2226 59802	(19)7.2965 75282	(19)7.3711 38193
8	(22)4.8475 75389	(22)4.9044 79523	(22)4.9619 67284	(22)5.0200 43794	(22)5.0787 14215
9	(25)3.3205 89142	(25)3.3644 72953	(25)3.4088 71524	(25)3.4537 90130	(25)3.4992 34094
10	(28)2.2746 03562	(28)2.3080 28446	(28)2.3418 94737	(28)2.3762 07610	(28)2.4109 72291
24	(68)1.1391 31118	(68)1.1797 19551	(68)1.2216 91886	(68)1.2650 93189	(68)1.3099 69927
1/2	(1)2.6172 50466	(1)2.6191 60171	(1)2.6210 68484	(1)2.6229 75410	(1)2.6248 80950
1/3	8.8151 59819	8.8194 47349	8.8237 30714	8.8280 09925	8.8322 84991
1/4	5.1159 07022	5.1177 73120	5.1196 37179	5.1214 99204	5.1233 59200
1/5	3.6909 49595	3.6920 26615	3.6931 02381	3.6941 76894	3.6952 50159

k	690	691	692	693	694
1	690	691	692	693	694
2	4 76100	4 77481	4 78864	4 80249	4 81636
3	3285 09000	3299 39371	3313 73888	3328 12557	3342 55384
4	(11)2.2667 12100	(11)2.2798 81054	(11)2.2931 07305	(11)2.3063 91020	(11)2.3197 32365
5	(14)1.5640 31349	(14)1.5753 97808	(14)1.5868 30255	(14)1.5983 28977	(14)1.6098 94261
6	(17)1.0791 81631	(17)1.0885 99885	(17)1.0980 86536	(17)1.1076 41981	(17)1.1172 66617
7	(19)7.4463 53253	(19)7.5222 25208	(19)7.5987 58832	(19)7.6759 58928	(19)7.7538 30324
8	(22)5.1379 83744	(22)5.1978 57619	(22)5.2583 44112	(22)5.3194 39537	(22)5.3811 58245
9	(25)3.5452 08784	(25)3.5917 19614	(25)3.6387 72050	(25)3.6863 71599	(25)3.7345 23822
10	(28)2.4461 94061	(28)2.4818 78254	(28)2.5180 30258	(28)2.5546 55518	(28)2.5917 59533
24	(68)1.3563 70007	(68)1.4043 42816	(68)1.4539 39271	(68)1.5052 11857	(68)1.5582 14678
1/2	(1)2.6267 85107	(1)2.6286 87886	(1)2.6305 89288	(1)2.6324 89316	(1)2.6343 87974
1/3	8.8365 55922	8.8408 22729	8.8450 85422	8.8493 44010	8.8535 98503
1/4	5.1252 17173	5.1270 73128	5.1289 27069	5.1307 79001	5.1326 28931
1/5	3.6963 22179	3.6973 92956	3.6984 62494	3.6995 30796	3.7005 97866

k	695	696	697	698	699
1	695	696	697	698	699
2	4 83025	4 84416	4 85809	4 87204	4 88601
3	3357 02375	3371 53536	3386 08873	3400 68392	3415 32099
4	(11)2.3331 31506	(11)2.3465 88611	(11)2.3601 03845	(11)2.3736 77376	(11)2.3873 09372
5	(14)1.6215 26397	(14)1.6332 25673	(14)1.6449 92380	(14)1.6568 26809	(14)1.6687 29251
6	(17)1.1269 60846	(17)1.1367 25068	(17)1.1465 59689	(17)1.1564 65112	(17)1.1664 41746
7	(19)7.8323 77878	(19)7.9116 06476	(19)7.9915 21031	(19)8.0721 26484	(19)8.1534 27808
8	(22)5.4435 02625	(22)5.5064 78107	(22)5.5700 90158	(22)5.6343 44286	(22)5.6992 46038
9	(25)3.7832 34325	(25)3.8325 08763	(25)3.8823 52840	(25)3.9327 72312	(25)3.9837 72980
10	(28)2.6293 47856	(28)2.6674 26099	(28)2.7059 99930	(28)2.7450 75074	(28)2.7846 57313
24	(68)1.6130 03502	(68)1.6696 35809	(68)1.7281 70846	(68)1.7886 69670	(68)1.8511 95210
1/2	(1)2.6362 85265	(1)2.6381 81192	(1)2.6400 75756	(1)2.6419 68963	(1)2.6438 60813
1/3	8.8578 48911	8.8620 95243	8.8663 37511	8.8705 75722	8.8748 09888
1/4	5.1344 76863	5.1363 22801	5.1381 66751	5.1400 08719	5.1418 48708
1/5	3.7016 63707	3.7027 28321	3.7037 91713	3.7048 53884	3.7059 14839

TABLE II.

POWERS AND ROOTS n^k

k					
1	700	701	702	703	704
2	4 90000	4 91401	4 92804	4 94209	4 95616
3	3430 00000	3444 72101	3459 48408	3474 28927	3489 13664
4	(11)2.4010 0000	(11)2.4147 49428	(11)2.4285 57824	(11)2.4424 25357	(11)2.4563 52195
5	(14)1.6807 00000	(14)1.6927 39349	(14)1.7048 47593	(14)1.7170 25026	(14)1.7292 71945
6	(17)1.1764 90000	(17)1.1866 10284	(17)1.1968 03010	(17)1.2070 68593	(17)1.2174 07449
7	(19)8.2354 30000	(19)8.3181 38089	(19)8.4015 57130	(19)8.4856 92210	(19)8.5705 48443
8	(22)5.7648 01000	(22)5.8310 14800	(22)5.8978 93105	(22)5.9654 41624	(22)6.0336 66104
9	(25)4.0353 60700	(25)4.0875 41375	(25)4.1403 20960	(25)4.1937 05461	(25)4.2477 00937
10	(28)2.8247 52490	(28)2.8653 66504	(28)2.9065 05314	(28)2.9481 74939	(28)2.9903 81460
24	(68)1.9158 12314	(68)1.9825 87808	(68)2.0515 90555	(68)2.1228 91511	(68)2.1965 63787
1/2	(1)2.6457 51311	(1)2.6476 40459	(1)2.6495 28260	(1)2.6514 14717	(1)2.6532 99832
1/3	8.8790 40017	8.8832 66120	8.8874 88205	8.8917 06283	8.8959 20362
1/4	5.1436 86724	5.1455 22771	5.1473 56856	5.1491 88981	5.1510 19154
1/5	3.7069 74581	3.7080 33112	3.7090 90435	3.7101 46554	3.7112 01473

k					
1	705	706	707	708	709
2	4 97025	4 98436	4 99849	5 01264	5 02681
3	3504 02625	3518 95816	3533 93243	3548 94912	3564 00829
4	(11)2.4703 38506	(11)2.4843 84461	(11)2.4984 90228	(11)2.5126 55977	(11)2.5268 81878
5	(14)1.7415 88647	(14)1.7539 75429	(14)1.7664 32591	(14)1.7789 60432	(14)1.7915 59251
6	(17)1.2278 19996	(17)1.2383 06653	(17)1.2488 67842	(17)1.2595 03986	(17)1.2702 15509
7	(19)8.6561 30972	(19)8.7424 44971	(19)8.8294 95643	(19)8.9172 88218	(19)9.0058 27960
8	(22)6.1025 72335	(22)6.1721 66150	(22)6.2424 53419	(22)6.3134 40059	(22)6.3851 32023
9	(25)4.3023 13497	(25)4.3575 49302	(25)4.4134 14568	(25)4.4699 15561	(25)4.5270 58605
10	(28)3.0331 31015	(28)3.0764 29807	(28)3.1202 84099	(28)3.1647 00218	(28)3.2096 84551
24	(68)2.2726 82709	(68)2.3513 25887	(68)2.4325 73275	(68)2.5165 07242	(68)2.6032 12640
1/2	(1)2.6551 83609	(1)2.6570 66051	(1)2.6589 47160	(1)2.6608 26939	(1)2.6627 05391
1/3	8.9001 30453	8.9043 36564	8.9085 38706	8.9127 36887	8.9169 31117
1/4	5.1528 47377	5.1546 73657	5.1564 97998	5.1583 20404	5.1601 40881
1/5	3.7122 55193	3.7133 07718	3.7143 59051	3.7154 09195	3.7164 58153

k					
1	710	711	712	713	714
2	5 04100	5 05521	5 06944	5 08369	5 09796
3	3579 11000	3594 25431	3609 44128	3624 67097	3639 94344
4	(11)2.5411 68100	(11)2.5555 14814	(11)2.5699 22191	(11)2.5843 90402	(11)2.5989 19616
5	(14)1.8042 29351	(14)1.8169 71033	(14)1.8297 84600	(14)1.8426 70356	(14)1.8556 28606
6	(17)1.2810 02839	(17)1.2918 66404	(17)1.3028 06635	(17)1.3138 23964	(17)1.3249 18825
7	(19)9.0951 20158	(19)9.1851 70136	(19)9.2759 83244	(19)9.3675 64864	(19)9.4599 20408
8	(22)6.4575 35312	(22)6.5306 55967	(22)6.6045 00070	(22)6.6790 73748	(22)6.7543 83171
9	(25)4.5848 50072	(25)4.6432 96392	(25)4.7024 04050	(25)4.7621 79582	(25)4.8226 29584
10	(28)3.2552 43551	(28)3.3013 83735	(28)3.3481 11683	(28)3.3954 34042	(28)3.4433 57523
24	(68)2.6927 76876	(68)2.7852 89985	(68)2.8808 44702	(68)2.9795 36544	(68)3.0814 63889
1/2	(1)2.6645 82519	(1)2.6664 58325	(1)2.6683 32813	(1)2.6702 05985	(1)2.6720 77843
1/3	8.9211 21404	8.9253 07760	8.9294 90191	8.9336 68708	8.9378 43321
1/4	5.1619 59433	5.1637 76065	5.1655 90782	5.1674 03588	5.1692 14489
1/5	3.7175 05928	3.7185 52523	3.7195 97942	3.7206 42186	3.7216 85260

k					
1	715	716	717	718	719
2	5 11225	5 12656	5 14089	5 15524	5 16961
3	3655 25875	3670 61696	3686 01813	3701 46232	3716 94959
4	(11)2.6135 10006	(11)2.6281 61743	(11)2.6428 74999	(11)2.6576 49946	(11)2.6724 86755
5	(14)1.8686 59654	(14)1.8817 63808	(14)1.8949 41374	(14)1.9081 92661	(14)1.9215 17777
6	(17)1.3360 91653	(17)1.3473 42887	(17)1.3586 72965	(17)1.3700 82331	(17)1.3815 71425
7	(19)9.5530 55319	(19)9.6469 75069	(19)9.7416 85162	(19)9.8371 91134	(19)9.9334 98549
8	(22)6.8304 34553	(22)6.9072 34149	(22)6.9847 88261	(22)7.0631 03234	(22)7.1421 85457
9	(25)4.8837 60705	(25)4.9455 79651	(25)5.0080 93183	(25)5.0713 08122	(25)5.1352 31343
10	(28)3.4918 88904	(28)3.5410 35030	(28)3.5908 02813	(28)3.6411 99232	(28)3.6922 31336
24	(68)3.1867 28051	(68)3.2954 33372	(68)3.4076 87302	(68)3.5236 00491	(68)3.6432 86875
1/2	(1)2.6739 48391	(1)2.6758 17632	(1)2.6776 85568	(1)2.6795 52201	(1)2.6814 17536
1/3	8.9420 14037	8.9461 80866	8.9503 43817	8.9545 02899	8.9586 58122
1/4	5.1710 23488	5.1728 30591	5.1746 35801	5.1764 39125	5.1782 40566
1/5	3.7227 27165	3.7237 67905	3.7248 07483	3.7258 45902	3.7268 83164

k					
1	720	721	722	723	724
2	5 18400	5 19841	5 21284	5 22729	5 24176
3	3732 48000	3748 05361	3763 67048	3779 33067	3795 03424
4	(11)2.6873 85600	(11)2.7023 46653	(11)2.7173 70087	(11)2.7324 56074	(11)2.7476 04790
5	(14)1.9349 17632	(14)1.9483 91937	(14)1.9619 41202	(14)1.9755 65742	(14)1.9892 65868
6	(17)1.3931 40695	(17)1.4047 90586	(17)1.4165 21548	(17)1.4283 34031	(17)1.4402 28488
7	(20)1.0030 61300	(20)1.0128 54013	(20)1.0227 28558	(20)1.0326 85505	(20)1.0427 25426
8	(22)7.2220 41363	(22)7.3026 77432	(22)7.3841 00187	(22)7.4663 16199	(22)7.5493 32081
9	(25)5.1998 69781	(25)5.2652 30428	(25)5.3313 20335	(25)5.3981 44612	(25)5.4657 16426
10	(28)3.7439 06243	(28)3.7962 31139	(28)3.8492 13282	(28)3.9028 60000	(28)3.9571 78693
24	(68)3.7668 63772	(68)3.8944 51981	(68)4.0261 75870	(68)4.1621 63488	(68)4.3025 46659
1/2	(1)2.6832 81573	(1)2.6851 44316	(1)2.6870 05769	(1)2.6888 65932	(1)2.6907 24809
1/3	8.9628 09493	8.9669 57022	8.9711 00718	8.9752 40590	8.9793 76646
1/4	5.1800 40128	5.1818 37817	5.1836 33637	5.1854 27593	5.1872 19688
1/5	3.7279 19273	3.7289 54232	3.7299 88042	3.7310 20708	3.7320 52232

TABLE II.

Table II 191

POWERS AND ROOTS n^k

k	725	726	727	728	729
1	725	726	727	728	729
2	5 25625	5 27076	5 28529	5 29984	5 31441
3	3810 78125	3826 57176	3842 40583	3858 28352	3874 20489
4	(11)2.7628 16406	(11)2.7780 91098	(11)2.7934 29038	(11)2.8088 30403	(11)2.8242 95365
5	(14)2.0030 41895	(14)2.0168 94137	(14)2.0308 22911	(14)2.0448 28533	(14)2.0589 11321
6	(17)1.4522 05374	(17)1.4642 65143	(17)1.4764 08256	(17)1.4886 35172	(17)1.5009 46353
7	(20)1.0528 48896	(20)1.0630 56494	(20)1.0733 48802	(20)1.0837 26405	(20)1.0941 89891
8	(22)7.6331 54495	(22)7.7177 90147	(22)7.8032 45793	(22)7.8895 28230	(22)7.9766 44308
9	(25)5.5340 37009	(25)5.6031 15647	(25)5.6729 59691	(25)5.7435 76552	(25)5.8149 73700
10	(28)4.0121 76831	(28)4.0678 61960	(28)4.1242 41696	(28)4.1813 23730	(28)4.2391 15828
24	(68)4.4474 61095	(68)4.5970 46501	(68)4.7514 46686	(68)4.9108 09683	(68)5.0752 87861
1/2	(1)2.6925 82404	(1)2.6944 38717	(1)2.6962 93753	(1)2.6981 47513	(1)2.7000 00000
1/3	8.9835 08896	8.9876 37347	8.9917 62009	8.9958 82891	9.0000 00000
1/4	5.1890 09928	5.1907 98317	5.1925 84860	5.1943 69560	5.1961 52423
1/5	3.7330 82616	3.7341 11864	3.7351 39979	3.7361 66963	3.7371 92819

k	730	731	732	733	734
1	730	731	732	733	734
2	5 32900	5 34361	5 35824	5 37289	5 38756
3	3890 17000	3906 17891	3922 23168	3938 32837	3954 46904
4	(11)2.8398 24100	(11)2.8554 16783	(11)2.8710 73590	(11)2.8867 94695	(11)2.9025 80275
5	(14)2.0730 71593	(14)2.0873 09669	(14)2.1016 25868	(14)2.1160 20512	(14)2.1304 93922
6	(17)1.5133 42263	(17)1.5258 23368	(17)1.5383 90135	(17)1.5510 43035	(17)1.5637 82539
7	(20)1.1047 39852	(20)1.1153 76882	(20)1.1261 01579	(20)1.1369 14545	(20)1.1478 16384
8	(22)8.0646 00919	(22)8.1534 05006	(22)8.2430 63558	(22)8.3335 83612	(22)8.4249 72255
9	(25)5.8871 58671	(25)5.9601 39059	(25)6.0339 22524	(25)6.1085 16788	(25)6.1839 29635
10	(28)4.2976 25830	(28)4.3568 61652	(28)4.4168 31288	(28)4.4775 42805	(28)4.5390 04352
24	(68)5.2450 38047	(68)5.4202 21655	(68)5.6010 04807	(68)5.7875 58467	(68)5.9800 58576
1/2	(1)2.7018 51217	(1)2.7037 01167	(1)2.7055 49852	(1)2.7073 97274	(1)2.7092 43437
1/3	9.0041 13346	9.0082 22937	9.0123 28782	9.0164 30890	9.0205 29268
1/4	5.1979 33452	5.1997 12653	5.2014 90029	5.2032 65584	5.2050 39324
1/5	3.7382 17550	3.7392 41158	3.7402 63647	3.7412 85019	3.7423 05277

k	735	736	737	738	739
1	735	736	737	738	739
2	5 40225	5 41696	5 43169	5 44644	5 46121
3	3970 65375	3986 88256	4003 15553	4019 47272	4035 83419
4	(11)2.9184 30506	(11)2.9343 45564	(11)2.9503 25626	(11)2.9663 70867	(11)2.9824 81466
5	(14)2.1450 46422	(14)2.1596 78335	(14)2.1743 89986	(14)2.1891 81700	(14)2.2040 53804
6	(17)1.5766 09120	(17)1.5895 23255	(17)1.6025 25420	(17)1.6156 16095	(17)1.6287 95761
7	(20)1.1588 07703	(20)1.1698 89115	(20)1.1810 61234	(20)1.1923 24678	(20)1.2036 80067
8	(22)8.5172 36620	(22)8.6103 83890	(22)8.7044 21297	(22)8.7993 56123	(22)8.8951 95697
9	(25)6.2601 68916	(25)6.3372 42543	(25)6.4151 58496	(25)6.4939 24819	(25)6.5735 49620
10	(28)4.6012 24153	(28)4.6642 10512	(28)4.7279 71812	(28)4.7925 16516	(28)4.8578 53170
24	(68)6.1786 86185	(68)6.3836 27605	(68)6.5950 74542	(68)6.8132 24254	(68)7.0382 79698
1/2	(1)2.7110 88342	(1)2.7129 31993	(1)2.7147 74392	(1)2.7166 15541	(1)2.7184 55444
1/3	9.0246 23926	9.0287 14871	9.0328 02112	9.0368 85658	9.0409 65517
1/4	5.2068 11253	5.2085 81374	5.2103 49693	5.2121 16213	5.2138 80938
1/5	3.7433 24423	3.7443 42461	3.7453 59393	3.7463 75222	3.7473 89950

k	740	741	742	743	744
1	740	741	742	743	744
2	5 47600	5 49081	5 50564	5 52049	5 53536
3	4052 24000	4068 69021	4085 18488	4101 72407	4118 30784
4	(11)2.9986 57600	(11)3.0148 99446	(11)3.0312 07181	(11)3.0475 80984	(11)3.0640 21033
5	(14)2.2190 06624	(14)2.2340 40489	(14)2.2491 55728	(14)2.2643 52671	(14)2.2796 31649
6	(17)1.6420 64902	(17)1.6554 24002	(17)1.6688 73550	(17)1.6824 14035	(17)1.6960 45947
7	(20)1.2151 28027	(20)1.2266 69186	(20)1.2383 04174	(20)1.2500 33628	(20)1.2618 58184
8	(22)8.9919 47402	(22)9.0896 18667	(22)9.1882 16974	(22)9.2877 49854	(22)9.3882 24890
9	(25)6.6540 41078	(25)6.7354 07432	(25)6.8176 56995	(25)6.9007 98142	(25)6.9848 39318
10	(28)4.9239 90397	(28)4.9909 36907	(28)5.0587 01490	(28)5.1272 93019	(28)5.1967 20453
24	(68)7.2704 49690	(68)7.5099 49065	(68)7.7569 98844	(68)8.0118 26396	(68)8.2746 65623
1/2	(1)2.7202 94102	(1)2.7221 31518	(1)2.7239 67694	(1)2.7258 02634	(1)2.7276 36339
1/3	9.0450 41696	9.0491 14206	9.0531 83053	9.0572 48245	9.0613 09792
1/4	5.2156 43874	5.2174 05023	5.2191 64391	5.2209 21982	5.2226 77799
1/5	3.7484 03580	3.7494 16115	3.7504 27557	3.7514 37909	3.7524 47174

k	745	746	747	748	749
1	745	746	747	748	749
2	5 55025	5 56516	5 58009	5 59504	5 61001
3	4134 93625	4151 60936	4168 32723	4185 08992	4201 89749
4	(11)3.0805 27506	(11)3.0971 00583	(11)3.1137 40441	(11)3.1304 47260	(11)3.1472 21220
5	(14)2.2949 92992	(14)2.3104 37035	(14)2.3259 64109	(14)2.3415 74551	(14)2.3572 68694
6	(17)1.7097 69779	(17)1.7235 86028	(17)1.7374 95190	(17)1.7514 97764	(17)1.7655 94252
7	(20)1.2737 78485	(20)1.2857 95177	(20)1.2979 08907	(20)1.3101 20327	(20)1.3224 30094
8	(22)9.4896 49717	(22)9.5920 32018	(22)9.6953 79533	(22)9.7997 00049	(22)9.9050 01408
9	(25)7.0697 89039	(25)7.1556 55886	(25)7.2424 48511	(25)7.3301 75636	(25)7.4188 46054
10	(28)5.2669 92834	(28)5.3381 19291	(28)5.4101 09038	(28)5.4829 71376	(28)5.5567 15695
24	(68)8.5457 57129	(68)8.8253 48404	(68)9.1136 94019	(68)9.4110 15807	(68)9.7177 03069
1/2	(1)2.7294 68813	(1)2.7313 00057	(1)2.7331 30074	(1)2.7349 58866	(1)2.7367 86437
1/3	9.0653 67701	9.0694 21981	9.0734 72639	9.0775 19683	9.0815 63122
1/4	5.2244 31847	5.2261 84131	5.2279 34653	5.2296 83419	5.2314 30432
1/5	3.7534 55355	3.7544 62453	3.7554 68472	3.7564 73415	3.7574 77282

TABLE II.

Table II

POWERS AND ROOTS n^k

k					
1	750	751	752	753	754
2	5 62500	5 64001	5 65504	5 67009	5 68516
3	4218 75000	4235 64751	4252 59008	4269 57777	4286 61064
4	(11)3.1640 62500	(11)3.1809 71280	(11)3.1979 47740	(11)3.2149 92061	(11)3.2321 04423
5	(14)2.3730 46875	(14)2.3889 09431	(14)2.4048 56701	(14)2.4208 89022	(14)2.4370 06735
6	(17)1.7797 85156	(17)1.7940 70983	(17)1.8084 52239	(17)1.8229 29433	(17)1.8375 03078
7	(20)1.3348 38867	(20)1.3473 47308	(20)1.3599 56084	(20)1.3726 65863	(20)1.3854 77321
8	(23)1.0011 29150	(23)1.0118 57828	(23)1.0226 86975	(23)1.0336 17395	(23)1.0446 49900
9	(25)7.5084 68628	(25)7.5990 52291	(25)7.6906 06051	(25)7.7831 38985	(25)7.8766 60245
10	(28)5.6313 51471	(28)5.7068 88271	(28)5.7833 35750	(28)5.8607 03656	(28)5.9390 01825
24	(69)1.0033 91278	(69)1.0359 96977	(69)1.0696 16698	(69)1.1042 80565	(69)1.1400 19555
1/2	(1)2.7386 12788	(1)2.7404 37921	(1)2.7422 61840	(1)2.7440 84547	(1)2.7459 06044
1/3	9.0856 02964	9.0896 39217	9.0936 71888	9.0977 00985	9.1017 26517
1/4	5.2331 75697	5.2349 19217	5.2366 60997	5.2384 01041	5.2401 39353
1/5	3.7584 80079	3.7594 81806	3.7604 82467	3.7614 82064	3.7624 80599

k					
1	755	756	757	758	759
2	5 70025	5 71536	5 73049	5 74564	5 76081
3	4303 68875	4320 81216	4337 98093	4355 19512	4372 45479
4	(11)3.2492 85006	(11)3.2665 33993	(11)3.2838 51564	(11)3.3012 37901	(11)3.3186 93186
5	(14)2.4532 10180	(14)2.4694 99699	(14)2.4858 75634	(14)2.5023 38329	(14)2.5188 88128
6	(17)1.8521 73686	(17)1.8669 41772	(17)1.8818 07855	(17)1.8967 72453	(17)1.9118 36089
7	(20)1.3983 91133	(20)1.4114 07980	(20)1.4245 28546	(20)1.4377 53520	(20)1.4510 83592
8	(23)1.0557 85305	(23)1.0670 24433	(23)1.0783 68109	(23)1.0898 17168	(23)1.1013 72446
9	(25)7.9711 79054	(25)8.0667 04711	(25)8.1632 46588	(25)8.2608 14132	(25)8.3594 16865
10	(28)6.0182 40186	(28)6.0984 28762	(28)6.1795 77667	(28)6.2616 97112	(28)6.3447 97401
24	(69)1.1768 65520	(69)1.2148 51214	(69)1.2540 10313	(69)1.2943 77441	(69)1.3359 88198
1/2	(1)2.7477 26333	(1)2.7495 45417	(1)2.7513 63298	(1)2.7531 79980	(1)2.7549 95463
1/3	9.1057 48491	9.1097 66916	9.1137 81798	9.1177 93146	9.1218 00968
1/4	5.2418 75936	5.2436 10795	5.2453 43934	5.2470 75356	5.2488 05067
1/5	3.7634 78075	3.7644 74495	3.7654 69862	3.7664 64176	3.7674 57442

k					
1	760	761	762	763	764
2	5 77600	5 79121	5 80644	5 82169	5 83696
3	4389 76000	4407 11081	4424 50728	4441 94947	4459 43744
4	(11)3.3362 17600	(11)3.3538 11326	(11)3.3714 74547	(11)3.3892 07446	(11)3.4070 10204
5	(14)2.5355 25376	(14)2.5522 50419	(14)2.5690 63605	(14)2.5859 65281	(14)2.6029 55796
6	(17)1.9269 99286	(17)1.9422 62569	(17)1.9576 26467	(17)1.9730 91509	(17)1.9886 58228
7	(20)1.4645 19457	(20)1.4780 61815	(20)1.4917 11368	(20)1.5054 68822	(20)1.5193 34886
8	(23)1.1130 34787	(23)1.1248 05041	(23)1.1366 84062	(23)1.1486 72711	(23)1.1607 71853
9	(25)8.4590 64385	(25)8.5597 66364	(25)8.6615 32555	(25)8.7643 72784	(25)8.8682 96958
10	(28)6.4288 88932	(28)6.5139 82203	(28)6.6000 87807	(28)6.6872 16435	(28)6.7753 78876
24	(69)1.3788 79182	(69)1.4230 88020	(69)1.4686 53390	(69)1.5156 15056	(69)1.5640 13890
1/2	(1)2.7568 09750	(1)2.7586 22845	(1)2.7604 34748	(1)2.7622 45463	(1)2.7640 54992
1/3	9.1258 05271	9.1298 06063	9.1338 03351	9.1377 97144	9.1417 87449
1/4	5.2505 33069	5.2522 59366	5.2539 83963	5.2557 06863	5.2574 28071
1/5	3.7684 49662	3.7694 40838	3.7704 30972	3.7714 20068	3.7724 08126

k					
1	765	766	767	768	769
2	5 85225	5 86756	5 88289	5 89824	5 91361
3	4476 97125	4494 55096	4512 17663	4529 84832	4547 56609
4	(11)3.4248 83006	(11)3.4428 26035	(11)3.4608 39475	(11)3.4789 23510	(11)3.4970 78323
5	(14)2.6200 35500	(14)2.6372 04743	(14)2.6544 63877	(14)2.6718 13255	(14)2.6892 53231
6	(17)2.0043 27157	(17)2.0200 98833	(17)2.0359 73794	(17)2.0519 52580	(17)2.0680 35734
7	(20)1.5333 10275	(20)1.5473 95706	(20)1.5615 91900	(20)1.5758 99582	(20)1.5903 19480
8	(23)1.1729 82361	(23)1.1853 05111	(23)1.1977 40987	(23)1.2102 90879	(23)1.2229 55680
9	(25)8.9733 15059	(25)9.0794 37150	(25)9.1866 73373	(25)9.2950 33948	(25)9.4045 29178
10	(28)6.8645 86020	(28)6.9548 48857	(28)7.0461 78477	(28)7.1385 86072	(28)7.2320 82938
24	(69)1.6138 91907	(69)1.6652 92289	(69)1.7182 59425	(69)1.7728 38934	(69)1.8290 77701
1/2	(1)2.7658 63337	(1)2.7676 70501	(1)2.7694 76485	(1)2.7712 81292	(1)2.7730 84925
1/3	9.1457 74274	9.1497 57625	9.1537 37512	9.1577 13940	9.1616 86919
1/4	5.2591 47590	5.2608 65424	5.2625 81576	5.2642 96052	5.2660 08854
1/5	3.7733 95151	3.7743 81144	3.7753 66108	3.7763 50045	3.7773 32958

k					
1	770	771	772	773	774
2	5 92900	5 94441	5 95984	5 97529	5 99076
3	4565 33000	4583 14011	4600 99648	4618 89917	4636 84824
4	(11)3.5153 04100	(11)3.5336 01025	(11)3.5519 69283	(11)3.5704 09058	(11)3.5889 20538
5	(14)2.7067 84157	(14)2.7244 06390	(14)2.7421 20286	(14)2.7599 26202	(14)2.7778 24496
6	(17)2.0842 23801	(17)2.1005 17327	(17)2.1169 16861	(17)2.1334 22954	(17)2.1500 36160
7	(20)1.6048 52327	(20)1.6194 98859	(20)1.6342 59817	(20)1.6491 35944	(20)1.6641 27988
8	(23)1.2357 36292	(23)1.2486 33620	(23)1.2616 48578	(23)1.2747 82084	(23)1.2880 35063
9	(25)9.5151 69445	(25)9.6269 65212	(25)9.7399 27025	(25)9.8540 65513	(25)9.9693 91385
10	(28)7.3266 80473	(28)7.4223 90179	(28)7.5192 23664	(28)7.6171 92641	(28)7.7163 08932
24	(69)1.8870 23915	(69)1.9467 27094	(69)2.0082 38127	(69)2.0716 09310	(69)2.1368 94378
1/2	(1)2.7748 87385	(1)2.7766 88675	(1)2.7784 88798	(1)2.7802 87755	(1)2.7820 85549
1/3	9.1656 56454	9.1696 22555	9.1735 85227	9.1775 44479	9.1815 00317
1/4	5.2677 19986	5.2694 29452	5.2711 37257	5.2728 43403	5.2745 47894
1/5	3.7783 14849	3.7792 95720	3.7802 75573	3.7812 54412	3.7822 32239

TABLE II.

Table II   POWERS AND ROOTS n^k 193

k	775	776	777	778	779
1	775	776	777	778	779
2	6 00625	6 02176	6 03729	6 05284	6 06841
3	4654 84375	4672 88576	4690 97433	4709 10952	4727 29139
4	(11)3.6075 03906	(11)3.6261 59350	(11)3.6448 87054	(11)3.6636 87207	(11)3.6825 59993
5	(14)2.7958 15527	(14)2.8138 99655	(14)2.8320 77241	(14)2.8503 48647	(14)2.8687 14234
6	(17)2.1667 57034	(17)2.1835 86133	(17)2.2005 24016	(17)2.2175 71247	(17)2.2347 28389
7	(20)1.6792 36701	(20)1.6944 62839	(20)1.7098 07161	(20)1.7252 70430	(20)1.7408 53415
8	(23)1.3014 08443	(23)1.3149 03163	(23)1.3285 20164	(23)1.3422 60395	(23)1.3561 24810
9	(26)1.0085 91544	(26)1.0203 64854	(26)1.0322 60167	(26)1.0442 78587	(26)1.0564 21227
10	(28)7.8165 84463	(28)7.9180 31271	(28)8.0206 61501	(28)8.1244 87408	(28)8.2295 21359
24	(69)2.2041 48547	(69)2.2734 28553	(69)2.3447 92689	(69)2.4183 00846	(69)2.4940 14558
1/2	(1)2.7838 82181	(1)2.7856 77655	(1)2.7874 71973	(1)2.7892 65136	(1)2.7910 57147
1/3	9.1854 52750	9.1894 01784	9.1933 47428	9.1972 89687	9.2012 28569
1/4	5.2762 50735	5.2779 51928	5.2796 51478	5.2813 49388	5.2830 45663
1/5	3.7832 09055	3.7841 84864	3.7851 59667	3.7861 33467	3.7871 06266

k	780	781	782	783	784
1	780	781	782	783	784
2	6 08400	6 09961	6 11524	6 13089	6 14656
3	4745 52000	4763 79541	4782 11768	4800 48687	4818 90304
4	(11)3.7015 05600	(11)3.7205 24215	(11)3.7396 16026	(11)3.7587 81219	(11)3.7780 19983
5	(14)2.8871 74368	(14)2.9057 29412	(14)2.9243 79732	(14)2.9431 25695	(14)2.9619 67667
6	(17)2.2519 96007	(17)2.2693 74671	(17)2.2868 64951	(17)2.3044 67419	(17)2.3221 82651
7	(20)1.7565 56885	(20)1.7723 81618	(20)1.7883 28391	(20)1.8043 97989	(20)1.8205 91198
8	(23)1.3701 14371	(23)1.3842 30044	(23)1.3984 72802	(23)1.4128 43625	(23)1.4273 43499
9	(26)1.0686 89209	(26)1.0810 83664	(26)1.0936 05731	(26)1.1062 56559	(26)1.1190 37304
10	(28)8.3357 75831	(28)8.4432 63416	(28)8.5519 96818	(28)8.6619 88854	(28)8.7732 52460
24	(69)2.5719 97041	(69)2.6523 13239	(69)2.7350 29868	(69)2.8202 15463	(69)2.9079 40422
1/2	(1)2.7928 48009	(1)2.7946 37722	(1)2.7964 26291	(1)2.7982 13716	(1)2.8000 00000
1/3	9.2051 64083	9.2090 96233	9.2130 25029	9.2169 50477	9.2208 72584
1/4	5.2847 40305	5.2864 33318	5.2881 24706	5.2898 14473	5.2915 02622
1/5	3.7880 78066	3.7890 48871	3.7900 18681	3.7909 87500	3.7919 55329

k	785	786	787	788	789
1	785	786	787	788	789
2	6 16225	6 17796	6 19369	6 20944	6 22521
3	4837 36625	4855 87656	4874 43403	4893 03872	4911 69069
4	(11)3.7973 32506	(11)3.8167 18976	(11)3.8361 79582	(11)3.8557 14511	(11)3.8753 23954
5	(14)2.9809 06017	(14)2.9999 41115	(14)3.0190 73331	(14)3.0383 03035	(14)3.0576 30600
6	(17)2.3400 11224	(17)2.3579 53717	(17)2.3760 10711	(17)2.3941 82792	(17)2.4124 70543
7	(20)1.8369 08811	(20)1.8533 51621	(20)1.8699 20430	(20)1.8866 16040	(20)1.9034 39259
8	(23)1.4419 73416	(23)1.4567 34374	(23)1.4716 27378	(23)1.4866 53439	(23)1.5018 13575
9	(26)1.1319 49132	(26)1.1449 93218	(26)1.1581 70747	(26)1.1714 82910	(26)1.1849 30911
10	(28)8.8858 00685	(28)8.9996 46695	(28)9.1148 03776	(28)9.2312 85332	(28)9.3491 04886
24	(69)2.9982 77060	(69)3.0912 99652	(69)3.1870 84488	(69)3.2857 09926	(69)3.3872 56439
1/2	(1)2.8017 85145	(1)2.8035 69154	(1)2.8053 52028	(1)2.8071 33770	(1)2.8089 14381
1/3	9.2247 91357	9.2287 06804	9.2326 18931	9.2365 27746	9.2404 33255
1/4	5.2931 89157	5.2948 74081	5.2965 57399	5.2982 39113	5.2999 19227
1/5	3.7929 22172	3.7938 88029	3.7948 52904	3.7958 16799	3.7967 79716

k	790	791	792	793	794
1	790	791	792	793	794
2	6 24100	6 25681	6 27264	6 28849	6 30436
3	4930 39000	4949 13671	4967 93088	4986 77257	5005 66184
4	(11)3.8950 08100	(11)3.9147 67138	(11)3.9346 01257	(11)3.9545 10648	(11)3.9744 95501
5	(14)3.0770 56399	(14)3.0965 80806	(14)3.1162 04196	(14)3.1359 26944	(14)3.1557 49428
6	(17)2.4308 74555	(17)2.4493 95417	(17)2.4680 33723	(17)2.4867 90066	(17)2.5056 65046
7	(20)1.9203 90899	(20)1.9374 71775	(20)1.9546 82708	(20)1.9720 24523	(20)1.9894 98046
8	(23)1.5171 08810	(23)1.5325 40174	(23)1.5481 08705	(23)1.5638 15447	(23)1.5796 61449
9	(26)1.1985 15960	(26)1.2122 39278	(26)1.2261 02094	(26)1.2401 05649	(26)1.2542 51190
10	(28)9.4682 76083	(28)9.5888 12687	(28)9.7107 28588	(28)9.8340 37797	(28)9.9587 54451
24	(69)3.4918 06676	(69)3.5994 45514	(69)3.7102 60118	(69)3.8243 39997	(69)3.9417 77065
1/2	(1)2.8106 93865	(1)2.8124 72222	(1)2.8142 49456	(1)2.8160 25568	(1)2.8178 00561
1/3	9.2443 35465	9.2482 34384	9.2521 30018	9.2560 22375	9.2599 11460
1/4	5.3015 97745	5.3032 74670	5.3049 50005	5.3066 23755	5.3082 95923
1/5	3.7977 41656	3.7987 02623	3.7996 62619	3.8006 21646	3.8015 79705

k	795	796	797	798	799
1	795	796	797	798	799
2	6 32025	6 33616	6 35209	6 36804	6 38401
3	5024 59875	5043 58336	5062 61573	5081 69592	5100 82399
4	(11)3.9945 56006	(11)4.0146 92355	(11)4.0349 04737	(11)4.0551 93344	(11)4.0755 58368
5	(14)3.1756 72025	(14)3.1956 95114	(14)3.2158 19075	(14)3.2360 44289	(14)3.2563 71136
6	(17)2.5246 59260	(17)2.5437 73311	(17)2.5630 07803	(17)2.5823 63342	(17)2.6018 40538
7	(20)2.0071 04112	(20)2.0248 43555	(20)2.0427 17219	(20)2.0607 25947	(20)2.0788 70590
8	(23)1.5956 47769	(23)1.6117 75470	(23)1.6280 45624	(23)1.6444 59306	(23)1.6610 17601
9	(26)1.2685 39976	(26)1.2829 73274	(26)1.2975 52362	(26)1.3122 78526	(26)1.3271 53063
10	(29)1.0084 89281	(29)1.0212 46726	(29)1.0341 49232	(29)1.0471 98264	(29)1.0603 95298
24	(69)4.0626 65702	(69)4.1871 02820	(69)4.3151 87922	(69)4.4470 23172	(69)4.5827 13463
1/2	(1)2.8195 74436	(1)2.8213 47196	(1)2.8231 18843	(1)2.8248 89378	(1)2.8266 58805
1/3	9.2637 97282	9.2676 79846	9.2715 59160	9.2754 35230	9.2793 08064
1/4	5.3099 66512	5.3116 35526	5.3133 02968	5.3149 68841	5.3166 33150
1/5	3.8025 36800	3.8034 92932	3.8044 48104	3.8054 02317	3.8063 55574

TABLE II.

POWERS AND ROOTS n^k

k	800	801	802	803	804
1	800	801	802	803	804
2	6 40000	6 41601	6 43204	6 44809	6 46416
3	5120 00000	5139 22401	5158 49608	5177 81627	5197 18464
4	(11)4.0960 00000	(11)4.1165 18432	(11)4.1371 13856	(11)4.1577 86465	(11)4.1785 36451
5	(14)3.2768 00000	(14)3.2973 31264	(14)3.3179 65313	(14)3.3387 02531	(14)3.3595 43306
6	(17)2.6214 40000	(17)2.6411 62342	(17)2.6610 08181	(17)2.6809 78133	(17)2.7010 72818
7	(20)2.0971 52000	(20)2.1155 71036	(20)2.1341 28561	(20)2.1528 25440	(20)2.1716 62546
8	(23)1.6777 21600	(23)1.6945 72400	(23)1.7115 71106	(23)1.7287 18829	(23)1.7460 16687
9	(26)1.3421 77280	(26)1.3573 52492	(26)1.3726 80027	(26)1.3881 61219	(26)1.4037 97416
10	(29)1.0737 41824	(29)1.0872 39346	(29)1.1008 89382	(29)1.1146 93459	(29)1.1286 53123
24	(69)4.7223 66483	(69)4.8660 92789	(69)5.0140 05879	(69)5.1662 22264	(69)5.3228 61548
1/2	(1)2.8284 27125	(1)2.8301 94340	(1)2.8319 60452	(1)2.8337 25463	(1)2.8354 89376
1/3	9.2831 77667	9.2870 44047	9.2909 07211	9.2947 67164	9.2986 23915
1/4	5.3182 95897	5.3199 57086	5.3216 16720	5.3232 74803	5.3249 31338
1/5	3.8073 07877	3.8082 59229	3.8092 09631	3.8101 59085	3.8111 07593

k	805	806	807	808	809
1	805	806	807	808	809
2	6 48025	6 49636	6 51249	6 52864	6 54481
3	5216 60125	5236 06616	5255 57943	5275 14112	5294 75129
4	(11)4.1993 64006	(11)4.2202 69325	(11)4.2412 52600	(11)4.2623 14025	(11)4.2834 53794
5	(14)3.3804 88025	(14)3.4015 37076	(14)3.4226 90848	(14)3.4439 49732	(14)3.4653 14119
6	(17)2.7212 92860	(17)2.7416 38883	(17)2.7621 11515	(17)2.7827 11384	(17)2.8034 39122
7	(20)2.1906 40752	(20)2.2097 60940	(20)2.2290 23992	(20)2.2484 30798	(20)2.2679 82250
8	(23)1.7634 65806	(23)1.7810 67318	(23)1.7988 22362	(23)1.8167 32085	(23)1.8347 97640
9	(26)1.4195 89974	(26)1.4355 40258	(26)1.4516 49646	(26)1.4679 19524	(26)1.4843 51291
10	(29)1.1427 69929	(29)1.1570 45448	(29)1.1714 81264	(29)1.1860 78976	(29)1.2008 40194
24	(69)5.4840 46503	(69)5.6499 03151	(69)5.8205 60843	(69)5.9961 52346	(69)6.1768 13927
1/2	(1)2.8372 52192	(1)2.8390 13913	(1)2.8407 74542	(1)2.8425 34081	(1)2.8442 92531
1/3	9.3024 77468	9.3063 27832	9.3101 75012	9.3140 19016	9.3178 59849
1/4	5.3265 86329	5.3282 39778	5.3298 91690	5.3315 42067	5.3331 90912
1/5	3.8120 55159	3.8130 01783	3.8139 47468	3.8148 92216	3.8158 36029

k	810	811	812	813	814
1	810	811	812	813	814
2	6 56100	6 57721	6 59344	6 60969	6 62596
3	5314 41000	5334 11731	5353 87328	5373 67797	5393 53144
4	(11)4.3046 72100	(11)4.3259 69138	(11)4.3473 45103	(11)4.3688 00190	(11)4.3903 34592
5	(14)3.4867 84401	(14)3.5083 60971	(14)3.5300 44224	(14)3.5518 34554	(14)3.5737 32358
6	(17)2.8242 95365	(17)2.8452 80748	(17)2.8663 95910	(17)2.8876 41493	(17)2.9090 18139
7	(20)2.2876 79245	(20)2.3075 22686	(20)2.3275 13479	(20)2.3476 52533	(20)2.3679 40765
8	(23)1.8530 20189	(23)1.8714 00899	(23)1.8899 40945	(23)1.9086 41510	(23)1.9275 03783
9	(26)1.5009 46353	(26)1.5177 06129	(26)1.5346 32047	(26)1.5517 25547	(26)1.5689 88079
10	(29)1.2157 66546	(29)1.2308 59670	(29)1.2461 21222	(29)1.2615 52870	(29)1.2771 56297
24	(69)6.3626 85441	(69)6.5539 10420	(69)6.7506 36166	(69)6.9530 13847	(69)7.1611 98588
1/2	(1)2.8460 49894	(1)2.8478 06173	(1)2.8495 61370	(1)2.8513 15486	(1)2.8530 68524
1/3	9.3216 97518	9.3255 32030	9.3293 63391	9.3331 91608	9.3370 16687
1/4	5.3348 38230	5.3364 84023	5.3381 28295	5.3397 71049	5.3414 12288
1/5	3.8167 78910	3.8177 20859	3.8186 61880	3.8196 01974	3.8205 41144

k	815	816	817	818	819
1	815	816	817	818	819
2	6 64225	6 65856	6 67489	6 69124	6 70761
3	5413 43375	5433 38496	5453 38513	5473 43432	5493 53259
4	(11)4.4119 48506	(11)4.4336 42127	(11)4.4554 15651	(11)4.4772 69274	(11)4.4992 03191
5	(14)3.5957 38033	(14)3.6178 51976	(14)3.6400 74587	(14)3.6624 06266	(14)3.6848 47414
6	(17)2.9305 26497	(17)2.9521 67212	(17)2.9739 40938	(17)2.9958 48326	(17)3.0178 90032
7	(20)2.3883 79095	(20)2.4089 68445	(20)2.4297 09746	(20)2.4506 03930	(20)2.4716 51936
8	(23)1.9465 28962	(23)1.9657 18251	(23)1.9850 72863	(23)2.0045 94015	(23)2.0242 82936
9	(26)1.5864 21104	(26)1.6040 26093	(26)1.6218 04529	(26)1.6397 57904	(26)1.6578 87724
10	(29)1.2929 33200	(29)1.3088 85292	(29)1.3250 14300	(29)1.3413 21966	(29)1.3578 10046
24	(69)7.3753 49576	(69)7.5956 30157	(69)7.8222 07941	(69)8.0552 54907	(69)8.2949 47511
1/2	(1)2.8548 20485	(1)2.8565 71371	(1)2.8583 21186	(1)2.8600 69929	(1)2.8618 17604
1/3	9.3408 38634	9.3446 57457	9.3484 73160	9.3522 85752	9.3560 95237
1/4	5.3430 52016	5.3446 90236	5.3463 26950	5.3479 62163	5.3495 95877
1/5	3.8214 79391	3.8224 16717	3.8233 53125	3.8242 88616	3.8252 23193

k	820	821	822	823	824
1	820	821	822	823	824
2	6 72400	6 74041	6 75684	6 77329	6 78976
3	5513 68000	5533 87661	5554 12248	5574 41767	5594 76224
4	(11)4.5212 17600	(11)4.5433 12697	(11)4.5654 88679	(11)4.5877 45742	(11)4.6100 84086
5	(14)3.7073 98432	(14)3.7300 59724	(14)3.7528 31694	(14)3.7757 14746	(14)3.7987 09287
6	(17)3.0400 66714	(17)3.0623 79033	(17)3.0848 27652	(17)3.1074 13236	(17)3.1301 36452
7	(20)2.4928 54706	(20)2.5142 13186	(20)2.5357 28330	(20)2.5574 01093	(20)2.5792 32437
8	(23)2.0441 40859	(23)2.0641 69026	(23)2.0843 68687	(23)2.1047 41100	(23)2.1252 87528
9	(26)1.6761 95504	(26)1.6946 82770	(26)1.7133 51061	(26)1.7322 01925	(26)1.7512 36923
10	(29)1.3744 80313	(29)1.3913 39355	(29)1.4083 74572	(29)1.4256 02184	(29)1.4430 19224
24	(69)8.5414 66801	(69)8.7949 98523	(69)9.0557 33244	(69)9.3238 66467	(69)9.5995 98755
1/2	(1)2.8635 64213	(1)2.8653 09756	(1)2.8670 54237	(1)2.8687 97658	(1)2.8705 40019
1/3	9.3599 01623	9.3637 04916	9.3675 05121	9.3713 02245	9.3750 96295
1/4	5.3512 28095	5.3528 58822	5.3544 88059	5.3561 15810	5.3577 42079
1/5	3.8261 56858	3.8270 89612	3.8280 21458	3.8289 52397	3.8298 82432

TABLE II.

k	825	826	827	828	829
1	825	826	827	828	829
2	6 80625	6 82276	6 83929	6 85584	6 87241
3	5615 15625	5635 59976	5656 09283	5676 63552	5697 22789
4	(11)4.6325 03906	(11)4.6550 05402	(11)4.6775 88770	(11)4.7002 54211	(11)4.7230 01921
5	(14)3.8218 15723	(14)3.8450 34462	(14)3.8683 65913	(14)3.8918 10486	(14)3.9153 68592
6	(17)3.1529 97971	(17)3.1759 98465	(17)3.1991 38610	(17)3.2224 19083	(17)3.2458 40563
7	(20)2.6012 23326	(20)2.6233 74732	(20)2.6456 87631	(20)2.6681 63000	(20)2.6908 01827
8	(23)2.1460 09244	(23)2.1669 07529	(23)2.1879 83671	(23)2.2092 38964	(23)2.2306 74714
9	(26)1.7704 57626	(26)1.7898 65619	(26)1.8094 62496	(26)1.8292 49863	(26)1.8492 29338
10	(29)1.4606 27542	(29)1.4784 29001	(29)1.4964 25484	(29)1.5146 18886	(29)1.5330 11121
24	(69)9.8831 35853	(70)1.0174 68882	(70)1.0474 47415	(70)1.0782 71392	(70)1.1099 63591
1/2	(1)2.8722 81323	(1)2.8740 21573	(1)2.8757 60769	(1)2.8774 98914	(1)2.8792 36010
1/3	9.3788 87277	9.3826 75196	9.3864 60060	9.3902 41873	9.3940 20643
1/4	5.3593 66869	5.3609 90182	5.3626 12021	5.3642 32391	5.3658 51293
1/5	3.8308 11564	3.8317 39795	3.8326 67128	3.8335 93565	3.8345 19107

k	830	831	832	833	834
1	830	831	832	833	834
2	6 88900	6 90561	6 92224	6 93889	6 95556
3	5717 87000	5738 56191	5759 30368	5780 09537	5800 93704
4	(11)4.7458 32100	(11)4.7687 44947	(11)4.7917 40662	(11)4.8148 19443	(11)4.8379 81491
5	(14)3.9390 40643	(14)3.9628 27051	(14)3.9867 28231	(14)4.0107 44596	(14)4.0348 76564
6	(17)3.2694 03734	(17)3.2931 09279	(17)3.3169 57888	(17)3.3409 50249	(17)3.3650 87054
7	(20)2.7136 05099	(20)2.7365 73811	(20)2.7597 08963	(20)2.7830 11557	(20)2.8064 82603
8	(23)2.2522 92232	(23)2.2740 92837	(23)2.2960 77857	(23)2.3182 48627	(23)2.3406 06491
9	(26)1.8694 02553	(26)1.8897 71148	(26)1.9103 36777	(26)1.9311 01106	(26)1.9520 65814
10	(29)1.5516 04119	(29)1.5703 99824	(29)1.5894 00198	(29)1.6086 07222	(29)1.6280 22889
24	(70)1.1425 47375	(70)1.1760 46709	(70)1.2104 86167	(70)1.2458 90957	(70)1.2822 86929
1/2	(1)2.8809 72058	(1)2.8827 07061	(1)2.8844 41020	(1)2.8861 73938	(1)2.8879 05816
1/3	9.3977 96375	9.4015 69076	9.4053 38751	9.4091 05407	9.4128 69049
1/4	5.3674 68731	5.3690 84709	5.3706 99229	5.3723 12294	5.3739 23907
1/5	3.8354 43756	3.8363 67514	3.8372 90383	3.8382 12366	3.8391 33463

k	835	836	837	838	839
1	835	836	837	838	839
2	6 97225	6 98896	7 00569	7 02244	7 03921
3	5821 82875	5842 77056	5863 76253	5884 80472	5905 89719
4	(11)4.8612 27006	(11)4.8845 56188	(11)4.9079 69238	(11)4.9314 66355	(11)4.9550 47742
5	(14)4.0591 24550	(14)4.0834 88973	(14)4.1079 70252	(14)4.1325 68806	(14)4.1572 85056
6	(17)3.3893 68999	(17)3.4137 96782	(17)3.4383 71101	(17)3.4630 92659	(17)3.4879 62162
7	(20)2.8301 23115	(20)2.8539 34109	(20)2.8779 16611	(20)2.9020 71648	(20)2.9264 00254
8	(23)2.3631 52801	(23)2.3858 88916	(23)2.4088 16204	(23)2.4319 36041	(23)2.4552 49813
9	(26)1.9732 32589	(26)1.9946 03133	(26)2.0161 79163	(26)2.0379 62403	(26)2.0599 54593
10	(29)1.6476 49211	(29)1.6674 88220	(29)1.6875 41959	(29)1.7078 12493	(29)1.7283 01904
24	(70)1.3197 00592	(70)1.3581 59133	(70)1.3976 90431	(70)1.4383 23072	(70)1.4800 86372
1/2	(1)2.8896 36655	(1)2.8913 66459	(1)2.8930 95228	(1)2.8948 22965	(1)2.8965 49672
1/3	9.4166 29685	9.4203 87319	9.4241 41957	9.4278 93606	9.4316 42272
1/4	5.3755 34071	5.3771 42790	5.3787 50067	5.3803 55904	5.3819 60304
1/5	3.8400 53677	3.8409 73010	3.8418 91464	3.8428 09040	3.8437 25741

k	840	841	842	843	844
1	840	841	842	843	844
2	7 05600	7 07281	7 08964	7 10649	7 12336
3	5927 04000	5948 23321	5969 47688	5990 77107	6012 11584
4	(11)4.9787 13600	(11)5.0024 64130	(11)5.0262 99533	(11)5.0502 20012	(11)5.0742 25769
5	(14)4.1821 19424	(14)4.2070 72333	(14)4.2321 44207	(14)4.2573 35470	(14)4.2826 46549
6	(17)3.5129 80316	(17)3.5381 47832	(17)3.5634 65422	(17)3.5889 33801	(17)3.6145 53687
7	(20)2.9509 03466	(20)2.9755 82327	(20)3.0004 37885	(20)3.0254 71195	(20)3.0506 83312
8	(23)2.4787 58911	(23)2.5024 64737	(23)2.5263 68700	(23)2.5504 72217	(23)2.5747 76715
9	(26)2.0821 57485	(26)2.1045 72844	(26)2.1272 02445	(26)2.1500 48079	(26)2.1731 11548
10	(29)1.7490 12288	(29)1.7699 45762	(29)1.7911 04459	(29)1.8124 90531	(29)1.8341 06146
24	(70)1.5230 10388	(70)1.5671 25939	(70)1.6124 64626	(70)1.6590 58848	(70)1.7069 41821
1/2	(1)2.8982 75349	(1)2.9000 00000	(1)2.9017 23626	(1)2.9034 46228	(1)2.9051 67809
1/3	9.4353 87961	9.4391 30677	9.4428 70428	9.4466 07220	9.4503 41057
1/4	5.3835 63271	5.3851 64807	5.3867 64916	5.3883 63600	5.3899 60862
1/5	3.8446 41568	3.8455 56523	3.8464 70609	3.8473 83826	3.8482 96177

k	845	846	847	848	849
1	845	846	847	848	849
2	7 14025	7 15716	7 17409	7 19104	7 20801
3	6033 51125	6054 95736	6076 45423	6098 00192	6119 60049
4	(11)5.0983 17006	(11)5.1224 93927	(11)5.1467 56733	(11)5.1711 05628	(11)5.1955 40816
5	(14)4.3080 77870	(14)4.3336 29862	(14)4.3593 02953	(14)4.3850 97573	(14)4.4110 14153
6	(17)3.6403 25800	(17)3.6662 50863	(17)3.6923 29601	(17)3.7185 62742	(17)3.7449 51016
7	(20)3.0760 75301	(20)3.1016 48230	(20)3.1274 03172	(20)3.1533 41205	(20)3.1794 63412
8	(23)2.5992 83630	(23)2.6239 94403	(23)2.6489 10487	(23)2.6740 33342	(23)2.6993 64437
9	(26)2.1963 94667	(26)2.2198 99265	(26)2.2436 27182	(26)2.2675 80274	(26)2.2917 60407
10	(29)1.8559 53494	(29)1.8780 34778	(29)1.9003 52223	(29)1.9229 08072	(29)1.9457 04586
24	(70)1.7561 47601	(70)1.8067 11101	(70)1.8586 68111	(70)1.9120 55324	(70)1.9669 10351
1/2	(1)2.9068 88371	(1)2.9086 07914	(1)2.9103 26442	(1)2.9120 43956	(1)2.9137 60457
1/3	9.4540 71946	9.4577 99893	9.4615 24903	9.4652 46982	9.4689 66137
1/4	5.3915 56705	5.3931 51133	5.3947 44148	5.3963 35753	5.3979 25951
1/5	3.8492 07664	3.8501 18288	3.8510 28051	3.8519 36956	3.8528 45003

TABLE II.

POWERS AND ROOTS n^k

Table II

k	850	851	852	853	854
1	850	851	852	853	854
2	7 22500	7 24201	7 25904	7 27609	7 29316
3	6141 25000	6162 95051	6184 70208	6206 50477	6228 35864
4	(11)5.2200 62500	(11)5.2446 70884	(11)5.2693 66172	(11)5.2941 48569	(11)5.3190 18279
5	(14)4.4370 53125	(14)4.4632 14922	(14)4.4894 99979	(14)4.5159 08729	(14)4.5424 41610
6	(17)3.7714 95156	(17)3.7981 95899	(17)3.8250 53982	(17)3.8520 70146	(17)3.8792 45135
7	(20)3.2057 70883	(20)3.2322 64710	(20)3.2589 45993	(20)3.2858 15835	(20)3.3128 75345
8	(23)2.7249 05250	(23)2.7506 57268	(23)2.7766 21986	(23)2.8028 00907	(23)2.8291 95545
9	(26)2.3161 69463	(26)2.3408 09335	(26)2.3656 81932	(26)2.3907 89174	(26)2.4161 32995
10	(29)1.9687 44043	(29)1.9920 28744	(29)2.0155 61006	(29)2.0393 43165	(29)2.0633 77578
24	(70)2.0232 71747	(70)2.0811 79034	(70)2.1406 72719	(70)2.2017 94325	(70)2.2645 86409
1/2	(1)2.9154 75947	(1)2.9171 90429	(1)2.9189 03904	(1)2.9206 16373	(1)2.9223 27839
1/3	9.4726 82372	9.4763 95693	9.4801 06107	9.4838 13619	9.4875 18234
1/4	5.3995 14744	5.4011 02137	5.4026 88131	5.4042 72729	5.4058 55935
1/5	3.8537 52195	3.8546 58534	3.8555 64021	3.8564 68659	3.8573 72448

k	855	856	857	858	859
1	855	856	857	858	859
2	7 31025	7 32736	7 34449	7 36164	7 37881
3	6250 26375	6272 22016	6294 22793	6316 28712	6338 39779
4	(11)5.3439 75506	(11)5.3690 20457	(11)5.3941 53336	(11)5.4193 74349	(11)5.4446 83702
5	(14)4.5690 99058	(14)4.5958 81511	(14)4.6227 89409	(14)4.6498 23191	(14)4.6769 83300
6	(17)3.9065 79694	(17)3.9340 74574	(17)3.9617 30523	(17)3.9895 48298	(17)4.0175 28654
7	(20)3.3401 25639	(20)3.3675 67835	(20)3.3952 03059	(20)3.4230 32440	(20)3.4510 57114
8	(23)2.8558 07421	(23)2.8826 38067	(23)2.9096 89021	(23)2.9369 61833	(23)2.9644 58061
9	(26)2.4417 15345	(26)2.4675 38185	(26)2.4936 03491	(26)2.5199 13253	(26)2.5464 69474
10	(29)2.0876 66620	(29)2.1122 12686	(29)2.1370 18192	(29)2.1620 85571	(29)2.1874 17279
24	(/0)2.3290 92589	(70)2.3953 57569	(70)2.4634 27165	(70)2.5333 48329	(70)2.6051 69182
1/2	(1)2.9240 38303	(1)2.9257 47768	(1)2.9274 56234	(1)2.9291 63703	(1)2.9308 70178
1/3	9.4912 19958	9.4949 18797	9.4986 14756	9.5023 07842	9.5059 98059
1/4	5.4074 37751	5.4090 18180	5.4105 97225	5.4121 74889	5.4137 51174
1/5	3.8582 75391	3.8591 77490	3.8600 78746	3.8609 79161	3.8618 78737

k	860	861	862	863	864
1	860	861	862	863	864
2	7 39600	7 41321	7 43044	7 44769	7 46496
3	6360 56000	6382 77381	6405 03928	6427 35647	6449 72544
4	(11)5.4700 81600	(11)5.4955 68250	(11)5.5211 43859	(11)5.5468 08634	(11)5.5725 62780
5	(14)4.7042 70176	(14)4.7316 84264	(14)4.7592 26007	(14)4.7868 95851	(14)4.8146 94242
6	(17)4.0456 72351	(17)4.0739 80151	(17)4.1024 52818	(17)4.1310 91119	(17)4.1598 95825
7	(20)3.4792 78222	(20)3.5076 96910	(20)3.5363 14329	(20)3.5651 31636	(20)3.5941 49993
8	(23)2.9921 79271	(23)3.0201 27039	(23)3.0483 02952	(23)3.0767 08602	(23)3.1053 45594
9	(26)2.5732 74173	(26)2.6003 29381	(26)2.6276 37144	(26)2.6551 99523	(26)2.6830 18593
10	(29)2.2130 15789	(29)2.2388 83597	(29)2.2650 23218	(29)2.2914 37189	(29)2.3181 28064
24	(70)2.6789 39031	(70)2.7547 08410	(70)2.8325 29097	(70)2.9124 54150	(70)2.9945 37938
1/2	(1)2.9325 75660	(1)2.9342 80150	(1)2.9359 83651	(1)2.9376 86164	(1)2.9393 87691
1/3	9.5096 85413	9.5133 69910	9.5170 51555	9.5207 30354	9.5244 06312
1/4	5.4153 26084	5.4168 99621	5.4184 71787	5.4200 42587	5.4216 12022
1/5	3.8627 77475	3.8636 75378	3.8645 72447	3.8654 68684	3.8663 64090

k	865	866	867	868	869
1	865	866	867	868	869
2	7 48225	7 49956	7 51689	7 53424	7 55161
3	6472 14625	6494 61896	6517 14363	6539 72032	6562 34909
4	(11)5.5984 06506	(11)5.6243 40019	(11)5.6503 63527	(11)5.6764 77238	(11)5.7026 81359
5	(14)4.8426 21628	(14)4.8706 78457	(14)4.8988 65178	(14)4.9271 82242	(14)4.9556 30101
6	(17)4.1888 67708	(17)4.2180 07544	(17)4.2473 16109	(17)4.2767 94186	(17)4.3064 42558
7	(20)3.6233 70568	(20)3.6527 94533	(20)3.6824 23067	(20)3.7122 57354	(20)3.7422 98583
8	(23)3.1342 15541	(23)3.1633 20065	(23)3.1926 60799	(23)3.2222 39383	(23)3.2520 57468
9	(26)2.7110 96443	(26)2.7394 35177	(26)2.7680 36913	(26)2.7969 03785	(26)2.8260 37940
10	(29)2.3450 98423	(29)2.3723 50863	(29)2.3998 88003	(29)2.4277 12485	(29)2.4558 26970
24	(70)3.0788 36164	(70)3.1654 05907	(70)3.2543 05644	(70)3.3455 95291	(70)3.4393 36231
1/2	(1)2.9410 88234	(1)2.9427 87794	(1)2.9444 86373	(1)2.9461 83973	(1)2.9478 80595
1/3	9.5280 79435	9.5317 49727	9.5354 17196	9.5390 81845	9.5427 43681
1/4	5.4231 80095	5.4247 46809	5.4263 12167	5.4278 76171	5.4294 38824
1/5	3.8672 58668	3.8681 52418	3.8690 45344	3.8699 37445	3.8708 28725

k	870	871	872	873	874
1	870	871	872	873	874
2	7 56900	7 58641	7 60384	7 62129	7 63876
3	6585 03000	6607 76311	6630 54848	6653 38617	6676 27624
4	(11)5.7289 76100	(11)5.7553 61669	(11)5.7818 38275	(11)5.8084 06126	(11)5.8350 65434
5	(14)4.9842 09207	(14)5.0129 20014	(14)5.0417 62975	(14)5.0707 38548	(14)5.0998 47189
6	(17)4.3362 62010	(17)4.3662 53332	(17)4.3964 17315	(17)4.4267 54753	(17)4.4572 66443
7	(20)3.7725 47949	(20)3.8030 06652	(20)3.8336 75898	(20)3.8645 56899	(20)3.8956 50871
8	(23)3.2821 16715	(23)3.3124 18794	(23)3.3429 65383	(23)3.3737 58173	(23)3.4047 98862
9	(26)2.8554 41542	(26)2.8851 16769	(26)2.9150 65814	(26)2.9452 90885	(26)2.9757 94205
10	(29)2.4842 34142	(29)2.5129 36706	(29)2.5419 37390	(29)2.5712 38943	(29)2.6008 44135
24	(70)3.5355 91351	(70)3.6344 25075	(70)3.7359 03403	(70)3.8400 93943	(70)3.9470 65953
1/2	(1)2.9495 76241	(1)2.9512 70913	(1)2.9529 64612	(1)2.9546 57341	(1)2.9563 49100
1/3	9.5464 02709	9.5500 58934	9.5537 12362	9.5573 62998	9.5610 10846
1/4	5.4310 00130	5.4325 60090	5.4341 18707	5.4356 75984	5.4372 31924
1/5	3.8717 19185	3.8726 08827	3.8734 97651	3.8743 85661	3.8752 72857

TABLE II.

k					
1	875	876	877	878	879
2	7 65625	7 67376	7 69129	7 70884	7 72641
3	6699 21875	6722 21376	6745 26133	6768 36152	6791 51439
4	(11)5.8618 16406	(11)5.8886 59254	(11)5.9155 94186	(11)5.9426 21415	(11)5.9697 41149
5	(14)5.1290 89355	(14)5.1584 65506	(14)5.1879 76101	(14)5.2176 21602	(14)5.2474 02470
6	(17)4.4879 53186	(17)4.5188 15784	(17)4.5498 55041	(17)4.5810 71767	(17)4.6124 66771
7	(20)3.9269 59038	(20)3.9584 82626	(20)3.9902 22871	(20)4.0221 81011	(20)4.0543 58292
8	(23)3.4360 89158	(23)3.4676 30781	(23)3.4994 25458	(23)3.5314 74928	(23)3.5637 80938
9	(26)3.0065 78013	(26)3.0376 44564	(26)3.0689 96127	(26)3.1006 34987	(26)3.1325 63445
10	(29)2.6307 55762	(29)2.6609 76638	(29)2.6915 09603	(29)2.7223 57518	(29)2.7535 23268
24	(70)4.0568 90376	(70)4.1696 39882	(70)4.2853 88904	(70)4.4042 13682	(70)4.5261 92303
1/2	(1)2.9580 39892	(1)2.9597 29717	(1)2.9614 18579	(1)2.9631 06478	(1)2.9647 93416
1/3	9.5646 55914	9.5682 98205	9.5719 37725	9.5755 74480	9.5792 08475
1/4	5.4387 86530	5.4403 39803	5.4418 91747	5.4434 42365	5.4449 91658
1/5	3.8761 59242	3.8770 44816	3.8779 29583	3.8788 13542	3.8796 96696

k					
1	880	881	882	883	884
2	7 74400	7 76161	7 77924	7 79689	7 81456
3	6814 72000	6837 97841	6861 28968	6884 65387	6908 07104
4	(11)5.9969 53600	(11)6.0242 58979	(11)6.0516 57498	(11)6.0791 49367	(11)6.1067 34799
5	(14)5.2773 19168	(14)5.3073 72161	(14)5.3375 61913	(14)5.3678 88891	(14)5.3983 53563
6	(17)4.6440 40868	(17)4.6757 94874	(17)4.7077 29607	(17)4.7398 45891	(17)4.7721 44549
7	(20)4.0867 55964	(20)4.1193 75284	(20)4.1522 17514	(20)4.1852 83922	(20)4.2185 75782
8	(23)3.5963 45248	(23)3.6291 69625	(23)3.6622 55847	(23)3.6956 05703	(23)3.7292 20991
9	(26)3.1647 83818	(26)3.1972 98440	(26)3.2301 09657	(26)3.2632 19836	(26)3.2966 31356
10	(29)2.7850 09760	(29)2.8168 19925	(29)2.8489 56718	(29)2.8814 23115	(29)2.9142 22119
24	(70)4.6514 04745	(70)4.7799 32920	(70)4.9118 60716	(70)5.0472 74047	(70)5.1862 60897
1/2	(1)2.9664 79395	(1)2.9681 64416	(1)2.9698 48481	(1)2.9715 31592	(1)2.9732 13749
1/3	9.5828 39714	9.5864 68204	9.5900 93948	9.5937 16954	9.5973 37224
1/4	5.4465 39631	5.4480 86284	5.4496 31621	5.4511 75645	5.4527 18358
1/5	3.8805 79047	3.8814 60596	3.8823 41346	3.8832 21296	3.8841 00450

k					
1	885	886	887	888	889
2	7 83225	7 84996	7 86769	7 88544	7 90321
3	6931 54125	6955 06456	6978 64103	7002 27072	7025 95369
4	(11)6.1344 14006	(11)6.1621 87200	(11)6.1900 54594	(11)6.2180 16399	(11)6.2460 72830
5	(14)5.4289 56396	(14)5.4596 97859	(14)5.4905 78425	(14)5.5215 98563	(14)5.5527 58746
6	(17)4.8046 26410	(17)4.8372 92303	(17)4.8701 43063	(17)4.9031 79524	(17)4.9364 02525
7	(20)4.2520 94373	(20)4.2858 40981	(20)4.3198 16896	(20)4.3540 23417	(20)4.3884 61845
8	(23)3.7631 03520	(23)3.7972 55109	(23)3.8316 77587	(23)3.8663 72794	(23)3.9013 42580
9	(26)3.3303 46615	(26)3.3643 68027	(26)3.3986 98020	(26)3.4333 39041	(26)3.4682 93554
10	(29)2.9473 56754	(29)2.9808 30072	(29)3.0146 45144	(29)3.0488 05069	(29)3.0833 12969
24	(70)5.3289 11365	(70)5.4753 17719	(70)5.6255 74442	(70)5.7797 78281	(70)5.9380 28303
1/2	(1)2.9748 94956	(1)2.9765 75213	(1)2.9782 54522	(1)2.9799 32885	(1)2.9816 10303
1/3	9.6009 54766	9.6045 69584	9.6081 81682	9.6117 91067	9.6153 97744
1/4	5.4542 59763	5.4557 99862	5.4573 38658	5.4588 76153	5.4604 12350
1/5	3.8849 78808	3.8858 56373	3.8867 33146	3.8876 09128	3.8884 84321

k					
1	890	891	892	893	894
2	7 92100	7 93881	7 95664	7 97449	7 99236
3	7049 69000	7073 47971	7097 32288	7121 21957	7145 16984
4	(11)6.2742 24100	(11)6.3024 70422	(11)6.3308 12009	(11)6.3592 49076	(11)6.3877 81837
5	(14)5.5840 59449	(14)5.6155 01146	(14)5.6470 84312	(14)5.6788 09425	(14)5.7106 76962
6	(17)4.9698 12910	(17)5.0034 11521	(17)5.0371 99206	(17)5.0711 76816	(17)5.1053 45204
7	(20)4.4231 33490	(20)4.4580 39665	(20)4.4931 81692	(20)4.5285 60897	(20)4.5641 78613
8	(23)3.9365 88806	(23)3.9721 13342	(23)4.0079 18069	(23)4.0440 04881	(23)4.0803 75680
9	(26)3.5035 64037	(26)3.5391 52987	(26)3.5750 62918	(26)3.6112 96359	(26)3.6478 55858
10	(29)3.1181 71993	(29)3.1533 85312	(29)3.1889 56118	(29)3.2248 87648	(29)3.2611 83137
24	(70)6.1004 25945	(70)6.2670 75070	(70)6.4380 82017	(70)6.6135 55666	(70)6.7936 07487
1/2	(1)2.9832 86778	(1)2.9849 62311	(1)2.9866 36905	(1)2.9883 10559	(1)2.9899 83278
1/3	9.6190 01716	9.6226 02990	9.6262 01570	9.6297 97462	9.6333 90671
1/4	5.4619 47252	5.4634 80860	5.4650 13179	5.4665 44210	5.4680 73955
1/5	3.8893 58728	3.8902 32348	3.8911 05185	3.8919 77239	3.8928 48512

k					
1	895	896	897	898	899
2	8 01025	8 02816	8 04609	8 06404	8 08201
3	7169 17375	7193 23136	7217 34273	7241 50792	7265 72699
4	(11)6.4164 10506	(11)6.4451 35299	(11)6.4739 56429	(11)6.5028 74112	(11)6.5318 88564
5	(14)5.7426 87403	(14)5.7748 41228	(14)5.8071 38917	(14)5.8395 80953	(14)5.8721 67819
6	(17)5.1397 05226	(17)5.1742 57740	(17)5.2090 03608	(17)5.2439 43696	(17)5.2790 78869
7	(20)4.6000 36177	(20)4.6361 34935	(20)4.6724 76237	(20)4.7090 61439	(20)4.7458 91904
8	(23)4.1170 32378	(23)4.1539 76902	(23)4.1912 11184	(23)4.2287 37172	(23)4.2665 56821
9	(26)3.6847 43979	(26)3.7219 63304	(26)3.7595 16432	(26)3.7974 05980	(26)3.8356 34582
10	(29)3.2978 45861	(29)3.3348 79120	(29)3.3722 86240	(29)3.4100 70570	(29)3.4482 35490
24	(70)6.9783 51604	(70)7.1679 04854	(70)7.3623 86846	(70)7.5619 20026	(70)7.7666 29743
1/2	(1)2.9916 55060	(1)2.9933 25909	(1)2.9949 95826	(1)2.9966 64813	(1)2.9983 32870
1/3	9.6369 81200	9.6405 69057	9.6441 54244	9.6477 36769	9.6513 16634
1/4	5.4696 02417	5.4711 29599	5.4726 55504	5.4741 80133	5.4757 03489
1/5	3.8937 19006	3.8945 88722	3.8954 57662	3.8963 25828	3.8971 93220

TABLE II.

POWERS AND ROOTS n^k

Table II

k	900	901	902	903	904
1	900	901	902	903	904
2	8 10000	8 11801	8 13604	8 15409	8 17216
3	7290 00000	7314 32701	7338 70808	7363 14327	7387 63264
4	(11)6.5610 00000	(11)6.5902 08636	(11)6.6195 14688	(11)6.6489 18373	(11)6.6784 19907
5	(14)5.9049 00000	(14)5.9377 77981	(14)5.9708 02249	(14)6.0039 73291	(14)6.0372 91596
6	(17)5.3144 10000	(17)5.3499 37961	(17)5.3856 63628	(17)5.4215 87881	(17)5.4577 11602
7	(20)4.7829 69000	(20)4.8202 94103	(20)4.8578 68593	(20)4.8956 93857	(20)4.9337 71289
8	(23)4.3046 72100	(23)4.3430 84987	(23)4.3817 97471	(23)4.4208 11553	(23)4.4601 29245
9	(26)3.8742 04890	(26)3.9131 19573	(26)3.9523 81319	(26)3.9919 92832	(26)4.0319 56837
10	(29)3.4867 84401	(29)3.5257 20735	(29)3.5650 47949	(29)3.6047 69527	(29)3.6448 88981
24	(70)7.9766 44308	(70)8.1920 95066	(70)8.4131 16465	(70)8.6398 46120	(70)8.8724 24888
1/2	(1)3.0000 00000	(1)3.0016 66204	(1)3.0033 31484	(1)3.0049 95840	(1)3.0066 59276
1/3	9.6548 93846	9.6584 68409	9.6620 40328	9.6656 09608	9.6691 76254
1/4	5.4772 25575	5.4787 46393	5.4802 65946	5.4817 84235	5.4833 01264
1/5	3.8980 59841	3.8989 25692	3.8997 90774	3.9006 55089	3.9015 18640

k	905	906	907	908	909
1	905	906	907	908	909
2	8 19025	8 20836	8 22649	8 24464	8 26281
3	7412 17625	7436 77416	7461 42643	7486 13312	7510 89429
4	(11)6.7080 19506	(11)6.7377 17389	(11)6.7675 13772	(11)6.7974 08873	(11)6.8274 02910
5	(14)6.0707 57653	(14)6.1043 71954	(14)6.1381 34991	(14)6.1720 47257	(14)6.2061 09245
6	(17)5.4940 35676	(17)5.5305 60991	(17)5.5672 88437	(17)5.6042 18909	(17)5.6413 53304
7	(20)4.9721 02287	(20)5.0106 88258	(20)5.0495 30612	(20)5.0886 30769	(20)5.1279 90153
8	(23)4.4997 52570	(23)4.5396 83561	(23)4.5799 24265	(23)4.6204 76739	(23)4.6613 43049
9	(26)4.0722 76076	(26)4.1129 53307	(26)4.1539 91309	(26)4.1953 92879	(26)4.2371 60832
10	(29)3.6854 09848	(29)3.7263 35696	(29)3.7676 70117	(29)3.8094 16734	(29)3.8515 79196
24	(70)9.1109 96943	(70)9.3557 09844	(70)9.6067 14616	(70)9.8641 65825	(71)1.0128 22166
1/2	(1)3.0083 21791	(1)3.0099 83389	(1)3.0116 44069	(1)3.0133 03835	(1)3.0149 62686
1/3	9.6727 40271	9.6763 01663	9.6798 60436	9.6834 16593	9.6869 70141
1/4	5.4848 17035	5.4863 31551	5.4878 44813	5.4893 56824	5.4908 67587
1/5	3.9023 81426	3.9032 43449	3.9041 04712	3.9049 65216	3.9058 24962

k	910	911	912	913	914
1	910	911	912	913	914
2	8 28100	8 29921	8 31744	8 33569	8 35396
3	7535 71000	7560 58031	7585 50528	7610 48497	7635 51944
4	(11)6.8574 96100	(11)6.8876 88662	(11)6.9179 80815	(11)6.9483 72778	(11)6.9788 64768
5	(14)6.2403 21451	(14)6.2746 84371	(14)6.3091 98504	(14)6.3438 64346	(14)6.3786 82398
6	(17)5.6786 92520	(17)5.7162 37462	(17)5.7539 89035	(17)5.7919 48148	(17)5.8301 15712
7	(20)5.1676 10194	(20)5.2074 92328	(20)5.2476 38000	(20)5.2880 48659	(20)5.3287 25761
8	(23)4.7025 25276	(23)4.7440 25511	(23)4.7858 45856	(23)4.8279 88426	(23)4.8704 55345
9	(26)4.2792 98001	(26)4.3218 07241	(26)4.3646 91421	(26)4.4079 53433	(26)4.4515 96186
10	(29)3.8941 61181	(29)3.9371 66396	(29)3.9805 98576	(29)4.0244 61484	(29)4.0687 58914
24	(71)1.0399 04400	(71)1.0676 79852	(71)1.0961 65476	(71)1.1253 78622	(71)1.1553 37042
1/2	(1)3.0166 20626	(1)3.0182 77655	(1)3.0199 33774	(1)3.0215 88986	(1)3.0232 43292
1/3	9.6905 21083	9.6940 69425	9.6976 15172	9.7011 58327	9.7046 98896
1/4	5.4923 77104	5.4938 85378	5.4953 92410	5.4968 98203	5.4984 02760
1/5	3.9066 83951	3.9075 42186	3.9083 99668	3.9092 56397	3.9101 12376

k	915	916	917	918	919
1	915	916	917	918	919
2	8 37225	8 39056	8 40889	8 42724	8 44561
3	7660 60875	7685 75296	7710 95213	7736 20632	7761 51559
4	(11)7.0094 57006	(11)7.0401 49711	(11)7.0709 43103	(11)7.1018 37402	(11)7.1328 32827
5	(14)6.4136 53161	(14)6.4487 77136	(14)6.4840 54826	(14)6.5194 86735	(14)6.5550 73368
6	(17)5.8684 92642	(17)5.9070 79856	(17)5.9458 78275	(17)5.9848 88823	(17)6.0241 12425
7	(20)5.3696 70767	(20)5.4108 85148	(20)5.4523 70378	(20)5.4941 27939	(20)5.5361 59319
8	(23)4.9132 48752	(23)4.9563 70796	(23)4.9998 23637	(23)5.0436 09448	(23)5.0877 30414
9	(26)4.4956 22608	(26)4.5400 35649	(26)4.5848 38275	(26)4.6300 33473	(26)4.6756 24251
10	(29)4.1134 94687	(29)4.1586 72654	(29)4.2042 96698	(29)4.2503 70729	(29)4.2968 98686
24	(71)1.1860 58902	(71)1.2175 62793	(71)1.2498 67732	(71)1.2829 93183	(71)1.3169 59057
1/2	(1)3.0248 96692	(1)3.0265 49190	(1)3.0282 00786	(1)3.0298 51482	(1)3.0315 01278
1/3	9.7082 36884	9.7117 72294	9.7153 05133	9.7188 35404	9.7223 63112
1/4	5.4999 06083	5.5014 08174	5.5029 09036	5.5044 08671	5.5059 07081
1/5	3.9109 67606	3.9118 22089	3.9126 75826	3.9135 28819	3.9143 81068

k	920	921	922	923	924
1	920	921	922	923	924
2	8 46400	8 48241	8 50084	8 51929	8 53776
3	7786 88000	7812 29961	7837 77448	7863 30467	7888 89024
4	(11)7.1639 29600	(11)7.1951 27941	(11)7.2264 28071	(11)7.2578 30210	(11)7.2893 34582
5	(14)6.5908 15232	(14)6.6267 12833	(14)6.6627 66681	(14)6.6989 77284	(14)6.7353 45154
6	(17)6.0635 50013	(17)6.1032 02520	(17)6.1430 70880	(17)6.1831 56033	(17)6.2234 58922
7	(20)5.5784 66012	(20)5.6210 49521	(20)5.6639 11351	(20)5.7070 53019	(20)5.7504 76044
8	(23)5.1321 88731	(23)5.1769 86608	(23)5.2221 26266	(23)5.2676 09936	(23)5.3134 39864
9	(26)4.7216 13633	(26)4.7680 04666	(26)4.8148 00417	(26)4.8620 03971	(26)4.9096 18435
10	(29)4.3438 84542	(29)4.3913 32298	(29)4.4392 45985	(29)4.4876 29665	(29)4.5364 87434
24	(71)1.3517 85726	(71)1.3874 94035	(71)1.4241 05308	(71)1.4616 41363	(71)1.5001 24518
1/2	(1)3.0331 50178	(1)3.0347 98181	(1)3.0364 45290	(1)3.0380 91506	(1)3.0397 36831
1/3	9.7258 88262	9.7294 10859	9.7329 30906	9.7364 48410	9.7399 63373
1/4	5.5074 04268	5.5089 00236	5.5103 94986	5.5118 88520	5.5133 80842
1/5	3.9152 32576	3.9160 83344	3.9169 33373	3.9177 82664	3.9186 31220

TABLE II.

k	925	926	927	928	929
1	925	926	927	928	929
2	8 55625	8 57476	8 59329	8 61184	8 63041
3	7914 53125	7940 22776	7965 97983	7991 78752	8017 65089
4	(11)7.3209 41406	(11)7.3526 50906	(11)7.3844 63302	(11)7.4163 78819	(11)7.4483 97677
5	(14)6.7718 70801	(14)6.8085 54739	(14)6.8453 97481	(14)6.8823 99544	(14)6.9195 61442
6	(17)6.2639 80491	(17)6.3047 21688	(17)6.3456 83465	(17)6.3868 66776	(17)6.4282 72579
7	(20)5.7941 81954	(20)5.8381 72283	(20)5.8824 48572	(20)5.9270 12369	(20)5.9718 65226
8	(23)5.3596 18307	(23)5.4061 47534	(23)5.4530 29826	(23)5.5002 67478	(23)5.5478 62795
9	(26)4.9576 46934	(26)5.0060 92617	(26)5.0549 58649	(26)5.1042 48220	(26)5.1539 64537
10	(29)4.5858 23414	(29)4.6356 41763	(29)4.6859 46668	(29)4.7367 42348	(29)4.7880 33055
24	(71)1.5395 77607	(71)1.5800 23988	(71)1.6214 87554	(71)1.6639 92748	(71)1.7075 64573
1/2	(1)3.0413 81265	(1)3.0430 24811	(1)3.0446 67470	(1)3.0463 09242	(1)3.0479 50131
1/3	9.7434 75802	9.7469 85700	9.7504 93072	9.7539 97922	9.7575 00256
1/4	5.5148 71952	5.5163 61854	5.5178 50550	5.5193 38042	5.5208 24332
1/5	3.9194 79042	3.9203 26131	3.9211 72488	3.9220 18115	3.9228 63013

k	930	931	932	933	934
1	930	931	932	933	934
2	8 64900	8 66761	8 68624	8 70489	8 72356
3	8043 57000	8069 54491	8095 57568	8121 66237	8147 80504
4	(11)7.4805 20100	(11)7.5127 46311	(11)7.5450 76534	(11)7.5775 10991	(11)7.6100 49907
5	(14)6.9568 83693	(14)6.9943 66816	(14)7.0320 11329	(14)7.0698 17755	(14)7.1077 86613
6	(17)6.4699 01834	(17)6.5117 55505	(17)6.5538 34559	(17)6.5961 39965	(17)6.6386 72697
7	(20)6.0170 08706	(20)6.0624 44376	(20)6.1081 73809	(20)6.1541 98588	(20)6.2005 20299
8	(23)5.5958 18097	(23)5.6441 35714	(23)5.6928 17990	(23)5.7418 67282	(23)5.7912 85959
9	(26)5.2041 10830	(26)5.2546 90349	(26)5.3057 06367	(26)5.3571 62174	(26)5.4090 61086
10	(29)4.8398 23072	(29)4.8921 16715	(29)4.9449 18334	(29)4.9982 32309	(29)5.0520 63054
24	(71)1.7522 28603	(71)1.7980 10997	(71)1.8449 38512	(71)1.8930 38514	(71)1.9423 38996
1/2	(1)3.0495 90136	(1)3.0512 29260	(1)3.0528 67504	(1)3.0545 04870	(1)3.0561 41358
1/3	9.7610 00077	9.7644 97390	9.7679 92199	9.7714 84510	9.7749 74326
1/4	5.5223 09423	5.5237 93317	5.5252 76015	5.5267 57521	5.5282 37837
1/5	3.9237 07185	3.9245 50630	3.9253 93351	3.9262 35348	3.9270 76625

k	935	936	937	938	939
1	935	936	937	938	939
2	8 74225	8 76096	8 77969	8 79844	8 81721
3	8174 00375	8200 25856	8226 56953	8252 93672	8279 36019
4	(11)7.6426 93506	(11)7.6754 42012	(11)7.7082 95650	(11)7.7412 54643	(11)7.7743 19218
5	(14)7.1459 18428	(14)7.1842 13723	(14)7.2226 73024	(14)7.2612 96855	(14)7.3000 85746
6	(17)6.6814 33731	(17)6.7244 24045	(17)6.7676 44623	(17)6.8110 96450	(17)6.8547 80516
7	(20)6.2471 40538	(20)6.2940 60906	(20)6.3412 83012	(20)6.3888 08471	(20)6.4366 38904
8	(·23)5.8410 76403	(23)5.8912 41008	(23)5.9417 82182	(23)5.9927 02455	(23)6.0440 03931
9	(26)5.4614 06437	(26)5.5142 01584	(26)5.5674 49905	(26)5.6211 54800	(26)5.6753 19691
10	(29)5.1064 15018	(29)5.1612 92682	(29)5.2167 00561	(29)5.2726 43202	(29)5.3291 25190
24	(71)1.9928 68584	(71)2.0446 56558	(71)2.0977 32860	(71)2.1521 28115	(71)2.2078 73640
1/2	(1)3.0577 76970	(1)3.0594 11708	(1)3.0610 45573	(1)3.0626 78566	(1)3.0643 10689
1/3	9.7784 61652	9.7819 46493	9.7854 28852	9.7889 08735	9.7923 86145
1/4	5.5297 16964	5.5311 94905	5.5326 71663	5.5341 47239	5.5356 21636
1/5	3.9279 17180	3.9287 57017	3.9295 96137	3.9304 34540	3.9312 72229

k	940	941	942	943	944
1	940	941	942	943	944
2	8 83600	8 85481	8 87364	8 89249	8 91136
3	8305 84000	8332 37621	8358 96888	8385 61807	8412 32384
4	(11)7.8074 89600	(11)7.8407 66014	(11)7.8741 48685	(11)7.9076 37840	(11)7.9412 33705
5	(14)7.3390 40224	(14)7.3781 60819	(14)7.4174 48061	(14)7.4569 02483	(14)7.4965 24617
6	(17)6.8986 97811	(17)6.9428 49330	(17)6.9872 36074	(17)7.0318 59042	(17)7.0767 19239
7	(20)6.4847 75942	(20)6.5332 21220	(20)6.5819 76381	(20)6.6310 43076	(20)6.6804 22962
8	(23)6.0956 89385	(23)6.1477 61168	(23)6.2002 21751	(23)6.2530 73621	(23)6.3063 19276
9	(26)5.7299 48022	(26)5.7850 43259	(26)5.8406 08890	(26)5.8966 48424	(26)5.9531 65396
10	(29)5.3861 51141	(29)5.4437 25707	(29)5.5018 53574	(29)5.5605 39464	(29)5.6197 88134
24	(71)2.2650 01461	(71)2.3235 44328	(71)2.3835 35733	(71)2.4450 09921	(71)2.5080 01911
1/2	(1)3.0659 41943	(1)3.0675 72330	(1)3.0692 01851	(1)3.0708 30507	(1)3.0724 58299
1/3	9.7958 61087	9.7993 33566	9.8028 03585	9.8062 71149	9.8097 36263
1/4	5.5370 94855	5.5385 66899	5.5400 37771	5.5415 07472	5.5429 76005
1/5	3.9321 09204	3.9329 45467	3.9337 81020	3.9346 15863	3.9354 49998

k	945	946	947	948	949
1	945	946	947	948	949
2	8 93025	8 94916	8 96809	8 98704	9 00601
3	8439 08625	8465 90536	8492 78123	8519 71392	8546 70349
4	(11)7.9749 36506	(11)8.0087 46471	(11)8.0426 63825	(11)8.0766 88796	(11)8.1108 21612
5	(14)7.5363 14998	(14)7.5762 74161	(14)7.6164 02642	(14)7.6567 00979	(14)7.6971 69710
6	(17)7.1218 17673	(17)7.1671 55356	(17)7.2127 33302	(17)7.2585 52528	(17)7.3046 14055
7	(20)6.7301 17701	(20)6.7801 28967	(20)6.8304 58437	(20)6.8811 07796	(20)6.9320 78738
8	(23)6.3599 61228	(23)6.4140 02003	(23)6.4684 44140	(23)6.5232 90191	(23)6.5785 42722
9	(26)6.0101 63360	(26)6.0676 45895	(26)6.1256 16600	(26)6.1840 79101	(26)6.2430 37043
10	(29)5.6796 04376	(29)5.7399 93016	(29)5.8009 58921	(29)5.8625 06988	(29)5.9246 42154
24	(71)2.5725 47511	(71)2.6386 83331	(71)2.7064 46809	(71)2.7758 76218	(71)2.8470 10693
1/2	(1)3.0740 85230	(1)3.0757 11300	(1)3.0773 36511	(1)3.0789 60864	(1)3.0805 84360
1/3	9.8131 98931	9.8166 59156	9.8201 16944	9.8235 72299	9.8270 25224
1/4	5.5444 43371	5.5459 09574	5.5473 74614	5.5488 38494	5.5503 01217
1/5	3.9362 83427	3.9371 16151	3.9379 48170	3.9387 79487	3.9396 10103

TABLE II.

Table II

POWERS AND ROOTS n^k

k	950	951	952	953	954
1	950	951	952	953	954
2	9 02500	9 04401	9 06304	9 08209	9 10116
3	8573 75000	8600 85351	8628 01408	8655 23177	8682 50664
4	(11)8.1450 62500	(11)8.1794 11688	(11)8.2138 69404	(11)8.2484 35877	(11)8.2831 11335
5	(14)7.7378 09375	(14)7.7786 20515	(14)7.8196 03673	(14)7.8607 59391	(14)7.9020 88213
6	(17)7.3509 18906	(17)7.3974 68110	(17)7.4442 62696	(17)7.4913 03699	(17)7.5385 92155
7	(20)6.9833 72961	(20)7.0349 92173	(20)7.0869 38087	(20)7.1392 12425	(20)7.1918 16916
8	(23)6.6342 04313	(23)6.6902 77556	(23)6.7467 65059	(23)6.8036 69441	(23)6.8609 93338
9	(26)6.3024 94097	(26)6.3624 53956	(26)6.4229 20336	(26)6.4838 96978	(26)6.5453 87645
10	(29)5.9873 69392	(29)6.0506 93712	(29)6.1146 20160	(29)6.1791 53820	(29)6.2442 99813
24	(71)2.9198 90243	(71)2.9945 55775	(71)3.0710 49109	(71)3.1494 12996	(71)3.2296 91146
1/2	(1)3.0822 07001	(1)3.0838 28789	(1)3.0854 49724	(1)3.0870 69808	(1)3.0886 89042
1/3	9.8304 75725	9.8339 23805	9.8373 69469	9.8408 12721	9.8442 53565
1/4	5.5517 62784	5.5532 23198	5.5546 82461	5.5561 40574	5.5575 97541
1/5	3.9404 40019	3.9412 69236	3.9420 97756	3.9429 25580	3.9437 52709

k	955	956	957	958	959
1	955	956	957	958	959
2	9 12025	9 13936	9 15849	9 17764	9 19681
3	8709 83875	8737 22816	8764 67493	8792 17912	8819 74079
4	(11)8.3178 96006	(11)8.3527 90121	(11)8.3877 93908	(11)8.4229 07597	(11)8.4581 31418
5	(14)7.9435 90686	(14)7.9852 67356	(14)8.0271 18770	(14)8.0691 45478	(14)8.1113 48029
6	(17)7.5861 29105	(17)7.6339 15592	(17)7.6819 52663	(17)7.7302 41368	(17)7.7787 82760
7	(20)7.2447 53295	(20)7.2980 23306	(20)7.3516 28698	(20)7.4055 71230	(20)7.4598 52667
8	(23)6.9187 39397	(23)6.9769 10280	(23)7.0355 08664	(23)7.0945 37239	(23)7.1539 98708
9	(26)6.6073 96124	(26)6.6699 26228	(26)6.7329 81792	(26)6.7965 66675	(26)6.8606 84761
10	(29)6.3100 63299	(29)6.3764 49474	(29)6.4434 63575	(29)6.5111 10874	(29)6.5793 96686
24	(71)3.3119 28238	(71)3.3961 69948	(71)3.4824 62966	(71)3.5708 55021	(71)3.6613 94899
1/2	(1)3.0903 07428	(1)3.0919 24967	(1)3.0935 41660	(1)3.0951 57508	(1)3.0967 72513
1/3	9.8476 92005	9.8511 28046	9.8545 61691	9.8579 92945	9.8614 21813
1/4	5.5590 53362	5.5605 08040	5.5619 61578	5.5634 13977	5.5648 65240
1/5	3.9445 79145	3.9454 04889	3.9462 29943	3.9470 54307	3.9478 77983

k	960	961	962	963	964
1	960	961	962	963	964
2	9 21600	9 23521	9 25444	9 27369	9 29296
3	8847 36000	8875 03681	8902 77128	8930 56347	8958 41344
4	(11)8.4934 65600	(11)8.5289 10374	(11)8.5644 65971	(11)8.6001 32622	(11)8.6359 10556
5	(14)8.1537 26976	(14)8.1962 82870	(14)8.2390 16264	(14)8.2819 27715	(14)8.3250 17776
6	(17)7.8275 77897	(17)7.8766 27838	(17)7.9259 33646	(17)7.9754 96389	(17)8.0253 17136
7	(20)7.5144 74781	(20)7.5694 39352	(20)7.6247 48168	(20)7.6804 03023	(20)7.7364 05719
8	(23)7.2138 95790	(23)7.2742 31217	(23)7.3350 07737	(23)7.3962 28111	(23)7.4578 95113
9	(26)6.9253 39958	(26)6.9905 36200	(26)7.0562 77443	(26)7.1225 66751	(26)7.1894 10889
10	(29)6.6483 26360	(29)6.7179 05288	(29)6.7881 38901	(29)6.8590 32667	(29)6.9305 92097
24	(71)3.7541 32467	(71)3.8491 18699	(71)3.9464 05693	(71)4.0460 46699	(71)4.1480 96142
1/2	(1)3.0983 86677	(1)3.1000 00000	(1)3.1016 12484	(1)3.1032 24130	(1)3.1048 34939
1/3	9.8648 48297	9.8682 72403	9.8716 94135	9.8751 13495	9.8785 30490
1/4	5.5663 15367	5.5677 64363	5.5692 12228	5.5706 58964	5.5721 04575
1/5	3.9487 00972	3.9495 23275	3.9503 44894	3.9511 65831	3.9519 86085

k	965	966	967	968	969
1	965	966	967	968	969
2	9 31225	9 33156	9 35089	9 37024	9 38961
3	8986 32125	9014 28696	9042 31063	9070 39232	9098 53209
4	(11)8.6718 00006	(11)8.7078 01203	(11)8.7439 14379	(11)8.7801 39766	(11)8.8164 77595
5	(14)8.3682 87006	(14)8.4117 35962	(14)8.4553 65205	(14)8.4991 75293	(14)8.5431 66790
6	(17)8.0753 96961	(17)8.1257 36940	(17)8.1763 38153	(17)8.2272 01684	(17)8.2783 28619
7	(20)7.7927 58067	(20)7.8494 61884	(20)7.9065 18994	(20)7.9639 31230	(20)8.0217 00432
8	(23)7.5200 11535	(23)7.5825 80180	(23)7.6456 03867	(23)7.7090 85431	(23)7.7730 27719
9	(26)7.2568 11131	(26)7.3247 72454	(26)7.3932 98939	(26)7.4623 94697	(26)7.5320 63859
10	(29)7.0028 22742	(29)7.0757 30190	(29)7.1493 20074	(29)7.2235 98067	(29)7.2985 69880
24	(71)4.2526 09649	(71)4.3596 44069	(71)4.4692 57504	(71)4.5815 09331	(71)4.6964 60232
1/2	(1)3.1064 44913	(1)3.1080 54054	(1)3.1096 62361	(1)3.1112 69837	(1)3.1128 76483
1/3	9.8819 45122	9.8853 57396	9.8887 67316	9.8921 74886	9.8955 80110
1/4	5.5735 49061	5.5749 92425	5.5764 34668	5.5778 75794	5.5793 15803
1/5	3.9528 05659	3.9536 24554	3.9544 42771	3.9552 60312	3.9560 77177

k	970	971	972	973	974
1	970	971	972	973	974
2	9 40900	9 42841	9 44784	9 46729	9 48676
3	9126 73000	9154 98611	9183 30048	9211 67317	9240 10424
4	(11)8.8529 28100	(11)8.8894 91513	(11)8.9261 68067	(11)8.9629 57994	(11)8.9998 61530
5	(14)8.5873 40257	(14)8.6316 96259	(14)8.6762 35361	(14)8.7209 58129	(14)8.7658 65130
6	(17)8.3297 20049	(17)8.3813 77067	(17)8.4333 00771	(17)8.4854 92259	(17)8.5379 52637
7	(20)8.0798 28448	(20)8.1383 17132	(20)8.1971 68349	(20)8.2563 83968	(20)8.3159 65868
8	(23)7.8374 33594	(23)7.9023 05936	(23)7.9676 47635	(23)8.0334 61601	(23)8.0997 50755
9	(26)7.6023 10587	(26)7.6731 39063	(26)7.7445 53501	(26)7.8165 58138	(26)7.8891 57236
10	(29)7.3742 41269	(29)7.4506 18031	(29)7.5277 06003	(29)7.6055 11068	(29)7.6840 39148
24	(71)4.8141 72219	(71)4.9347 08664	(71)5.0581 34323	(71)5.1845 15371	(71)5.3139 19427
1/2	(1)3.1144 82300	(1)3.1160 87290	(1)3.1176 91454	(1)3.1192 94792	(1)3.1208 97307
1/3	9.8989 82992	9.9023 83537	9.9057 81747	9.9091 77627	9.9125 71181
1/4	5.5807 54698	5.5821 92482	5.5836 29155	5.5850 64719	5.5864 99178
1/5	3.9568 93368	3.9577 08886	3.9585 23732	3.9593 37908	3.9601 51415

TABLE II.

Table II   **POWERS AND ROOTS** n^k 201

k	975	976	977	978	979
1	975	976	977	978	979
2	9 50625	9 52576	9 54529	9 56484	9 58441
3	9268 59375	9297 14176	9325 74833	9354 41352	9383 13739
4	(11)9.0368 78906	(11)9.0740 10358	(11)9.1112 56118	(11)9.1486 16423	(11)9.1860 91505
5	(14)8.8109 56934	(14)8.8562 34109	(14)8.9016 97228	(14)8.9473 46861	(14)8.9931 83583
6	(17)8.5906 83010	(17)8.6436 84491	(17)8.6969 58191	(17)8.7505 05230	(17)8.8043 26728
7	(20)8.3759 15935	(20)8.4362 36063	(20)8.4969 28153	(20)8.5579 94115	(20)8.6194 35867
8	(23)8.1665 18037	(23)8.2337 66397	(23)8.3014 98806	(23)8.3697 18245	(23)8.4384 27713
9	(26)7.9623 55086	(26)8.0361 56004	(26)8.1105 64333	(26)8.1855 84443	(26)8.2612 20731
10	(29)7.7632 96209	(29)7.8432 88260	(29)7.9240 21353	(29)8.0055 01586	(29)8.0877 35096
24	(71)5.4464 15584	(71)5.5820 74443	(71)5.7209 68141	(71)5.8631 70383	(71)6.0087 56477
1/2	(1)3.1224 98999	(1)3.1240 99870	(1)3.1256 99922	(1)3.1272 99154	(1)3.1288 97569
1/3	9.9159 62413	9.9193 51328	9.9227 37928	9.9261 22218	9.9295 04202
1/4	5.5879 32533	5.5893 64785	5.5907 95938	5.5922 25992	5.5936 54950
1/5	3.9609 64254	3.9617 76427	3.9625 87934	3.9633 98776	3.9642 08956

k	980	981	982	983	984
1	980	981	982	983	984
2	9 60400	9 62361	9 64324	9 66289	9 68256
3	9411 92000	9440 76141	9469 66168	9498 62087	9527 63904
4	(11)9.2236 81600	(11)9.2613 86943	(11)9.2992 07770	(11)9.3371 44315	(11)9.3751 96815
5	(14)9.0392 07968	(14)9.0854 20591	(14)9.1318 22030	(14)9.1784 12862	(14)9.2251 93666
6	(17)8.8584 23809	(17)8.9127 97600	(17)8.9674 49233	(17)9.0223 79843	(17)9.0775 90568
7	(20)8.6812 55332	(20)8.7434 54446	(20)8.8060 35147	(20)8.8689 99386	(20)8.9323 49119
8	(23)8.5076 30226	(23)8.5773 28811	(23)8.6475 26515	(23)8.7182 26396	(23)8.7894 31533
9	(26)8.3374 77621	(26)8.4143 59564	(26)8.4918 71037	(26)8.5700 16548	(26)8.6488 00628
10	(29)8.1707 28069	(29)8.2544 86732	(29)8.3390 17359	(29)8.4243 26266	(29)8.5104 19818
24	(71)6.1578 03365	(71)6.3103 89657	(71)6.4665 95666	(71)6.6265 03443	(71)6.7901 96812
1/2	(1)3.1304 95168	(1)3.1320 91953	(1)3.1336 87923	(1)3.1352 83081	(1)3.1368 77428
1/3	9.9328 83884	9.9362 61267	9.9396 36356	9.9430 09155	9.9463 79667
1/4	5.5950 82813	5.5965 09584	5.5979 35265	5.5993 59857	5.6007 83363
1/5	3.9650 18474	3.9658 27331	3.9666 35529	3.9674 43069	3.9682 49952

k	985	986	987	988	989
1	985	986	987	988	989
2	9 70225	9 72196	9 74169	9 76144	9 78121
3	9556 71625	9585 85256	9615 04803	9644 30272	9673 61669
4	(11)9.4133 65506	(11)9.4516 50624	(11)9.4900 52406	(11)9.5285 71087	(11)9.5672 06906
5	(14)9.2721 65024	(14)9.3193 27515	(14)9.3666 81724	(14)9.4142 28234	(14)9.4619 67630
6	(17)9.1330 82548	(17)9.1888 56930	(17)9.2449 14862	(17)9.3012 57495	(17)9.3578 85987
7	(20)8.9960 86310	(20)9.0602 12933	(20)9.1247 30969	(20)9.1896 42406	(20)9.2549 49241
8	(23)8.8611 45015	(23)8.9333 69952	(23)9.0061 09466	(23)9.0793 66697	(23)9.1531 44799
9	(26)8.7282 27840	(26)8.8083 02773	(26)8.8890 30043	(26)8.9704 14296	(26)9.0524 60206
10	(29)8.5973 04423	(29)8.6849 86534	(29)8.7734 72653	(29)8.8627 69325	(29)8.9528 83144
24	(71)6.9577 61406	(71)7.1292 84708	(71)7.3048 56083	(71)7.4845 66822	(71)7.6685 10178
1/2	(1)3.1384 70965	(1)3.1400 63694	(1)3.1416 55614	(1)3.1432 46729	(1)3.1448 37039
1/3	9.9497 47896	9.9531 13846	9.9564 77521	9.9598 38925	9.9631 98061
1/4	5.6022 05785	5.6036 27123	5.6050 47381	5.6064 66560	5.6078 84662
1/5	3.9690 56179	3.9698 61752	3.9706 66671	3.9714 70939	3.9722 74555

k	990	991	992	993	994
1	990	991	992	993	994
2	9 80100	9 82081	9 84064	9 86049	9 88036
3	9702 99000	9732 42271	9761 91488	9791 46657	9821 07784
4	(11)9.6059 60100	(11)9.6448 30906	(11)9.6838 19561	(11)9.7229 26304	(11)9.7621 51373
5	(14)9.5099 00499	(14)9.5580 27427	(14)9.6063 49004	(14)9.6548 65820	(14)9.7035 78465
6	(17)9.4148 01494	(17)9.4720 05181	(17)9.5294 98212	(17)9.5872 81759	(17)9.6453 56994
7	(20)9.3206 53479	(20)9.3867 57134	(20)9.4532 62227	(20)9.5201 70787	(20)9.5874 84852
8	(23)9.2274 46944	(23)9.3022 76320	(23)9.3776 36129	(23)9.4535 29591	(23)9.5299 59943
9	(26)9.1351 72475	(26)9.2185 55833	(26)9.3026 15040	(26)9.3873 54884	(26)9.4727 80183
10	(29)9.0438 20750	(29)9.1355 88830	(29)9.2281 94120	(29)9.3216 43400	(29)9.4159 43502
24	(71)7.8567 81408	(71)8.0494 77813	(71)8.2466 98779	(71)8.4485 45822	(71)8.6551 22630
1/2	(1)3.1464 26545	(1)3.1480 15248	(1)3.1496 03150	(1)3.1511 90251	(1)3.1527 76554
1/3	9.9665 54934	9.9699 09547	9.9732 61904	9.9766 12009	9.9799 59866
1/4	5.6093 01690	5.6107 17644	5.6121 32527	5.6135 46340	5.6149 59086
1/5	3.9730 77521	3.9738 79839	3.9746 81509	3.9754 82534	3.9762 82913

k	995	996	997	998	999
1	995	996	997	998	999
2	9 90025	9 92016	9 94009	9 96004	9 98001
3	9850 74875	9880 47936	9910 26973	9940 11992	9970 02999
4	(11)9.8014 95006	(11)9.8409 57443	(11)9.8805 38921	(11)9.9202 39680	(11)9.9600 59960
5	(14)9.7524 87531	(14)9.8015 93613	(14)9.8508 97304	(14)9.9003 99201	(14)9.9500 99900
6	(17)9.7037 25094	(17)9.7623 87238	(17)9.8213 44612	(17)9.8805 98402	(17)9.9401 49800
7	(20)9.6552 06468	(20)9.7233 37689	(20)9.7918 80578	(20)9.8608 37206	(20)9.9302 09650
8	(23)9.6069 30436	(23)9.6844 44339	(23)9.7625 04937	(23)9.8411 15531	(23)9.9202 79441
9	(26)9.5588 95784	(26)9.6457 06561	(26)9.7332 17422	(26)9.8214 33300	(26)9.9103 59161
10	(29)9.5111 01305	(29)9.6071 23735	(29)9.7040 17769	(29)9.8017 90434	(29)9.9004 48802
24	(71)8.8665 35105	(71)9.0828 91413	(71)9.3043 02025	(71)9.5308 79767	(71)9.7627 39866
1/2	(1)3.1543 62059	(1)3.1559 46768	(1)3.1575 30681	(1)3.1591 13800	(1)3.1606 96126
1/3	9.9833 05478	9.9866 48849	9.9899 89983	9.9933 28884	9.9966 65555
1/4	5.6163 70767	5.6177 81384	5.6191 90939	5.6205 99434	5.6220 06871
1/5	3.9770 82648	3.9778 81740	3.9786 80191	3.9794 78001	3.9802 75173

TABLE II.

Table III

RECIPROCALS, CIRCUMFERENCE AND AREA OF CIRCLES

As a matter of convenience, the values of 1000 × (1/n) are given in the table. To obtain the actual value of the reciprocal, shift the decimal point three places to the left.

Circumferences and areas of circles are given for the values of n as the diameter.

$n = $ dia	$1000\frac{1}{n}$	Circumference πn	Area $\frac{\pi n^2}{4}$	$n = $ dia	$1000\frac{1}{n}$	Circumference πn	Area $\frac{\pi n^2}{4}$
0	∞	0.000000	.0000000	50	20.00000	157.0796	1963.495
1	1000.000	3.141593	.7853982	51	19.60784	160.2212	2042.821
2	500.0000	6.283185	3.141593	52	19.23077	163.3628	2123.717
3	333.3333	9.424778	7.068583	53	18.86792	166.5044	2206.183
4	250.0000	12.56637	12.56637	54	18.51852	169.6460	2290.221
5	200.0000	15.70796	19.63495	55	18.18182	172.7876	2375.829
6	166.6667	18.84956	28.27433	56	17.85714	175.9292	2463.009
7	142.8571	21.99115	38.48451	57	17.54386	179.0708	2551.759
8	125.0000	25.13274	50.26548	58	17.24138	182.2124	2642.079
9	111.1111	28.27433	63.61725	59	16.94915	185.3540	2733.971
10	100.0000	31.41593	78.53982	60	16.66667	188.4956	2827.433
11	90.90909	34.55752	95.03318	61	16.39344	191.6372	2922.467
12	83.33333	37.69911	113.0973	62	16.12903	194.7787	3019.071
13	76.92308	40.84070	132.7323	63	15.87302	197.9203	3117.245
14	71.42857	43.98230	153.9380	64	15.62500	201.0619	3216.991
15	66.66667	47.12389	176.7146	65	15.38462	204.2035	3318.307
16	62.50000	50.26548	201.0619	66	15.15152	207.3451	3421.194
17	58.82353	53.40708	226.9801	67	14.92537	210.4867	3525.652
18	55.55556	56.54867	254.4690	68	14.70588	213.6283	3631.681
19	52.63158	59.69026	283.5287	69	14.49275	216.7699	3739.281
20	50.00000	62.83185	314.1593	70	14.28571	219.9115	3848.451
21	47.61905	65.97345	346.3606	71	14.08451	223.0531	3959.192
22	45.45455	69.11504	380.1327	72	13.88889	226.1947	4071.504
23	43.47826	72.25663	415.4756	73	13.69863	229.3363	4185.387
24	41.66667	75.39822	452.3893	74	13.51351	232.4779	4300.840
25	40.00000	78.53982	490.8739	75	13.33333	235.6194	4417.865
26	38.46154	81.68141	530.9292	76	13.15789	238.7610	4536.460
27	37.03704	84.82300	572.5553	77	12.98701	241.9026	4656.626
28	35.71429	87.96459	615.7522	78	12.82051	245.0442	4778.362
29	34.48276	91.10619	660.5199	79	12.65823	248.1858	4901.670
30	33.33333	94.24778	706.8583	80	12.50000	251.3274	5026.548
31	32.25806	97.38937	754.7676	81	12.34568	254.4690	5152.997
32	31.25000	100.5310	804.2477	82	12.19512	257.6106	5281.017
33	30.30303	103.6726	855.2986	83	12.04819	260.7522	5410.608
34	29.41176	106.8142	907.9203	84	11.90476	263.8938	5541.769
35	28.57143	109.9557	962.1128	85	11.76471	267.0354	5674.502
36	27.77778	113.0973	1017.876	86	11.62791	270.1770	5808.805
37	27.02703	116.2389	1075.210	87	11.49425	273.3186	5944.679
38	26.31579	119.3805	1134.115	88	11.36364	276.4602	6082.123
39	25.64103	122.5221	1194.591	89	11.23596	279.6017	6221.139
40	25.00000	125.6637	1256.637	90	11.11111	282.7433	6361.725
41	24.39024	128.8053	1320.254	91	10.98901	285.8849	6503.882
42	23.80952	131.9469	1385.442	92	10.86957	289.0265	6647.610
43	23.25581	135.0885	1452.201	93	10.75269	292.1681	6792.909
44	22.72727	138.2301	1520.531	94	10.63830	295.3097	6939.778
45	22.22222	141.3717	1590.431	95	10.52632	298.4513	7088.218
46	21.73913	144.5133	1661.903	96	10.41667	301.5929	7238.229
47	21.27660	147.6549	1734.945	97	10.30928	304.7345	7389.811
48	20.83333	150.7964	1809.557	98	10.20408	307.8761	7542.964
49	20.40816	153.9380	1885.741	99	10.10101	311.0177	7697.687
50	20.00000	157.0796	1963.495	100	10.00000	314.1593	7853.982

TABLE III.

From *Standard Mathematical Tables*, Eighteenth Edition, © 1970 by the Chemical Rubber Company, pp. 21–30.

Table   III   203

RECIPROCALS, CIRCUMFERENCE AND AREA OF CIRCLES (Continued)

$n = $ dia	$1000\dfrac{1}{n}$	Circumference πn	Area $\dfrac{\pi n^2}{4}$	$n = $ dia	$1000\dfrac{1}{n}$	Circumference πn	Area $\dfrac{\pi n^2}{4}$
100	10.00000	314.1593	7853.982	**150**	6.666 667	471.2389	17671.46
101	9.900 990	317.3009	8011.847	151	6.622 517	474.3805	17907.86
102	9.803 922	320.4425	8171.282	152	6.578 947	477.5221	18145.84
103	9.708 738	323.5840	8332.289	153	6.535 948	480.6637	18385.39
104	9.615 385	326.7256	8494.867	154	6.493 506	483.8053	18626.50
105	9.523 810	329.8672	8659.015	155	6.451 613	486.9469	18869.19
106	9.433 962	333.0088	8824.734	156	6.410 256	490.0885	19113.45
107	9.345 794	336.1504	8992.024	157	6.369 427	493.2300	19359.28
108	9.259 259	339.2920	9160.884	158	6.329 114	496.3716	19606.68
109	9.174 312	342.4336	9331.316	159	6.289 308	499.5132	19855.65
110	9.090 909	345.5752	9503.318	**160**	6.250 000	502.6548	20106.19
111	9.009 009	348.7168	9676.891	161	6.211 180	505.7964	20358.31
112	8.928 571	351.8584	9852.035	162	6.172 840	508.9380	20611.99
113	8.849 558	355.0000	10028.75	163	6.134 969	512.0796	20867.24
114	8.771 930	358.1416	10207.03	164	6.097 561	515.2212	21124.07
115	8.695 652	361.2832	10386.89	165	6.060 606	518.3628	21382.46
116	8.620 690	364.4247	10568.32	166	6.024 096	521.5044	21642.43
117	8.547 009	367.5663	10751.32	167	5.988 024	524.6460	21903.97
118	8.474 576	370.7079	10935.88	168	5.952 381	527.7876	22167.08
119	8.403 361	373.8495	11122.02	169	5.917 160	530.9292	22431.76
120	8.333 333	376.9911	11309.73	**170**	5.882 353	534.0708	22698.01
121	8.264 463	380.1327	11499.01	171	5.847 953	537.2123	22965.83
122	8.196 721	383.2743	11689.87	172	5.813 953	540.3539	23235.22
123	8.130 081	386.4159	11882.29	173	5.780 347	543.4955	23506.18
124	8.064 516	389.5575	12076.28	174	5.747 126	546.6371	23778.71
125	8.000 000	392.6991	12271.85	175	5.714 286	549.7787	24052.82
126	7.936 508	395.8407	12468.98	176	5.681 818	552.9203	24328.49
127	7.874 016	398.9823	12667.69	177	5.649 718	556.0619	24605.74
128	7.812 500	402.1239	12867.96	178	5.617 978	559.2035	24884.56
129	7.751 938	405.2655	13069.81	179	5.586 592	562.3451	25164.94
130	7.692 308	408.4070	13273.23	**180**	5.555 556	565.4867	25446.90
131	7.633 588	411.5486	13478.22	181	5.524 862	568.6283	25730.43
132	7.575 758	414.6902	13684.78	182	5.494 505	571.7699	26015.53
133	7.518 797	417.8318	13892.91	183	5.464 481	574.9115	26302.20
134	7.462 687	420.9734	14102.61	184	5.434 783	578.0530	26590.44
135	7.407 407	424.1150	14313.88	185	5.405 405	581.1946	26880.25
136	7.352 941	427.2566	14526.72	186	5.376 344	584.3362	27171.63
137	7.299 270	430.3982	14741.14	187	5.347 594	587.4778	27464.59
138	7.246 377	433.5398	14957.12	188	5.319 149	590.6194	27759.11
139	7.194 245	436.6814	15174.68	189	5.291 005	593.7610	28055.21
140	7.142 857	439.8230	15393.80	**190**	5.263 158	596.9026	28352.87
141	7.092 199	442.9646	15614.50	191	5.235 602	600.0442	28652.11
142	7.042 254	446.1062	15836.77	192	5.208 333	603.1858	28952.92
143	6.993 007	449.2477	16060.61	193	5.181 347	606.3274	29255.30
144	6.944 444	452.3893	16286.02	194	5.154 639	609.4690	29559.25
145	6.896 552	455.5309	16513.00	195	5.128 205	612.6106	29864.77
146	6.849 315	458.6725	16741.55	196	5.102 041	615.7522	30171.86
147	6.802 721	461.8141	16971.67	197	5.076 142	618.8938	30480.52
148	6.756 757	464.9557	17203.36	198	5.050 505	622.0353	30790.75
149	6.711 409	468.0973	17436.62	199	5.025 126	625.1769	31102.55
150	6.666 667	471.2389	17671.46	**200**	5.000 000	628.3185	31415.93

TABLE III.

RECIPROCALS, CIRCUMFERENCE AND AREA OF CIRCLES (Continued)

n = dia	$1000\frac{1}{n}$	Circum-ference πn	Area $\frac{\pi n^2}{4}$	n = dia	$1000\frac{1}{n}$	Circum-ference πn	Area $\frac{\pi n^2}{4}$
200	5.000 000	628.3185	31415.93	250	4.000 000	785.3982	49087.39
201	4.975 124	631.4601	31730.87	251	3.984 066	788.5398	49480.87
202	4.950 495	634.6017	32047.39	252	3.968 254	791.6813	49875.92
203	4.926 108	637.7433	32365.47	253	3.952 569	794.8229	50272.55
204	4.901 961	640.8849	32685.13	254	3.937 008	797.9645	50670.75
205	4.878 049	644.0265	33006.36	255	3.921 569	801.1061	51070.52
206	4.854 369	647.1681	33329.16	256	3.906 250	804.2477	51471.85
207	4.830 918	650.3097	33653.53	257	3.891 051	807.3893	51874.76
208	4.807 692	653.4513	33979.47	258	3.875 969	810.5309	52279.24
209	4.784 689	656.5929	34306.98	259	3.861 004	813.6725	52685.29
210	4.761 905	659.7345	34636.06	260	3.846 154	816.8141	53092.92
211	4.739 336	662.8760	34966.71	261	3.831 418	819.9557	53502.11
212	4.716 981	666.0176	35298.94	262	3.816 794	823.0973	53912.87
213	4.694 836	669.1592	35632.73	263	3.802 281	826.2389	54325.21
214	4.672 897	672.3008	35968.09	264	3.787 879	829.3805	54739.11
215	4.651 163	675.4424	36305.03	265	3.773 585	832.5221	55154.59
216	4.629 630	678.5840	36643.54	266	3.759 398	835.6636	55571.63
217	4.608 295	681.7256	36983.61	267	3.745 318	838.8052	55990.25
218	4.587 156	684.8672	37325.26	268	3.731 343	841.9468	56410.44
219	4.566 210	688.0088	37668.48	269	3.717 472	845.0884	56832.20
220	4.545 455	691.1504	38013.27	270	3.703 704	848.2300	57255.53
221	4.524 887	694.2920	38359.63	271	3.690 037	851.3716	57680.43
222	4.504 505	697.4336	38707.56	272	3.676 471	854.5132	58106.90
223	4.484 305	700.5752	39057.07	273	3.663 004	857.6548	58534.94
224	4.464 286	703.7168	39408.14	274	3.649 635	860.7964	58964.55
225	4.444 444	706.8583	39760.78	275	3.636 364	863.9380	59395.74
226	4.424 779	709.9999	40115.00	276	3.623 188	867.0796	59828.49
227	4.405 286	713.1415	40470.78	277	3.610 108	870.2212	60262.82
228	4.385 965	716.2831	40828.14	278	3.597 122	873.3628	60698.71
229	4.366 812	719.4247	41187.07	279	3.584 229	876.5044	61136.18
230	4.347 826	722.5663	41547.56	280	3.571 429	879.6459	61575.22
231	4.329 004	725.7079	41909.63	281	3.558 719	882.7875	62015.82
232	4.310 345	728.8495	42273.27	282	3.546 099	885.9291	62458.00
233	4.291 845	731.9911	42638.48	283	3.533 569	889.0707	62901.75
234	4.273 504	735.1327	43005.26	284	3.521 127	892.2123	63347.07
235	4.255 319	738.2743	43373.61	285	3.508 772	895.3539	63793.97
236	4.237 288	741.4159	43743.54	286	3.496 503	898.4955	64242.43
237	4.219 409	744.5575	44115.03	287	3.484 321	901.6371	64692.46
238	4.201 681	747.6991	44488.09	288	3.472 222	904.7787	65144.07
239	4.184 100	750.8406	44862.73	289	3.460 208	907.9203	65597.24
240	4.166 667	753.9822	45238.93	290	3.448 276	911.0619	66051.99
241	4.149 378	757.1238	45616.71	291	3.436 426	914.2035	66508.30
242	4.132 231	760.2654	45996.06	292	3.424 658	917.3451	66966.19
243	4.115 226	763.4070	46376.98	293	3.412 969	920.4866	67425.65
244	4.098 361	766.5486	46759.47	294	3.401 361	923.6282	67886.68
245	4.081 633	769.6902	47143.52	295	3.389 831	926.7698	68349.28
246	4.065 041	772.8318	47529.16	296	3.378 378	929.9114	68813.45
247	4.048 583	775.9734	47916.36	297	3.367 003	933.0530	69279.19
248	4.032 258	779.1150	48305.13	298	3.355 705	936.1946	69746.50
249	4.016 064	782.2566	48695.47	299	3.344 482	939.3362	70215.38
250	4.000 000	785.3982	49087.39	300	3.333 333	942.4778	70685.83

TABLE III.

Table III 205

RECIPROCALS, CIRCUMFERENCE AND AREA OF CIRCLES (Continued)

$n = $ dia	$1000\frac{1}{n}$	Circum-ference πn	Area $\frac{\pi n^2}{4}$	$n = $ dia	$1000\frac{1}{n}$	Circum-ference πn	Area $\frac{\pi n^2}{4}$
300	3.333 333	942.4778	70685.83	**350**	2.857 143	1099.557	96211.28
301	3.322 259	945.6194	71157.86	351	2.849 003	1102.699	96761.84
302	3.311 258	948.7610	71631.45	352	2.840 909	1105.841	97313.97
303	3.300 330	951.9026	72106.62	353	2.832 861	1108.982	97867.68
304	3.289 474	955.0442	72583.36	354	2.824 859	1112.124	98422.96
305	3.278 689	958.1858	73061.66	355	2.816 901	1115.265	98979.80
306	3.267 974	961.3274	73541.54	356	2.808 989	1118.407	99538.22
307	3.257 329	964.4689	74022.99	357	2.801 120	1121.549	100 098.2
308	3.246 753	967.6105	74506.01	358	2.793 296	1124.690	100 659.8
309	3.236 246	970.7521	74990.60	359	2.785 515	1127.832	101 222.9
310	3.225 806	973.8937	75476.76	**360**	2.777 778	1130.973	101 787.6
311	3.215 434	977.0353	75964.50	361	2.770 083	1134.155	102 353.9
312	3.205 128	980.1769	76453.80	362	2.762 431	1137.257	102 921.7
313	3.194 888	983.3185	76944.67	363	2.754 821	1140.398	103 491.1
314	3.184 713	986.4601	77437.12	364	2.747 253	1143.540	104 062.1
315	3.174 603	989.6017	77931.13	365	2.739 726	1146.681	104 634.7
316	3.164 557	992.7433	78426.72	366	2.732 240	1149.823	105 208.8
317	3.154 574	995.8849	78923.88	367	2.724 796	1152.965	105 784.5
318	3.144 654	999.0265	79422.60	368	2.717 391	1156.106	106 361.8
319	3.134 796	1002.168	79922.90	369	2.710 027	1159.248	106 940.6
320	3.125 000	1005.310	80424.77	**370**	2.702 703	1162.389	107 521.0
321	3.115 265	1008.451	80928.21	371	2.695 418	1165.531	108 103.0
322	3.105 590	1011.593	81433.22	372	2.688 172	1168.672	108 686.5
323	3.095 975	1014.734	81939.80	373	2.680 965	1171.814	109 271.7
324	3.086 420	1017.876	82447.96	374	2.673 797	1174.956	109 858.4
325	3.076 923	1021.018	82957.68	375	2.666 667	1178.097	110 446.6
326	3.067 485	1024.159	83468.98	376	2.659 574	1181.239	111 036.5
327	3.058 104	1027.301	83981.84	377	2.652 520	1184.380	111 627.9
328	3.048 780	1030.442	84496.28	378	2.645 503	1187.522	112 220.8
329	3.039 514	1033.584	85012.28	379	2.638 522	1190.664	112 815.4
330	3.030 303	1036.726	85529.86	**380**	2.631 579	1193.805	113 411.5
331	3.021 148	1039.867	86049.01	381	2.624 672	1196.947	114 009.2
332	3.012 048	1043.009	86569.73	382	2.617 801	1200.088	114 608.4
333	3.003 003	1046.150	87092.02	383	2.610 966	1203.230	115 209.3
334	2.994 012	1049.292	87615.88	384	2.604 167	1206.372	115 811.7
335	2.985 075	1052.434	88141.31	385	2.597 403	1209.513	116 415.6
336	2.976 190	1055.575	88668.31	386	2.590 674	1212.655	117 021.2
337	2.967 359	1058.717	89196.88	387	2.583 979	1215.796	117 628.3
338	2.958 580	1061.858	89727.03	388	2.577 320	1218.938	118 237.0
339	2.949 853	1065.000	90258.74	389	2.570 694	1222.080	118 847.2
340	2.941 176	1068.142	90792.03	**390**	2.564 103	1225.221	119 459.1
341	2.932 551	1071.283	91326.88	391	2.557 545	1228.363	120 072.5
342	2.923 977	1074.425	91863.31	392	2.551 020	1231.504	120 687.4
343	2.915 452	1077.566	92401.31	393	2.544 529	1234.646	121 304.0
344	2.906 977	1080.708	92940.88	394	2.538 071	1237.788	121 922.1
345	2.898 551	1083.849	93482.02	395	2.531 646	1240.929	122 541.7
346	2.890 173	1086.991	94024.73	396	2.525 253	1244.071	123 163.0
347	2.881 844	1090.133	94569.01	397	2.518 892	1247.212	123 785.8
348	2.873 563	1093.274	95114.86	398	2.512 563	1250.354	124 410.2
349	2.865 330	1096.416	95662.28	399	2.506 266	1253.495	125 036.2
350	2.857 143	1099.557	96211.28	**400**	2.500 000	1256.637	125 663.7

TABLE III.

RECIPROCALS, CIRCUMFERENCE AND AREA OF CIRCLES (Continued)

$n = $ dia	$1000\dfrac{1}{n}$	Circum-ference πn	Area $\dfrac{\pi n^2}{4}$	$n = $ dia	$1000\dfrac{1}{n}$	Circum-ference πn	Area $\dfrac{\pi n^2}{4}$
400	2.500 000	1256.637	125 663.7	**450**	2.222 222	1413.717	159 043.1
401	2.493 766	1259.779	126 292.8	451	2.217 295	1416.858	159 750.8
402	2.487 562	1262.920	126 923.5	452	2.212 389	1420.000	160 460.0
403	2.481 390	1266.062	127 555.7	453	2.207 506	1423.141	161 170.8
404	2.475 248	1269.203	128 189.5	454	2.202 643	1426.283	161 883.1
405	2.469 136	1272.345	128 824.9	455	2.197 802	1429.425	162 597.1
406	2.463 054	1275.487	129 461.9	456	2.192 982	1432.566	163 312.6
407	2.457 002	1278.628	130 100.4	457	2.188 184	1435.708	164 029.6
408	2.450 980	1281.770	130 740.5	458	2.183 406	1438.849	164 748.3
409	2.444 988	1284.911	131 382.2	459	2.178 649	1441.991	165 468.5
410	2.439 024	1288.053	132 025.4	**460**	2.173 913	1445.133	166 190.3
411	2.433 090	1291.195	132 670.2	461	2.169 197	1448.274	166 913.6
412	2.427 184	1294.336	133 316.6	462	2.164 502	1451.416	167 638.5
413	2.421 308	1297.478	133 964.6	463	2.159 827	1454.557	168 365.0
414	2.415 459	1300.619	134 614.1	464	2.155 172	1457.699	169 093.1
415	2.409 639	1303.761	135 265.2	465	2.150 538	1460.841	169 822.7
416	2.403 846	1306.903	135 917.9	466	2.145 923	1463.982	170 553.9
417	2.398 082	1310.044	136 572.1	467	2.141 328	1467.124	171 286.7
418	2.392 344	1313.186	137 227.9	468	2.136 752	1470.265	172 021.0
419	2.386 635	1316.327	137 885.3	469	2.132 196	1473.407	172 757.0
420	2.380 952	1319.469	138 544.2	**470**	2.127 660	1476.549	173 494.5
421	2.375 297	1322.611	139 204.8	471	2.123 142	1479.690	174 233.5
422	2.369 668	1325.752	139 866.8	472	2.118 644	1482.832	174 974.1
423	2.364 066	1328.894	140 530.5	473	2.114 165	1485.973	175 716.3
424	2.358 491	1332.035	141 195.7	474	2.109 705	1489.115	176 460.1
425	2.352 941	1335.177	141 862.5	475	2.105 263	1492.257	177 205.5
426	2.347 418	1338.318	142 530.9	476	2.100 840	1495.398	177 952.4
427	2.341 920	1341.460	143 200.9	477	2.096 436	1498.540	178 700.9
428	2.336 449	1344.602	143 872.4	478	2.092 050	1501.681	179 450.9
429	2.331 002	1347.743	144 545.5	479	2.087 683	1504.823	180 202.5
430	2.325 581	1350.885	145 220.1	**480**	2.083 333	1507.964	180 955.7
431	2.320 186	1354.026	145 896.3	481	2.079 002	1511.106	181 710.5
432	2.314 815	1357.168	146 574.1	482	2.074 689	1514.248	182 466.8
433	2.309 469	1360.310	147 253.5	483	2.070 393	1517.389	183 224.8
434	2.304 147	1363.451	147 934.5	484	2.066 116	1520.531	183 984.2
435	2.298 851	1366.593	148 617.0	485	2.061 856	1523.672	184 745.3
436	2.293 578	1369.734	149 301.0	486	2.057 613	1526.814	185 507.9
437	2.288 330	1372.876	149 986.7	487	2.053 388	1529.956	186 272.1
438	2.283 105	1376.018	150 673.9	488	2.049 180	1533.097	187 037.9
439	2.277 904	1379.159	151 362.7	489	2.044 990	1536.239	187 805.2
440	2.272 727	1382.301	152 053.1	**490**	2.040 816	1539.380	188 574.1
441	2.267 574	1385.442	152 745.0	491	2.036 660	1542.522	189 344.6
442	2.262 443	1388.584	153 438.5	492	2.032 520	1545.664	190 116.6
443	2.257 336	1391.726	154 133.6	493	2.028 398	1548.805	190 890.2
444	2.252 252	1394.867	154 830.3	494	2.024 291	1551.947	191 665.4
445	2.247 191	1398.009	155 528.5	495	2.020 202	1555.088	192 442.2
446	2.242 152	1401.150	156 228.3	496	2.016 129	1558.230	193 220.5
447	2.237 136	1404.292	156 929.6	497	2.012 072	1561.372	194 000.4
448	2.232 143	1407.434	157 632.6	498	2.008 032	1564.513	194 781.9
449	2.227 171	1410.575	158 337.1	499	2.004 008	1567.655	195 564.9
450	2.222 222	1413.717	159 043.1	**500**	2.000 000	1570.796	196 349.5

TABLE III.

Table   III 207

RECIPROCALS, CIRCUMFERENCE AND AREA OF CIRCLES (Continued)

n = dia	$1000\frac{1}{n}$	Circumference πn	Area $\frac{\pi n^2}{4}$	n = dia	$1000\frac{1}{n}$	Circumference πn	Area $\frac{\pi n^2}{4}$
500	2.000 000	1570.796	196 349.5	**550**	1.818 182	1727.876	237 582.9
501	1.996 008	1573.938	197 135.7	551	1.814 882	1731.018	238 447.7
502	1.992 032	1577.080	197 923.5	552	1.811 594	1734.159	239 314.0
503	1.988 072	1580.221	198 712.8	553	1.808 318	1737.301	240 181.8
504	1.984 127	1583.363	199 503.7	554	1.805 054	1740.442	241 051.3
505	1.980 198	1586.504	200 296.2	555	1.801 802	1743.584	241 922.3
506	1.976 285	1589.646	201 090.2	556	1.798 561	1746.726	242 794.8
507	1.972 387	1592.787	201 885.8	557	1.795 332	1749.867	243 669.0
508	1.968 504	1595.929	202 683.0	558	1.792 115	1753.009	244 544.7
509	1.964 637	1599.071	203 481.7	559	1.788 909	1756.150	245 422.0
510	1.960 784	1602.212	204 282.1	**560**	1.785 714	1759.292	246 300.9
511	1.956 947	1605.354	205 084.0	561	1.782 531	1762.433	247 181.3
512	1.953 125	1608.495	205 887.4	562	1.779 359	1765.575	248 063.3
513	1.949 318	1611.637	206 692.4	563	1.776 199	1768.717	248 946.9
514	1.945 525	1614.779	207 499.1	564	1.773 050	1771.858	249 832.0
515	1.941 748	1617.920	208 307.2	565	1.769 912	1775.000	250 718.7
516	1.937 984	1621.062	209 117.0	566	1.766 784	1778.141	251 607.0
517	1.934 236	1624.203	209 928.3	567	1.763 668	1781.283	252 496.9
518	1.930 502	1627.345	210 741.2	568	1.760 563	1784.425	253 388.3
519	1.926 782	1630.487	211 555.6	569	1.757 469	1787.566	254 281.3
520	1.923 077	1633.628	212 371.7	**570**	1.754 386	1790.708	255 175.9
521	1.919 386	1636.770	213 189.3	571	1.751 313	1793.849	256 072.0
522	1.915 709	1639.911	214 008.4	572	1.748 252	1796.991	256 969.7
523	1.912 046	1643.053	214 829.2	573	1.745 201	1800.133	257 869.0
524	1.908 397	1646.195	215 651.5	574	1.742 160	1803.274	258 769.8
525	1.904 762	1649.336	216 475.4	575	1.739 130	1806.416	259 672.3
526	1.901 141	1652.478	217 300.8	576	1.736 111	1809.557	260 576.3
527	1.897 533	1655.619	218 127.8	577	1.733 102	1812.699	261 481.8
528	1.893 939	1658.761	218 956.4	578	1.730 104	1815.841	262 389.0
529	1.890 359	1661.903	219 786.6	579	1.727 116	1818.982	263 297.7
530	1.886 792	1665.044	220 618.3	**580**	1.724 138	1822.124	264 207.9
531	1.883 239	1668.186	221 451.7	581	1.721 170	1825.265	265 119.8
532	1.879 699	1671.327	222 286.5	582	1.718 213	1828.407	266 033.2
533	1.876 173	1674.469	223 123.0	583	1.715 266	1831.549	266 948.2
534	1.872 659	1677.610	223 961.0	584	1.712 329	1834.690	267 864.8
535	1.869 159	1680.752	224 800.6	585	1.709 402	1837.832	268 782.9
536	1.865 672	1683.894	225 641.8	586	1.706 485	1840.973	269 702.6
537	1.862 197	1687.035	226 484.5	587	1.703 578	1844.115	270 623.9
538	1.858 736	1690.177	227 328.8	588	1.700 680	1847.256	271 546.7
539	1.855 288	1693.318	228 174.7	589	1.697 793	1850.398	272 471.1
540	1.851 852	1696.460	229 022.1	**590**	1.694 915	1853.540	273 397.1
541	1.848 429	1699.602	229 871.1	591	1.692 047	1856.681	274 324.7
542	1.845 018	1702.743	230 721.7	592	1.689 189	1859.823	275 253.8
543	1.841 621	1705.885	231 573.9	593	1.686 341	1862.964	276 184.5
544	1.838 235	1709.026	232 427.6	594	1.683 502	1866.106	277 116.7
545	1.834 862	1712.168	233 282.9	595	1.680 672	1869.248	278 050.6
546	1.831 502	1715.310	234 139.8	596	1.677 852	1872.389	278 986.0
547	1.828 154	1718.451	234 998.2	597	1.675 042	1875.531	279 923.0
548	1.824 818	1721.593	235 858.2	598	1.672 241	1878.672	280 861.5
549	1.821 494	1724.734	236 719.8	599	1.669 449	1881.814	281 801.6
550	1.818 182	1727.876	237.582.9	**600**	1.666 667	1884.956	282 743.3

TABLE III.

Table III

RECIPROCALS, CIRCUMFERENCE AND AREA OF CIRCLES (Continued)

n = dia	$1000\dfrac{1}{n}$	Circum-ference πn	Area $\dfrac{\pi n^2}{4}$	n = dia	$1000\dfrac{1}{n}$	Circum-ference πn	Area $\dfrac{\pi n^2}{4}$
600	1.666 667	1884.956	282 743.3	**650**	1.538 462	2042.035	331 830.7
601	1.663 894	1888.097	283 686.6	651	1.536 098	2045.177	332 852.5
602	1.661 130	1891.239	284 631.4	652	1.533 742	2048.318	333 875.9
603	1.658 375	1894.380	285 577.8	653	1.531 394	2051.460	334 900.8
604	1.655 629	1897.522	286 525.8	654	1.529 052	2054.602	335 927.4
605	1.652 893	1900.664	287 475.4	655	1.526 718	2057.743	336 955.4
606	1.650 165	1903.805	288 426.5	656	1.524 390	2060.885	337 985.1
607	1.647 446	1906.947	289 379.2	657	1.522 070	2064.026	339 016.3
608	1.644 737	1910.088	290 333.4	658	1.519 757	2067.168	340 049.1
609	1.642 036	1913.230	291 289.3	659	1.517 451	2070.310	341 083.5
610	1.639 344	1916.372	292 246.7	**660**	1.515 152	2073.451	342 119.4
611	1.636 661	1919.513	293 205.6	661	1.512 859	2076.593	343 157.0
612	1.633 987	1922.655	294 166.2	662	1.510 574	2079.734	344 196.0
613	1.631 321	1925.796	295 128.3	663	1.508 296	2082.876	345 236.7
614	1.628 664	1928.938	296 092.0	664	1.506 024	2086.018	346 278.9
615	1.626 016	1932.079	297 057.2	665	1.503 759	2089.159	347 322.7
616	1.623 377	1935.221	298 024.0	666	1.501 502	2092.301	348 368.1
617	1.620 746	1938.363	298 992.4	667	1.499 250	2095.442	349 415.0
618	1.618 123	1941.504	299 962.4	668	1.497 006	2098.584	350 463.5
619	1.615 509	1944.646	300 933.9	669	1.494 768	2101.725	351 513.6
620	1.612 903	1947.787	301 907.1	**670**	1.492 537	2104.867	352 565.2
621	1.610 306	1950.929	302 881.7	671	1.490 313	2108.009	353 618.5
622	1.607 717	1954.071	303 858.0	672	1.488 095	2111.150	354 673.2
623	1.605 136	1957.212	304 835.8	673	1.485 884	2114.292	355 729.6
624	1.602 564	1960.354	305 815.2	674	1.483 680	2117.433	356 787.5
625	1.600 000	1963.495	306 796.2	675	1.481 481	2120.575	357 847.0
626	1.597 444	1966.637	307 778.7	676	1.479 290	2123.717	358 908.1
627	1.594 896	1969.779	308 762.8	677	1.477 105	2126.858	359 970.8
628	1.592 357	1972.920	309 748.5	678	1.474 926	2130.000	361 035.0
629	1.589 825	1976.062	310 735.7	679	1.472 754	2133.141	362 100.8
630	1.587 302	1979.203	311 724.5	**680**	1.470 588	2136.283	363 168.1
631	1.584 786	1982.345	312 714.9	681	1.468 429	2139.425	364 237.0
632	1.582 278	1985.487	313 706.9	682	1.466 276	2142.566	365 307.5
633	1.579 779	1988.628	314 700.4	683	1.464 129	2145.708	366 379.6
634	1.577 287	1991.770	315 695.5	684	1.461 988	2148.849	367 453.2
635	1.574 803	1994.911	316 692.2	685	1.459 854	2151.991	368 528.5
636	1.572 327	1998.053	317 690.4	686	1.457 726	2155.133	369 605.2
637	1.569 859	2001.195	318 690.2	687	1.455 604	2158.274	370 683.6
638	1.567 398	2004.336	319 691.6	688	1.453 488	2161.416	371 763.5
639	1.564 945	2007.478	320 694.6	689	1.451 379	2164.557	372 845.0
640	1.562 500	2010.619	321 699.1	**690**	1.449 275	2167.699	373 928.1
641	1.560 062	2013.761	322 705.2	691	1.447 178	2170.841	375 012.7
642	1.557 632	2016.902	323 712.8	692	1.445 087	2173.982	376 098.9
643	1.555 210	2020.044	324 722.1	693	1.443 001	2177.124	377 186.7
644	1.552 795	2023.186	325 732.9	694	1.440 922	2180.265	378 276.0
645	1.550 388	2026.327	326 745.3	695	1.438 849	2183.407	379 366.9
646	1.547 988	2029.469	327 759.2	696	1.436 782	2186.548	380 459.4
647	1.545 595	2032.610	328 774.7	697	1.434 720	2189.690	381 553.5
648	1.543 210	2035.752	329 791.8	698	1.432 665	2192.832	382 649.1
649	1.540 832	2038.894	330 810.5	699	1.430 615	2195.973	383 746.3
650	1.538 462	2042.035	331 830.7	**700**	1.428 571	2199.115	384 845.1

TABLE III.

Table   III   *209*

RECIPROCALS, CIRCUMFERENCE AND AREA OF CIRCLES (Continued)

$n = $ dia	$1000\dfrac{1}{n}$	Circumference πn	Area $\dfrac{\pi n^2}{4}$	$n = $ dia	$1000\dfrac{1}{n}$	Circumference πn	Area $\dfrac{\pi n^2}{4}$
700	1.428 571	2199.115	384 845.1	**750**	1.333 333	2356.194	441 786.5
701	1.426 534	2202.256	385 945.4	751	1.331 558	2359.336	442 965.3
702	1.424 501	2205.398	387 047.4	752	1.329 787	2362.478	444 145.8
703	1.422 475	2208.540	388 150.8	753	1.328 021	2365.619	445 327.8
704	1.420 455	2211.681	389 255.9	754	1.326 260	2368.761	446 511.4
705	1.418 440	2214.823	390 362.5	755	1.324 503	2371.902	447 696.6
706	1.416 431	2217.964	391 470.7	756	1.322 751	2375.044	448 883.3
707	1.414 427	2221.106	392 580.5	757	1.321 004	2378.186	450 071.6
708	1.412 429	2224.248	393 691.8	758	1.319 261	2381.327	451 261.5
709	1.410 437	2227.389	394 804.7	759	1.317 523	2384.469	452 453.0
710	1.408 451	2230.531	395 919.2	**760**	1.315 789	2387.610	453 646.0
711	1.406 470	2233.672	397 035.3	761	1.314 060	2390.752	454 840.6
712	1.404 494	2236.814	398 152.9	762	1.312 336	2393.894	456 036.7
713	1.402 525	2239.956	399 272.1	763	1.310 616	2397.035	457 234.5
714	1.400 560	2243.097	400 392.8	764	1.308 901	2400.177	458 433.8
715	1.398 601	2246.239	401 515.2	765	1.307 190	2403.318	459 634.6
716	1.396 648	2249.380	402 639.1	766	1.305 483	2406.460	460 837.1
717	1.394 700	2252.522	403 764.6	767	1.303 781	2409.602	462 041.1
718	1.392 758	2255.664	404 891.6	768	1.302 083	2412.743	463 246.7
719	1.390 821	2258.805	406 020.2	769	1.300 390	2415.885	464 453.8
720	1.388 889	2261.947	407 150.4	**770**	1.298 701	2419.026	465 662.6
721	1.386 963	2265.088	408 282.2	771	1.297 017	2422.168	466 872.9
722	1.385 042	2268.230	409 415.5	772	1.295 337	2425.310	468 084.7
723	1.383 126	2271.371	410 550.4	773	1.293 661	2428.451	469 298.2
724	1.381 215	2274.513	411 686.9	774	1.291 990	2431.593	470 513.2
725	1.379 310	2277.655	412 824.9	775	1.290 323	2434.734	471 729.8
726	1.377 410	2280.796	413 964.5	776	1.288 660	2437.876	472 947.9
727	1.375 516	2283.938	415 105.7	777	1.287 001	2441.017	474 167.6
728	1.373 626	2287.079	416 248.5	778	1.285 347	2444.159	475 388.9
729	1.371 742	2290.221	417 392.8	779	1.283 697	2447.301	476 611.8
730	1.369 863	2293.363	418 538.7	**780**	1.282 051	2450.442	477 836.2
731	1.367 989	2296.504	419 686.1	781	1.280 410	2453.584	479 062.2
732	1.366 120	2299.646	420 835.2	782	1.278 772	2456.725	480 289.8
733	1.364 256	2302.787	421 985.8	783	1.277 139	2459.867	481 519.0
734	1.362 398	2305.929	423 138.0	784	1.275 510	2463.009	482 749.7
735	1.360 544	2309.071	424 291.7	785	1.273 885	2466.150	483 982.0
736	1.358 696	2312.212	425 447.0	786	1.272 265	2469.292	485 215.8
737	1.356 852	2315.354	426 603.9	787	1.270 648	2472.433	486 451.3
738	1.355 014	2318.495	427 762.4	788	1.269 036	2475.575	487 688.3
739	1.353 180	2321.637	428 922.4	789	1.267 427	2478.717	488 926.9
740	1.351 351	2324.779	430 084.0	**790**	1.265 823	2481.858	490 167.0
741	1.349 528	2327.920	431 247.2	791	1.264 223	2485.000	491 408.7
742	1.347 709	2331.062	432 412.0	792	1.262 626	2488.141	492 652.0
743	1.345 895	2334.203	433 578.3	793	1.261 034	2491.283	493 896.8
744	1.344 086	2337.345	434 746.2	794	1.259 446	2494.425	495 143.3
745	1.342 282	2340.487	435 915.6	795	1.257 862	2497.566	496 391.3
746	1.340 483	2343.628	437 086.6	796	1.256 281	2500.708	497 640.8
747	1.338 688	2346.770	438 259.2	797	1.254 705	2503.849	498 892.0
748	1.336 898	2349.911	439 433.4	798	1.253 133	2506.991	500 144.7
749	1.335 113	2353.053	440 609.2	799	1.251 564	2510.133	501 399.0
750	1.333 333	2356.194	441 786.5	**800**	1.250 000	2513.274	502 654.8

TABLE III.

Table *III*

RECIPROCALS, CIRCUMFERENCE AND AREA OF CIRCLES (Continued)

$n = $ dia	$1000\dfrac{1}{n}$	Circum-ference πn	Area $\dfrac{\pi n^2}{4}$	$n = $ dia	$1000\dfrac{1}{n}$	Circum-ference πn	Area $\dfrac{\pi n^2}{4}$
800	1.250 000	2513.274	502 654.8	**850**	1.176 471	2670.354	567 450.2
801	1.248 439	2516.416	503 912.2	851	1.175 088	2673.495	568 786.1
802	1.246 883	2519.557	505 171.2	852	1.173 709	2676.637	570 123.7
803	1.245 330	2522.699	506 431.8	853	1.172 333	2679.779	571 462.8
804	1.243 781	2525.840	507 693.9	854	1.170 960	2682.920	572 803.4
805	1.242 236	2528.982	508 957.6	855	1.169 591	2686.062	574 145.7
806	1.240 695	2532.124	510 222.9	856	1.168 224	2689.203	575 489.5
807	1.239 157	2535.265	511 489.8	857	1.166 861	2692.345	576 834.9
808	1.237 624	2538.407	512 758.2	858	1.165 501	2695.486	578 181.9
809	1.236 094	2541.548	514 028.2	859	1.164 144	2698.628	579 530.4
810	1.234 568	2544.690	515 299.7	**860**	1.162 791	2701.770	580 880.5
811	1.233 046	2547.832	516 572.9	861	1.161 440	2704.911	582 232.2
812	1.231 527	2550.973	517 847.6	862	1.160 093	2708.053	583 585.4
813	1.230 012	2554.115	519 123.8	863	1.158 749	2711.194	584 940.2
814	1.228 501	2557.256	520 401.7	864	1.157 407	2714.336	586 296.6
815	1.226 994	2560.398	521 681.1	865	1.156 069	2717.478	587 654.5
816	1.225 490	2563.540	522 962.1	866	1.154 734	2720.619	589 014.1
817	1.223 990	2566.681	524 244.6	867	1.153 403	2723.761	590 375.2
818	1.222 494	2569.823	525 528.8	868	1.152 074	2726.902	591 737.8
819	1.221 001	2572.964	526 814.5	869	1.150 748	2730.044	593 102.1
820	1.219 512	2576.106	528 101.7	**870**	1.149 425	2733.186	594 467.9
821	1.218 027	2579.248	529 390.6	871	1.148 106	2736.327	595 835.2
822	1.216 545	2582.389	530 681.0	872	1.146 789	2739.469	597 204.2
823	1.215 067	2585.531	531 973.0	873	1.145 475	2742.610	598 574.7
824	1.213 592	2588.672	533 266.5	874	1.144 165	2745.752	599 946.8
825	1.212 121	2591.814	534 561.6	875	1.142 857	2748.894	601 320.5
826	1.210 654	2594.956	535 858.3	876	1.141 553	2752.035	602 695.7
827	1.209 190	2598.097	537 156.6	877	1.140 251	2755.177	604 072.5
828	1.207 729	2601.239	538 456.4	878	1.138 952	2758.318	605 450.9
829	1.206 273	2604.380	539 757.8	879	1.137 656	2761.460	606 830.8
830	1.204 819	2607.522	541 060.8	**880**	1.136 364	2764.602	608 212.3
831	1.203 369	2610.663	542 065.3	881	1.135 074	2767.743	609 595.4
832	1.201 923	2613.805	543 671.5	882	1.133 787	2770.885	610 980.1
833	1.200 480	2616.947	544 979.1	883	1.132 503	2774.026	612 366.3
834	1.199 041	2620.088	546 288.4	884	1.131 222	2777.168	613 754.1
835	1.197 605	2623.230	547 599.2	885	1.129 944	2780.309	615 143.5
836	1.196 172	2626.371	548 911.6	886	1.128 668	2783.451	616 534.4
837	1.194 743	2629.513	550 225.6	887	1.127 396	2786.593	617 926.9
838	1.193 317	2632.655	551 541.1	888	1.126 126	2789.734	619 321.0
839	1.191 895	2635.796	552 858.3	889	1.124 859	2792.876	620 716.7
840	1.190 476	2638.938	554 176.9	**890**	1.123 596	2796.017	622 113.9
841	1.189 061	2642.079	555 497.2	891	1.122 334	2799.159	623 512.7
842	1.187 648	2645.221	556 819.0	892	1.121 076	2802.301	624 913.0
843	1.186 240	2648.363	558 142.4	893	1.119 821	2805.442	626 315.0
844	1.184 834	2651.504	559 467.4	894	1.118 568	2808.584	627 718.5
845	1.183 432	2654.646	560 793.9	895	1.117 318	2811.725	629 123.6
846	1.182 033	2657.787	562 122.0	896	1.116 071	2814.867	630 530.2
847	1.180 638	2660.929	563 451.7	897	1.114 827	2818.009	631 938.4
848	1.179 245	2664.071	564 783.0	898	1.113 586	2821.150	633 348.2
849	1.177 856	2667.212	566 115.8	899	1.112 347	2824.292	634 759.6
850	1.176 471	2670.354	567 450.2	**900**	1.111 111	2827.433	636 172.5

TABLE III.

Table III 211

RECIPROCALS, CIRCUMFERENCE AND AREA OF CIRCLES (Continued)

n = dia	$1000\dfrac{1}{n}$	Circum-ference πn	Area $\dfrac{\pi n^2}{4}$	n = dia	$1000\dfrac{1}{n}$	Circum-ference πn	Area $\dfrac{\pi n^2}{4}$
900	1.111 111	2827.433	636 172.5	**950**	1.052 632	2984.513	708 821.8
901	1.109 878	2830.575	637 587.0	951	1.051 525	2987.655	710 314.9
902	1.108 647	2833.717	639 003.1	952	1.050 420	2990.796	711 809.5
903	1.107 420	2836.858	640 420.7	953	1.049 318	2993.938	713 305.7
904	1.106 195	2840.000	641 839.9	954	1.048 218	2997.079	714 803.4
905	1.104 972	2843.141	643 260.7	955	1.047 120	3000.221	716 302.8
906	1.103 753	2846.283	644 683.1	956	1.046 025	3003.363	717 803.7
907	1.102 536	2849.425	646 107.0	957	1.044 932	3006.504	719 306.1
908	1.101 322	2852.566	647 532.5	958	1.043 841	3009.646	720 810.2
909	1.100 110	2855.708	648 959.6	959	1.042 753	3012.787	722 315.8
910	1.098 901	2858.849	650 388.2	**960**	1.041 667	3015.929	723 822.9
911	1.097 695	2861.991	651 818.4	961	1.040 583	3019.071	725 331.7
912	1.096 491	2865.133	653 250.2	962	1.039 501	3022.212	726 842.0
913	1.095 290	2868.274	654 683.6	963	1.038 422	3025.354	728 353.9
914	1.094 092	2871.416	656 118.5	964	1.037 344	3028.495	729 867.4
915	1.092 896	2874.557	657 555.0	965	1.036 269	3031.637	731 382.4
916	1.091 703	2877.699	658 993.0	966	1.035 197	3034.779	732 899.0
917	1.090 513	2880.840	660 432.7	967	1.034 126	3037.920	734 417.2
918	1.089 325	2883.982	661 873.9	968	1.033 058	3041.062	735 936.9
919	1.088 139	2887.124	663 316.7	969	1.031 992	3044.203	737 458.2
920	1.086 957	2890.265	664 761.0	**970**	1.030 928	3047.345	738 981.1
921	1.085 776	2893.407	666 206.9	971	1.029 866	3050.486	740 505.6
922	1.084 599	2896.548	667 654.4	972	1.028 807	3053.628	742 031.6
923	1.083 424	2899.690	669 103.5	973	1.027 749	3056.770	743 559.2
924	1.082 251	2902.832	670 554.1	974	1.026 694	3059.911	745 088.4
925	1.081 081	2905.973	672 006.3	975	1.025 641	3063.053	746 619.1
926	1.079 914	2909.115	673 460.1	976	1.024 590	3066.194	748 151.4
927	1.078 749	2912.256	674 915.4	977	1.023 541	3069.336	749 685.3
928	1.077 586	2915.398	676 372.3	978	1.022 495	3072.478	751 220.8
929	1.076 426	2918.540	677 830.8	979	1.021 450	3075.619	752 757.8
930	1.075 269	2921.681	679 290.9	**980**	1.020 408	3078.761	754 296.4
931	1.074 114	2924.823	680 752.5	981	1.019 368	3081.902	755 836.6
932	1.072 961	2927.964	682 215.7	982	1.018 330	3085.044	757 378.3
933	1.071 811	2931.106	683 680.5	983	1.017 294	3088.186	758 921.6
934	1.070 664	2934.248	685 146.8	984	1.016 260	3091.327	760 466.5
935	1.069 519	2937.389	686 614.7	985	1.015 228	3094.469	762 012.9
936	1.068 376	2940.531	688 084.2	986	1.014 199	3097.610	763 561.0
937	1.067 236	2943.672	689 555.2	987	1.013 171	3100.752	765 110.5
938	1.066 098	2946.814	691 027.9	988	1.012 146	3103.894	766 661.7
939	1.064 963	2949.956	692 502.1	989	1.011 122	3107.035	768 214.4
940	1.063 830	2953.097	693 977.8	**990**	1.010 101	3110.177	769 768.7
941	1.062 699	2956.239	695 455.2	991	1.009 082	3113.318	771 324.6
942	1.061 571	2959.380	696 934.1	992	1.008 065	3116.460	772 882.1
943	1.060 445	2962.522	698 414.5	993	1.007 049	3119.602	774 441.1
944	1.059 322	2965.663	699 896.6	994	1.006 036	3122.743	776 001.7
945	1.058 201	2968.805	701 380.2	995	1.005 025	3125.885	777 563.8
946	1.057 082	2971.947	702 865.4	996	1.004 016	3129.026	779 127.5
947	1.055 966	2975.088	704 352.1	997	1.003 009	3132.168	780 692.8
948	1.054 852	2978.230	705 840.5	998	1.002 004	3135.309	782 259.7
949	1.053 741	2981.371	707 330.4	999	1.001 001	3138.451	783 828.2
950	1.052 632	2984.513	708 821.8	**1000**	1.000 000	3141.593	785 398.2

TABLE III.

n	n!
1	1
2	2
3	6
4	24
5	120
6	720
7	5040
8	40320
9	362880
10	3628800
11	39916800
12	479001600
13	6227020800
14	87178291200
15	1307674368000
16	20922789888000
17	355687428096000
18	6402373705728000
19	121645100408832000
20	2432902008176640000

TABLE IV.   FACTORIALS

Table V   *213*

COMMON LOGARITHMS

x	$\log_{10} x$	x	$\log_{10} x$	x	$\log_{10} x$	x	$\log_{10} x$	x	$\log_{10} x$	x	$\log_{10} x$
100	00000 00000	150	17609 12591	200	30102 99957	250	39794 00087	300	47712 12547		
101	00432 13738	151	17897 69473	201	30319 60574	251	39967 37215	301	47856 64956		
102	00860 01718	152	18184 35879	202	30535 13694	252	40140 05408	302	48000 69430		
103	01283 72247	153	18469 14308	203	30749 60379	253	40312 05212	303	48144 26285		
104	01703 33393	154	18752 07208	204	30963 01674	254	40483 37166	304	48287 35836		
105	02118 92991	155	19033 16982	205	31175 38611	255	40654 01804	305	48429 98393		
106	02530 58653	156	19312 45984	206	31386 72204	256	40823 99653	306	48572 14265		
107	02938 37777	157	19589 96524	207	31597 03455	257	40993 31233	307	48713 83755		
108	03342 37555	158	19865 70870	208	31806 33350	258	41161 97060	308	48855 07165		
109	03742 64979	159	20139 71243	209	32014 62861	259	41329 97641	309	48995 84794		
110	04139 26852	160	20411 99827	210	32221 92947	260	41497 33480	310	49136 16938		
111	04532 29788	161	20682 58760	211	32428 24553	261	41664 05073	311	49276 03890		
112	04921 80227	162	20951 50145	212	32633 58609	262	41830 12913	312	49415 45940		
113	05307 84435	163	21218 76044	213	32837 96034	263	41995 57485	313	49554 43375		
114	05690 48513	164	21484 38480	214	33041 37733	264	42160 39269	314	49692 96481		
115	06069 78404	165	21748 39442	215	33243 84599	265	42324 58739	315	49831 05538		
116	06445 79892	166	22010 80880	216	33445 37512	266	42488 16366	316	49968 70826		
117	06818 58617	167	22271 64711	217	33645 97338	267	42651 12614	317	50105 92622		
118	07188 20073	168	22530 92817	218	33845 64936	268	42813 47940	318	50242 71200		
119	07554 69614	169	22788 67046	219	34044 41148	269	42975 22800	319	50379 06831		
120	07918 12460	170	23044 89214	220	34242 26808	270	43136 37642	320	50514 99783		
121	08278 53703	171	23299 61104	221	34439 22737	271	43296 92909	321	50650 50324		
122	08635 98307	172	23552 84469	222	34635 29745	272	43456 89040	322	50785 58717		
123	08990 51114	173	23804 61031	223	34830 48630	273	43616 26470	323	50920 25223		
124	09342 16852	174	24054 92483	224	35024 80183	274	43775 05628	324	51054 50102		
125	09691 00130	175	24303 80487	225	35218 25181	275	43933 26938	325	51188 33610		
126	10037 05451	176	24551 26678	226	35410 84391	276	44090 90821	326	51321 76001		
127	10380 37210	177	24797 32664	227	35602 58572	277	44247 97691	327	51454 77527		
128	10720 99696	178	25042 00023	228	35793 48470	278	44404 47959	328	51587 38437		
129	11058 97103	179	25285 30310	229	35983 54823	279	44560 42033	329	51719 58979		
130	11394 33523	180	25527 25051	230	36172 78360	280	44715 80313	330	51851 39399		
131	11727 12957	181	25767 85749	231	36361 19799	281	44870 63199	331	51982 79938		
132	12057 39312	182	26007 13880	232	36548 79849	282	45024 91083	332	52113 80837		
133	12385 16410	183	26245 10897	233	36735 59210	283	45178 64355	333	52244 42335		
134	12710 47984	184	26481 78230	234	36921 58574	284	45331 83400	334	52374 64668		
135	13033 37685	185	26717 17284	235	37106 78623	285	45484 48600	335	52504 48070		
136	13353 89084	186	26951 29442	236	37291 20030	286	45636 60331	336	52633 92774		
137	13672 05672	187	27184 16065	237	37474 83460	287	45788 18967	337	52762 99009		
138	13987 90864	188	27415 78493	238	37657 69571	288	45939 24878	338	52891 67003		
139	14301 48003	189	27646 18042	239	37839 79009	289	46089 78428	339	53019 96982		
140	14612 80357	190	27875 36010	240	38021 12417	290	46239 79979	340	53147 89170		
141	14921 91127	191	28103 33672	241	38201 70426	291	46389 29890	341	53275 43790		
142	15228 83444	192	28330 12287	242	38381 53660	292	46538 28514	342	53402 61061		
143	15533 60375	193	28555 73090	243	38560 62736	293	46686 76204	343	53529 41200		
144	15836 24921	194	28780 17299	244	38738 98263	294	46834 73304	344	53655 84426		
145	16136 80022	195	29003 46114	245	38916 60844	295	46982 20160	345	53781 90951		
146	16435 28558	196	29225 60714	246	39093 51071	296	47129 17111	346	53907 60988		
147	16731 73347	197	29446 62262	247	39269 69533	297	47275 64493	347	54032 94748		
148	17026 17154	198	29666 51903	248	39445 16808	298	47421 62641	348	54157 92439		
149	17318 62684	199	29885 30764	249	39619 93471	299	47567 11883	349	54282 54270		
150	17609 12591	200	30102 99957	250	39794 00087	300	47712 12547	350	54406 80444		

TABLE V.

TABLE V. COMMON LOGARITHMS
From *Handbook of Mathematical Functions With Formulas, Graphs, and Mathematical Tables*. National Bureau of Standards Applied Mathematics Series—55, 1964, pp. 95–99.

Table V

COMMON LOGARITHMS

x	$\log_{10} x$	x	$\log_{10} x$	x	$\log_{10} x$	x	$\log_{10} x$	x	$\log_{10} x$
350	54406 80444	400	60205 99913	450	65321 25138	500	69897 00043	550	74036 26895
351	54530 71165	401	60314 43726	451	65417 65419	501	69983 77259	551	74115 15989
352	54654 26635	402	60422 60531	452	65513 84348	502	70070 37171	552	74193 90777
353	54777 47054	403	60530 50461	453	65609 82020	503	70156 79851	553	74272 51313
354	54900 32620	404	60638 13651	454	65705 58529	504	70243 05364	554	74350 97647
355	55022 83531	405	60745 50232	455	65801 13967	505	70329 13781	555	74429 29831
356	55144 99980	406	60852 60336	456	65896 48427	506	70415 05168	556	74507 47916
357	55266 82161	407	60959 44092	457	65991 62001	507	70500 79593	557	74585 51952
358	55388 30266	408	61066 01631	458	66086 54780	508	70586 37123	558	74663 41989
359	55509 44486	409	61172 33080	459	66181 26855	509	70671 77823	559	74741 18079
360	55630 25008	410	61278 38567	460	66275 78317	510	70757 01761	560	74818 80270
361	55750 72019	411	61384 18219	461	66370 09254	511	70842 09001	561	74896 28613
362	55870 85705	412	61489 72160	462	66464 19756	512	70926 99610	562	74973 63156
363	55990 66250	413	61595 00517	463	66558 09910	513	71011 73651	563	75050 83949
364	56110 13836	414	61700 03411	464	66651 79806	514	71096 31190	564	75127 91040
365	56229 28645	415	61804 80967	465	66745 29529	515	71180 72290	565	75204 84478
366	56348 10854	416	61909 33306	466	66838 59167	516	71264 97016	566	75281 64312
367	56466 60643	417	62013 60550	467	66931 68806	517	71349 05431	567	75358 30589
368	56584 78187	418	62117 62818	468	67024 58531	518	71432 97597	568	75434 83357
369	56702 63662	419	62221 40230	469	67117 28427	519	71516 73578	569	75511 22664
370	56820 17241	420	62324 92904	470	67209 78579	520	71600 33436	570	75587 48557
371	56937 39096	421	62428 20958	471	67302 09071	521	71683 77233	571	75663 61082
372	57054 29399	422	62531 24510	472	67394 19986	522	71767 05030	572	75739 60288
373	57170 88318	423	62634 03674	473	67486 11407	523	71850 16889	573	75815 46220
374	57287 16022	424	62736 58566	474	67577 83417	524	71933 12870	574	75891 18924
375	57403 12677	425	62838 89301	475	67669 36096	525	72015 93034	575	75966 78447
376	57518 78449	426	62940 95991	476	67760 69527	526	72098 57442	576	76042 24834
377	57634 13502	427	63042 78750	477	67851 83790	527	72181 06152	577	76117 58132
378	57749 17998	428	63144 37690	478	67942 78966	528	72263 39225	578	76192 78384
379	57863 92100	429	63245 72922	479	68033 55134	529	72345 56720	579	76267 85637
380	57978 35966	430	63346 84556	480	68124 12374	530	72427 58696	580	76342 79936
381	58092 49757	431	63447 72702	481	68214 50764	531	72509 45211	581	76417 61324
382	58206 33629	432	63548 37468	482	68304 70382	532	72591 16323	582	76492 29846
383	58319 87740	433	63648 78964	483	68394 71308	533	72672 72090	583	76566 85548
384	58433 12244	434	63748 97295	484	68484 53616	534	72754 12570	584	76641 28471
385	58546 07295	435	63848 92570	485	68574 17386	535	72835 37820	585	76715 58661
386	58658 73047	436	63948 64893	486	68663 62693	536	72916 47897	586	76789 76160
387	58771 09650	437	64048 14370	487	68752 89612	537	72997 42857	587	76863 81012
388	58883 17256	438	64147 41105	488	68841 98220	538	73078 22757	588	76937 73261
389	58994 96013	439	64246 45202	489	68930 88591	539	73158 87652	589	77011 52948
390	59106 46070	440	64345 26765	490	69019 60800	540	73239 37598	590	77085 20116
391	59217 67574	441	64443 85895	491	69108 14921	541	73319 72651	591	77158 74809
392	59328 60670	442	64542 22693	492	69196 51028	542	73399 92865	592	77232 17067
393	59439 25504	443	64640 37262	493	69284 69193	543	73479 98296	593	77305 46934
394	59549 62218	444	64738 29701	494	69372 69489	544	73559 88997	594	77378 64450
395	59659 70956	445	64836 00110	495	69460 51989	545	73639 65023	595	77451 69657
396	59769 51859	446	64933 48587	496	69548 16765	546	73719 26427	596	77524 62597
397	59879 05068	447	65030 75231	497	69635 63887	547	73798 73263	597	77597 43311
398	59988 30721	448	65127 80140	498	69722 93428	548	73878 05585	598	77670 11840
399	60097 28957	449	65224 63410	499	69810 05456	549	73957 23445	599	77742 68224
400	60205 99913	450	65321 25138	500	69897 00043	550	74036 26895	600	77815 12504

TABLE V.

Table V   *215*

COMMON LOGARITHMS

x	$\log_{10} x$	x	$\log_{10} x$	x	$\log_{10} x$	x	$\log_{10} x$	x	$\log_{10} x$
600	77815 12504	650	81291 33566	700	84509 80400	750	87506 12634	800	90308 99870
601	77887 44720	651	81358 09886	701	84571 80180	751	87563 99370	801	90363 25161
602	77959 64913	652	81424 75957	702	84633 71121	752	87621 78406	802	90417 43683
603	78031 73121	653	81491 31813	703	84695 53250	753	87679 49762	803	90471 55453
604	78103 69386	654	81557 77483	704	84757 26591	754	87737 13459	804	90525 60487
605	78175 53747	655	81624 13000	705	84818 91170	755	87794 69516	805	90579 58804
606	78247 26242	656	81690 38394	706	84880 47011	756	87852 17955	806	90633 50418
607	78318 86911	657	81756 53696	707	84941 94138	757	87909 58795	807	90687 35347
608	78390 35793	658	81822 58936	708	85003 32577	758	87966 92056	808	90741 13608
609	78461 72926	659	81888 54146	709	85064 62352	759	88024 17759	809	90794 85216
610	78532 98350	660	81954 39355	710	85125 83487	760	88081 35923	810	90848 50189
611	78604 12102	661	82020 14595	711	85186 96007	761	88138 46568	811	90902 08542
612	78675 14221	662	82085 79894	712	85247 99936	762	88195 49713	812	90955 60292
613	78746 04745	663	82151 35284	713	85308 95299	763	88252 45380	813	91009 05456
614	78816 83711	664	82216 80794	714	85369 82118	764	88309 33586	814	91062 44049
615	78887 51158	665	82282 16453	715	85430 60418	765	88366 14352	815	91115 76087
616	78958 07122	666	82347 42292	716	85491 30223	766	88422 87696	816	91169 01588
617	79028 51640	667	82412 58339	717	85551 91557	767	88479 53639	817	91222 20565
618	79098 84751	668	82477 64625	718	85612 44442	768	88536 12200	818	91275 33037
619	79169 06490	669	82542 61178	719	85672 88904	769	88592 63398	819	91328 39018
620	79239 16895	670	82607 48027	720	85733 24964	770	88649 07252	820	91381 38524
621	79309 16002	671	82672 25202	721	85793 52647	771	88705 43781	821	91434 31571
622	79379 03847	672	82736 92731	722	85853 71976	772	88761 73003	822	91487 18175
623	79448 80467	673	82801 50642	723	85913 82973	773	88817 94939	823	91539 98352
624	79518 45897	674	82865 98965	724	85973 85662	774	88874 09607	824	91592 72117
625	79588 00173	675	82930 37728	725	86033 80066	775	88930 17025	825	91645 39485
626	79657 43332	676	82994 66959	726	86093 66207	776	88986 17213	826	91698 00473
627	79726 75408	677	83058 86687	727	86153 44109	777	89042 10188	827	91750 55096
628	79795 96437	678	83122 96939	728	86213 13793	778	89097 95970	828	91803 03368
629	79865 06454	679	83186 97743	729	86272 75283	779	89153 74577	829	91855 45306
630	79934 05495	680	83250 89127	730	86332 28601	780	89209 46027	830	91907 80924
631	80002 93592	681	83314 71119	731	86391 73770	781	89265 10339	831	91960 10238
632	80071 70783	682	83378 43747	732	86451 10811	782	89320 67531	832	92012 33263
633	80140 37100	683	83442 07037	733	86510 39746	783	89376 17621	833	92064 50014
634	80208 92579	684	83505 61017	734	86569 60599	784	89431 60627	834	92116 60506
635	80277 37253	685	83569 05715	735	86628 73391	785	89486 96567	835	92168 64755
636	80345 71156	686	83632 41157	736	86687 78143	786	89542 25460	836	92220 62774
637	80413 94323	687	83695 67371	737	86746 74879	787	89597 47324	837	92272 54580
638	80482 06787	688	83758 84382	738	86805 63618	788	89652 62175	838	92324 40186
639	80550 08582	689	83821 92219	739	86864 44384	789	89707 70032	839	92376 19608
640	80617 99740	690	83884 90907	740	86923 17197	790	89762 70913	840	92427 92861
641	80685 80295	691	83947 80474	741	86981 82080	791	89817 64835	841	92479 59958
642	80753 50281	692	84010 60945	742	87040 39053	792	89872 51816	842	92531 20915
643	80821 09729	693	84073 32346	743	87098 88138	793	89927 31873	843	92582 75746
644	80888 58674	694	84135 94705	744	87157 29355	794	89982 05024	844	92634 24466
645	80955 97146	695	84198 48046	745	87215 62727	795	90036 71287	845	92685 67089
646	81023 25180	696	84260 92396	746	87273 88275	796	90091 30677	846	92737 03630
647	81090 42807	697	84323 27781	747	87332 06018	797	90145 83214	847	92788 34103
648	81157 50059	698	84385 54226	748	87390 15979	798	90200 28914	848	92839 58523
649	81224 46968	699	84447 71757	749	87448 18177	799	90254 67793	849	92890 76902
650	81291 33566	700	84509 80400	750	87506 12634	800	90308 99870	850	92941 89257

TABLE V.

Table V

COMMON LOGARITHMS

x	$\log_{10} x$	x	$\log_{10} x$	x	$\log_{10} x$	x	$\log_{10} x$	x	$\log_{10} x$
850	92941 89257	900	95424 25094	950	97772 36053	1000	00000 00000	1050	02118 92991
851	92992 95601	901	95472 47910	951	97818 05169	1001	00043 40775	1051	02160 27160
852	93043 95948	902	95520 65375	952	97863 69484	1002	00086 77215	1052	02201 57398
853	93094 90312	903	95568 77503	953	97909 29006	1003	00130 09330	1053	02242 83712
854	93145 78707	904	95616 84305	954	97954 83747	1004	00173 37128	1054	02284 06109
855	93196 61147	905	95664 85792	955	98000 33716	1005	00216 60618	1055	02325 24596
856	93247 37647	906	95712 81977	956	98045 78923	1006	00259 79807	1056	02366 39182
857	93298 08219	907	95760 72871	957	98091 19378	1007	00302 94706	1057	02407 49873
858	93348 72878	908	95808 58485	958	98136 55091	1008	00346 05321	1058	02448 56677
859	93399 31638	909	95856 38832	959	98181 86072	1009	00389 11662	1059	02489 59601
860	93449 84512	910	95904 13923	960	98227 12330	1010	00432 13738	1060	02530 58653
861	93500 31515	911	95951 83770	961	98272 33877	1011	00475 11556	1061	02571 53839
862	93550 72658	912	95999 48383	962	98317 50720	1012	00518 05125	1062	02612 45167
863	93601 07957	913	96047 07775	963	98362 62871	1013	00560 94454	1063	02653 32645
864	93651 37425	914	96094 61957	964	98407 70339	1014	00603 79550	1064	02694 16280
865	93701 61075	915	96142 10941	965	98452 73133	1015	00646 60422	1065	02734 96078
866	93751 78920	916	96189 54737	966	98497 71264	1016	00689 37079	1066	02775 72047
867	93801 90975	917	96236 93357	967	98542 64741	1017	00732 09529	1067	02816 44194
868	93851 97252	918	96284 26812	968	98587 53573	1018	00774 77780	1068	02857 12527
869	93901 97764	919	96331 55114	969	98632 37771	1019	00817 41840	1069	02897 77052
870	93951 92526	920	96378 78273	970	98677 17343	1020	00860 01718	1070	02938 37777
871	94001 81550	921	96425 96302	971	98721 92299	1021	00902 57421	1071	02978 94708
872	94051 64849	922	96473 09211	972	98766 62649	1022	00945 08958	1072	03019 47854
873	94101 42437	923	96520 17010	973	98811 28403	1023	00987 56337	1073	03059 97220
874	94151 14326	924	96567 19712	974	98855 89569	1024	01029 99566	1074	03100 42814
875	94200 80530	925	96614 17327	975	98900 46157	1025	01072 38654	1075	03140 84643
876	94250 41062	926	96661 09867	976	98944 98177	1026	01114 73608	1076	03181 22713
877	94299 95934	927	96707 97341	977	98989 45637	1027	01157 04436	1077	03221 57033
878	94349 45159	928	96754 79762	978	99033 88548	1028	01199 31147	1078	03261 87609
879	94398 88751	929	96801 57140	979	99078 26918	1029	01241 53748	1079	03302 14447
880	94448 26722	930	96848 29486	980	99122 60757	1030	01283 72247	1080	03342 37555
881	94497 59084	931	96894 96810	981	99166 90074	1031	01325 86653	1081	03382 56940
882	94546 85851	932	96941 59124	982	99211 14878	1032	01367 96973	1082	03422 72608
883	94596 07036	933	96988 16437	983	99255 35178	1033	01410 03215	1083	03462 84566
884	94645 22650	934	97034 68762	984	99299 50984	1034	01452 05388	1084	03502 92822
885	94694 32707	935	97081 16109	985	99343 62305	1035	01494 03498	1085	03542 97382
886	94743 37219	936	97127 58487	986	99387 69149	1036	01535 97554	1086	03582 98253
887	94792 36198	937	97173 95909	987	99431 71527	1037	01577 87564	1087	03622 95441
888	94841 29658	938	97220 28384	988	99475 69446	1038	01619 73535	1088	03662 88954
889	94890 17610	939	97266 55923	989	99519 62916	1039	01661 55476	1089	03702 78798
890	94939 00066	940	97312 78536	990	99563 51946	1040	01703 33393	1090	03742 64979
891	94987 77040	941	97358 96234	991	99607 36545	1041	01745 07295	1091	03782 47506
892	95036 48544	942	97405 09028	992	99651 16722	1042	01786 77190	1092	03822 26384
893	95085 14589	943	97451 16927	993	99694 92485	1043	01828 43084	1093	03862 01619
894	95133 75188	944	97497 19943	994	99738 63844	1044	01870 04987	1094	03901 73220
895	95182 30353	945	97543 18085	995	99782 30807	1045	01911 62904	1095	03941 41192
896	95230 80097	946	97589 11364	996	99825 93384	1046	01953 16845	1096	03981 05541
897	95279 24430	947	97634 99790	997	99869 51583	1047	01994 66817	1097	04020 66276
898	95327 63367	948	97680 83373	998	99913 05413	1048	02036 12826	1098	04060 23401
899	95375 96917	949	97726 62124	999	99956 54882	1049	02077 54882	1099	04099 76924
900	95424 25094	950	97772 36053	1000	00000 00000	1050	02118 92991	1100	04139 26852

TABLE V.

Table   V   *217*

COMMON LOGARITHMS

x	$\log_{10} x$	x	$\log_{10} x$	x	$\log_{10} x$	x	$\log_{10} x$	x	$\log_{10} x$	x	$\log_{10} x$
1100	04139 26852	1150	06069 78404	1200	07918 12460	1250	09691 00130	1300	11394 33523
1101	04178 ·73190	1151	06107 53236	1201	07954 30074	1251	09725 73097	1301	11427 72966
1102	04218 15945	1152	06145 24791	1202	07990 44677	1252	09760 43289	1302	11461 09842
1103	04257 55124	1153	06182 93073	1203	08026 56273	1253	09795 10710	1303	11494 44157
1104	04296 90734	1154	06220 58088	1204	08062 64869	1254	09829 75365	1304	11527 75914
1105	04336 22780	1155	06258 19842	1205	08098 70469	1255	09864 37258	1305	11561 05117
1106	04375 51270	1156	06295 78341	1206	08134 73078	1256	09898 96394	1306	11594 31769
1107	04414 76209	1157	06333 33590	1207	08170 72701	1257	09933 52777	1307	11627 55876
1108	04453 97604	1158	06370 85594	1208	08206 69343	1258	09968 06411	1308	11660 77440
1109	04493 15461	1159	06408 34360	1209	08242 63009	1259	10002 57301	1309	11693 96466
1110	04532 29788	1160	06445 79892	1210	08278 53703	1260	10037 05451	1310	11727 12957
1111	04571 40589	1161	06483 22197	1211	08314 41431	1261	10071 50866	1311	11760 26917
1112	04610 47872	1162	06520 61281	1212	08350 26198	1262	10105 93549	1312	11793 38350
1113	04649 51643	1163	06557 97147	1213	08386 08009	1263	10140 33506	1313	11826 47261
1114	04688 51908	1164	06595 29803	1214	08421 86867	1264	10174 70739	1314	11859 53652
1115	04727 48674	1165	06632 59254	1215	08457 62779	1265	10209 05255	1315	11892 57528
1116	04766 41946	1166	06669 85504	1216	08493 35749	1266	10243 37057	1316	11925 58893
1117	04805 31731	1167	06707 08560	1217	08529 05782	1267	10277 66149	1317	11958 57750
1118	04844 18036	1168	06744 28428	1218	08564 72883	1268	10311 92535	1318	11991 54103
1119	04883 00865	1169	06781 45112	1219	08600 37056	1269	10346 16221	1319	12024 47955
1120	04921 80227	1170	06818 58617	1220	08635 98307	1270	10380 37210	1320	12057 39312
1121	04960 56126	1171	06855 68951	1221	08671 56639	1271	10414 55506	1321	12090 28176
1122	04999 28569	1172	06892 76117	1222	08707 12059	1272	10448 71113	1322	12123 14551
1123	05037 97563	1173	06929 80121	1223	08742 64570	1273	10482 84037	1323	12155 98442
1124	05076 63112	1174	06966 80969	1224	08778 14178	1274	10516 94280	1324	12188 79851
1125	05115 25224	1175	07003 78666	1225	08813 60887	1275	10551 01848	1325	12221 58783
1126	05153 83905	1176	07040 73217	1226	08849 04702	1276	10585 06744	1326	12254 35241
1127	05192 39160	1177	07077 64628	1227	08884 45627	1277	10619 08973	1327	12287 09229
1128	05230 90996	1178	07114 52905	1228	08919 83668	1278	10653 08538	1328	12319 80750
1129	05269 39419	1179	07151 38051	1229	08955 18829	1279	10687 05445	1329	12352 49809
1130	05307 84435	1180	07188 20073	1230	08990 51114	1280	10720 99696	1330	12385 16410
1131	05346 26049	1181	07224 98976	1231	09025 80529	1281	10754 91297	1331	12417 80555
1132	05384 64269	1182	07261 74765	1232	09061 07078	1282	10788 80252	1332	12450 42248
1133	05422 99099	1183	07298 47446	1233	09096 30766	1283	10822 66564	1333	12483 01494
1134	05461 30546	1184	07335 17024	1234	09131 51597	1284	10856 50237	1334	12515 58296
1135	05499 58615	1185	07371 83503	1235	09166 69576	1285	10890 31277	1335	12548 12657
1136	05537 83314	1186	07408 46890	1236	09201 84708	1286	10924 09686	1336	12580 64581
1137	05576 04647	1187	07445 07190	1237	09236 96996	1287	10957 85469	1337	12613 14073
1138	05614 22621	1188	07481 64406	1238	09272 06447	1288	10991 58630	1338	12645 61134
1139	05652 37241	1189	07518 18546	1239	09307 13064	1289	11025 29174	1339	12678 05770
1140	05690 48513	1190	07554 69614	1240	09342 16852	1290	11058 97103	1340	12710 47984
1141	05728 56444	1191	07591 17615	1241	09377 17815	1291	11092 62423	1341	12742 87779
1142	05766 61039	1192	07627 62554	1242	09412 15958	1292	11126 25137	1342	12775 25158
1143	05804 62304	1193	07664 04437	1243	09447 11286	1293	11159 85249	1343	12807 60127
1144	05842 60245	1194	07700 43268	1244	09482 03804	1294	11193 42763	1344	12839 92687
1145	05880 54867	1195	07736 79053	1245	09516 93514	1295	11226 97684	1345	12872 22843
1146	05918 46176	1196	07773 11797	1246	09551 80423	1296	11260 50015	1346	12904 50599
1147	05956 34179	1197	07809 41504	1247	09586 64535	1297	11293 99761	1347	12936 75957
1148	05994 18881	1198	07845 68181	1248	09621 45853	1298	11327 46925	1348	12968 98922
1149	06032 00287	1199	07881 91831	1249	09656 24384	1299	11360 91511	1349	13001 19497
1150	06069 78404	1200	07918 12460	1250	09691 00130	1300	11394 33523	1350	13033 37685

TABLE V.

NATURAL LOGARITHMS

Table VI

x	$\ln x$	x	$\ln x$	x	$\ln x$
0.000	$-\infty$	0.050	-2.99573 22735 539910	0.100	-2.30258 50929 940457
0.001	-6.90775 52789 821371	0.051	-2.97592 96462 578113	0.101	-2.29263 47621 408776
0.002	-6.21460 80984 221917	0.052	-2.95651 15604 007097	0.102	-2.28278 24656 978660
0.003	-5.80914 29903 140274	0.053	-2.93746 33654 300152	0.103	-2.27302 62907 525013
0.004	-5.52146 09178 622464	0.054	-2.91877 12324 178627	0.104	-2.26336 43798 407644
0.005	-5.29831 73665 480367	0.055	-2.90042 20937 496661	0.105	-2.25379 49288 246137
0.006	-5.11599 58097 540821	0.056	-2.88240 35882 469878	0.106	-2.24431 61848 700699
0.007	-4.96184 51299 268237	0.057	-2.86470 40111 475869	0.107	-2.23492 64445 202309
0.008	-4.82831 37373 023011	0.058	-2.84731 22684 357177	0.108	-2.22562 40518 579174
0.009	-4.71053 07016 459177	0.059	-2.83021 78350 764176	0.109	-2.21640 73967 529934
0.010	-4.60517 01859 880914	0.060	-2.81341 07167 600364	0.110	-2.20727 49131 897208
0.011	-4.50986 00061 837665	0.061	-2.79688 14148 088258	0.111	-2.19822 50776 698029
0.012	-4.42284 86291 941367	0.062	-2.78062 08939 370455	0.112	-2.18925 64076 870425
0.013	-4.34280 59215 206003	0.063	-2.76462 05525 906044	0.113	-2.18036 74602 697965
0.014	-4.26869 79493 668784	0.064	-2.74887 21956 224652	0.114	-2.17155 68305 876416
0.015	-4.19970 50778 799270	0.065	-2.73336 80090 864999	0.115	-2.16282 31506 188870
0.016	-4.13516 65567 423558	0.066	-2.71810 05369 557115	0.116	-2.15416 50878 757724
0.017	-4.07454 19349 259210	0.067	-2.70306 26595 911710	0.117	-2.14558 13441 843809
0.018	-4.01738 35210 859724	0.068	-2.68824 75738 060304	0.118	-2.13707 06545 164723
0.019	-3.96331 62998 156966	0.069	-2.67364 87743 848777	0.119	-2.12863 17858 706077
0.020	-3.91202 30054 281461	0.070	-2.65926 00369 327781	0.120	-2.12026 35362 000911
0.021	-3.86323 28412 587141	0.071	-2.64507 54019 408216	0.121	-2.11196 47333 853960
0.022	-3.81671 28256 238212	0.072	-2.63108 91599 660817	0.122	-2.10373 42342 488805
0.023	-3.77226 10630 529874	0.073	-2.61729 58378 337459	0.123	-2.09557 09236 097196
0.024	-3.72970 14486 341914	0.074	-2.60369 01857 779673	0.124	-2.08747 37133 771002
0.025	-3.68887 94541 139363	0.075	-2.59026 71654 458266	0.125	-2.07944 15416 798359
0.026	-3.64965 87409 606550	0.076	-2.57702 19386 958060	0.126	-2.07147 33720 306591
0.027	-3.61191 84129 778080	0.077	-2.56394 98571 284532	0.127	-2.06356 81925 235458
0.028	-3.57555 07688 069331	0.078	-2.55104 64522 925453	0.128	-2.05572 50150 625199
0.029	-3.54045 94489 956630	0.079	-2.53830 74265 151156	0.129	-2.04794 28746 204649
0.030	-3.50655 78973 199817	0.080	-2.52572 86443 082554	0.130	-2.04022 08285 265546
0.031	-3.47376 80744 969908	0.081	-2.51330 61243 096983	0.131	-2.03255 79557 809855
0.032	-3.44201 93761 824105	0.082	-2.50103 60317 178839	0.132	-2.02495 33563 957662
0.033	-3.41124 77175 156568	0.083	-2.48891 46711 855391	0.133	-2.01740 61507 603833
0.034	-3.38139 47543 659757	0.084	-2.47693 84801 388234	0.134	-2.00991 54790 312257
0.035	-3.35240 72174 927234	0.085	-2.46510 40224 918206	0.135	-2.00248 05005 437076
0.036	-3.32423 63405 260271	0.086	-2.45340 79827 286293	0.136	-1.99510 03932 460850
0.037	-3.29683 73663 379126	0.087	-2.44184 71603 275533	0.137	-1.98777 43531 540121
0.038	-3.27016 91192 557513	0.088	-2.43041 84645 039306	0.138	-1.98050 15938 249324
0.039	-3.24419 36328 524906	0.089	-2.41911 89092 499972	0.139	-1.97328 13458 514453
0.040	-3.21887 58248 682007	0.090	-2.40794 56086 518720	0.140	-1.96611 28563 728328
0.041	-3.19418 32122 778292	0.091	-2.39689 57724 652870	0.141	-1.95899 53886 039688
0.042	-3.17008 56606 987687	0.092	-2.38596 67019 330967	0.142	-1.95192 82213 808763
0.043	-3.14655 51632 885746	0.093	-2.37515 57858 288811	0.143	-1.94491 06487 222298
0.044	-3.12356 56450 638759	0.094	-2.36446 04967 121332	0.144	-1.93794 19794 061364
0.045	-3.10109 27892 118173	0.095	-2.35387 83873 815962	0.145	-1.93102 15365 615627
0.046	-3.07911 38824 930421	0.096	-2.34340 70875 143008	0.146	-1.92414 86572 738006
0.047	-3.05760 76772 720785	0.097	-2.33304 43004 787542	0.147	-1.91732 26922 034008
0.048	-3.03655 42680 742461	0.098	-2.32278 78003 115651	0.148	-1.91054 30052 180220
0.049	-3.01593 49808 715104	0.099	-2.31263 54288 475471	0.149	-1.90380 89730 366779
0.050	-2.99573 22735 539910	0.100	-2.30258 50929 940457	0.150	-1.89711 99848 858813

TABLE VI.

TABLE VI. NATURAL LOGARIHMS
From *Handbook of Mathematical Functions With Formulas, Graphs, and Mathematical Tables*. National Bureau of Standards Applied Mathematics Series—55, 1964, pp. 100–113.

Table VI 219

NATURAL LOGARITHMS

x	$\ln x$	x	$\ln x$	x	$\ln x$
0.150	-1.89711 99848 858813	0.200	-1.60943 79124 341004	0.250	-1.38629 43611 198906
0.151	-1.89047 54421 672127	0.201	-1.60445 03709 230613	0.251	-1.38230 23398 503532
0.152	-1.88387 47581 358607	0.202	-1.59948 75815 809323	0.252	-1.37832 61914 707137
0.153	-1.87731 73575 897016	0.203	-1.59454 92999 403497	0.253	-1.37436 57902 546168
0.154	-1.87080 26765 685079	0.204	-1.58963 52851 379207	0.254	-1.37042 10119 636005
0.155	-1.86433 01620 628904	0.205	-1.58474 52998 437289	0.255	-1.36649 17338 237109
0.156	-1.85789 92717 326000	0.206	-1.57987 91101 925560	0.256	-1.36257 78345 025746
0.157	-1.85150 94736 338290	0.207	-1.57503 64857 167680	0.257	-1.35867 91940 869173
0.158	-1.84516 02459 551702	0.208	-1.57021 71992 808191	0.258	-1.35479 56940 605196
0.159	-1.83885 10767 619055	0.209	-1.56542 10270 173260	0.259	-1.35092 72172 825993
0.160	-1.83258 14637 483101	0.210	-1.56064 77482 646684	0.260	-1.34707 36479 666093
0.161	-1.82635 09139 976741	0.211	-1.55589 71455 060706	0.261	-1.34323 48716 594436
0.162	-1.82015 89437 497530	0.212	-1.55116 90043 101246	0.262	-1.33941 07752 210402
0.163	-1.81400 50781 753747	0.213	-1.54646 31132 727119	0.263	-1.33560 12468 043725
0.164	-1.80788 88511 579386	0.214	-1.54177 92639 602856	0.264	-1.33180 61758 358209
0.165	-1.80180 98050 815564	0.215	-1.53711 72508 544743	0.265	-1.32802 54529 959148
0.166	-1.79576 74906 255938	0.216	-1.53247 68712 979720	0.266	-1.32425 89702 004380
0.167	-1.78976 14665 653819	0.217	-1.52785 79254 416775	0.267	-1.32050 66205 818875
0.168	-1.78379 12995 788781	0.218	-1.52326 02161 930480	0.268	-1.31676 82984 712804
0.169	-1.77785 65640 590636	0.219	-1.51868 35491 656362	0.269	-1.31304 38993 802979
0.170	-1.77195 68419 318753	0.220	-1.51412 77326 297755	0.270	-1.30933 33199 837623
0.171	-1.76609 17224 794772	0.221	-1.50959 25774 643842	0.271	-1.30563 64581 024362
0.172	-1.76026 08021 686840	0.222	-1.50507 78971 098576	0.272	-1.30195 32126 861397
0.173	-1.75446 36844 843581	0.223	-1.50058 35075 220183	0.273	-1.29828 34837 971773
0.174	-1.74869 99797 676080	0.224	-1.49610 92271 270972	0.274	-1.29462 71725 940668
0.175	-1.74296 93050 586230	0.225	-1.49165 48767 777169	0.275	-1.29098 41813 155658
0.176	-1.73727 12839 439853	0.226	-1.48722 02797 098512	0.276	-1.28735 44132 649871
0.177	-1.73160 55464 083079	0.227	-1.48280 52615 007344	0.277	-1.28373 77727 947986
0.178	-1.72597 17286 900519	0.228	-1.47840 96500 276963	0.278	-1.28013 41652 915000
0.179	-1.72036 94731 413821	0.229	-1.47403 32754 278974	0.279	-1.27654 34971 607714
0.180	-1.71479 84280 919267	0.230	-1.46967 59700 589417	0.280	-1.27296 56758 128874
0.181	-1.70925 82477 163113	0.231	-1.46533 75684 603435	0.281	-1.26940 06096 483913
0.182	-1.70374 85919 053417	0.232	-1.46101 79073 158271	0.282	-1.26584 82080 440235
0.183	-1.69826 91261 407161	0.233	-1.45671 68254 164365	0.283	-1.26230 83813 388994
0.184	-1.69281 95213 731514	0.234	-1.45243 41636 244356	0.284	-1.25878 10408 209310
0.185	-1.68739 94539 038122	0.235	-1.44816 97648 379781	0.285	-1.25526 60987 134865
0.186	-1.68200 86052 689358	0.236	-1.44392 34739 565270	0.286	-1.25176 34681 622845
0.187	-1.67664 66621 275504	0.237	-1.43969 51378 470059	0.287	-1.24827 30632 225159
0.188	-1.67131 33161 521878	0.238	-1.43548 46053 106624	0.288	-1.24479 47988 461911
0.189	-1.66600 82639 224947	0.239	-1.43129 17270 506264	0.289	-1.24132 85908 697049
0.190	-1.66073 12068 216509	0.240	-1.42711 63556 401457	0.290	-1.23787 43560 016173
0.191	-1.65548 18509 355072	0.241	-1.42295 83454 914821	0.291	-1.23443 20118 106445
0.192	-1.65025 99069 543555	0.242	-1.41881 75528 254507	0.292	-1.23100 14767 138553
0.193	-1.64506 50900 772515	0.243	-1.41469 38356 415886	0.293	-1.22758 26699 650697
0.194	-1.63989 71199 188089	0.244	-1.41058 70536 889352	0.294	-1.22417 55116 434554
0.195	-1.63475 57204 183903	0.245	-1.40649 70684 374101	0.295	-1.22077 99226 423172
0.196	-1.62964 06197 516198	0.246	-1.40242 37430 497742	0.296	-1.21739 58246 580767
0.197	-1.62455 15502 441485	0.247	-1.39836 69423 541599	0.297	-1.21402 31401 794374
0.198	-1.61948 82482 876018	0.248	-1.39432 65328 171549	0.298	-1.21066 17924 767326
0.199	-1.61445 04542 576447	0.249	-1.39030 23825 174294	0.299	-1.20731 17055 914506
0.200	-1.60943 79124 341004	0.250	-1.38629 43611 198906	0.300	-1.20397 28043 259360

TABLE VI.

Table VI

NATURAL LOGARITHMS

x	$\ln x$	x	$\ln x$	x	$\ln x$
0.300	−1.20397 28043 259360	0.350	−1.04982 21244 986777	0.400	−0.91629 07318 741551
0.301	−1.20064 50142 332613	0.351	−1.04696 90555 162712	0.401	−0.91379 38516 755679
0.302	−1.19732 82616 072674	0.352	−1.04412 41033 840400	0.402	−0.91130 31903 631160
0.303	−1.19402 24734 727679	0.353	−1.04128 72220 488403	0.403	−0.90881 87170 354541
0.304	−1.19072 75775 759154	0.354	−1.03845 83658 483626	0.404	−0.90634 04010 209870
0.305	−1.18744 35023 747254	0.355	−1.03563 74895 067213	0.405	−0.90386 82118 755979
0.306	−1.18417 01770 297563	0.356	−1.03282 45481 301066	0.406	−0.90140 21193 804044
0.307	−1.18090 75313 949399	0.357	−1.03001 94972 024980	0.407	−0.89894 20935 395421
0.308	−1.17765 54960 085626	0.358	−1.02722 22925 814367	0.408	−0.89648 81045 779754
0.309	−1.17441 40020 843916	0.359	−1.02443 28904 938582	0.409	−0.89404 01229 393353
0.310	−1.17118 29815 029451	0.360	−1.02165 12475 319814	0.410	−0.89159 81192 837836
0.311	−1.16796 23668 029029	0.361	−1.01887 73206 492561	0.411	−0.88916 20644 859024
0.312	−1.16475 20911 726547	0.362	−1.01611 10671 563660	0.412	−0.88673 19296 326107
0.313	−1.16155 20884 419838	0.363	−1.01335 24447 172863	0.413	−0.88430 76860 211043
0.314	−1.15836 22930 738837	0.364	−1.01060 14113 453964	0.414	−0.88188 93051 568227
0.315	−1.15518 26401 565040	0.365	−1.00785 79253 996455	0.415	−0.87947 67587 514388
0.316	−1.15201 30653 952249	0.366	−1.00512 19455 807708	0.416	−0.87707 00187 208738
0.317	−1.14885 35051 048564	0.367	−1.00239 34309 275668	0.417	−0.87466 90571 833356
0.318	−1.14570 38962 019602	0.368	−0.99967 23408 132061	0.418	−0.87227 38464 573807
0.319	−1.14256 41761 972925	0.369	−0.99695 86349 416099	0.419	−0.86988 43590 599993
0.320	−1.13943 42831 883648	0.370	−0.99425 22733 438669	0.420	−0.86750 05677 047231
0.321	−1.13631 41558 521212	0.371	−0.99155 32163 747019	0.421	−0.86512 24452 997556
0.322	−1.13320 37334 377287	0.372	−0.98886 14247 089905	0.422	−0.86274 99649 461252
0.323	−1.13010 29557 594805	0.373	−0.98617 68593 383215	0.423	−0.86038 30999 358591
0.324	−1.12701 17631 898077	0.374	−0.98349 94815 676051	0.424	−0.85802 18237 501793
0.325	−1.12393 00966 523996	0.375	−0.98082 92530 117262	0.425	−0.85566 61100 577202
0.326	−1.12085 78976 154294	0.376	−0.97816 61355 922425	0.426	−0.85331 59327 127666
0.327	−1.11779 51080 848837	0.377	−0.97551 00915 341263	0.427	−0.85097 12657 535125
0.328	−1.11474 16705 979933	0.378	−0.97286 10833 625494	0.428	−0.84863 20834 003403
0.329	−1.11169 75282 167652	0.379	−0.97021 90738 997107	0.429	−0.84629 83600 541201
0.330	−1.10866 26245 216111	0.380	−0.96758 40262 617056	0.430	−0.84397 00702 945289
0.331	−1.10563 69036 050742	0.381	−0.96495 59038 554361	0.431	−0.84164 71888 783893
0.332	−1.10262 03100 656485	0.382	−0.96233 46703 755619	0.432	−0.83932 96907 380267
0.333	−1.09961 27890 016932	0.383	−0.95972 02898 014911	0.433	−0.83701 75509 796472
0.334	−1.09661 42860 054366	0.384	−0.95711 27263 944102	0.434	−0.83471 07448 817322
0.335	−1.09362 47471 570706	0.385	−0.95451 19446 943528	0.435	−0.83240 92478 934530
0.336	−1.09064 41190 189328	0.386	−0.95191 79095 173062	0.436	−0.83011 30356 331027
0.337	−1.08767 23486 297753	0.387	−0.94933 05859 523552	0.437	−0.82782 20838 865469
0.338	−1.08470 93834 991183	0.388	−0.94674 99393 588636	0.438	−0.82553 63686 056909
0.339	−1.08175 51716 016868	0.389	−0.94417 59353 636908	0.439	−0.82325 58659 069657
0.340	−1.07880 96613 719300	0.390	−0.94160 85398 584449	0.440	−0.82098 05520 698302
0.341	−1.07587 28016 986203	0.391	−0.93904 77189 967713	0.441	−0.81871 04035 352911
0.342	−1.07294 45419 195319	0.392	−0.93649 34391 916745	0.442	−0.81644 53969 044389
0.343	−1.07002 48318 161971	0.393	−0.93394 56671 128758	0.443	−0.81418 55089 370014
0.344	−1.06711 36216 087387	0.394	−0.93140 43696 842032	0.444	−0.81193 07165 499123
0.345	−1.06421 08619 507773	0.395	−0.92886 95140 810152	0.445	−0.80968 09968 158968
0.346	−1.06131 65039 244128	0.396	−0.92634 10677 276565	0.446	−0.80743 63269 620730
0.347	−1.05843 04990 352779	0.397	−0.92381 89982 949466	0.447	−0.80519 66843 685682
0.348	−1.05555 27992 076627	0.398	−0.92130 32736 976993	0.448	−0.80296 20465 671519
0.349	−1.05268 33567 797099	0.399	−0.91879 38620 922736	0.449	−0.80073 23912 398828
0.350	−1.04982 21244 986777	0.400	−0.91629 07318 741551	0.450	−0.79850 76962 177716

TABLE VI.

Table VI 221

NATURAL LOGARITHMS

x	$\ln x$	x	$\ln x$	x	$\ln x$
0.450	-0.79850 76962 177716	0.500	-0.69314 71805 599453	0.550	-0.59783 70007 556204
0.451	-0.79628 79394 794587	0.501	-0.69114 91778 972723	0.551	-0.59602 04698 292226
0.452	-0.79407 30991 499059	0.502	-0.68915 51592 904079	0.552	-0.59420 72327 050417
0.453	-0.79186 31534 991030	0.503	-0.68716 51088 823978	0.553	-0.59239 72774 598023
0.454	-0.78965 80809 407891	0.504	-0.68517 90109 107684	0.554	-0.59059 05922 348532
0.455	-0.78745 78600 311866	0.505	-0.68319 68497 067772	0.555	-0.58878 71652 357025
0.456	-0.78526 24694 677510	0.506	-0.68121 86096 946715	0.556	-0.58698 69847 315547
0.457	-0.78307 18880 879324	0.507	-0.67924 42753 909539	0.557	-0.58519 00390 548530
0.458	-0.78088 60948 679521	0.508	-0.67727 38314 036552	0.558	-0.58339 63166 008261
0.459	-0.77870 50689 215919	0.509	-0.67530 72624 316143	0.559	-0.58160 58058 270379
0.460	-0.77652 87894 989964	0.510	-0.67334 45532 637656	0.560	-0.57981 84952 529421
0.461	-0.77435 72359 854885	0.511	-0.67138 56887 784326	0.561	-0.57803 43734 594407
0.462	-0.77219 03879 003982	0.512	-0.66943 06539 426293	0.562	-0.57625 34290 884460
0.463	-0.77002 82248 959030	0.513	-0.66747 94338 113675	0.563	-0.57447 56503 424467
0.464	-0.76787 07267 558818	0.514	-0.66553 20135 269719	0.564	-0.57270 10274 840782
0.465	-0.76571 78733 947807	0.515	-0.66358 83783 184009	0.565	-0.57092 95478 356961
0.466	-0.76356 96448 564912	0.516	-0.66164 85135 005743	0.566	-0.56916 12007 789541
0.467	-0.76142 60213 132397	0.517	-0.65971 24044 737079	0.567	-0.56739 59752 543850
0.468	-0.75928 69830 644903	0.518	-0.65778 00367 226540	0.568	-0.56563 38602 609857
0.469	-0.75715 25105 358577	0.519	-0.65585 13958 162484	0.569	-0.56387 48448 558061
0.470	-0.75502 25842 780328	0.520	-0.65392 64674 066640	0.570	-0.56211 89181 535412
0.471	-0.75289 71849 657193	0.521	-0.65200 52372 287701	0.571	-0.56036 60693 261268
0.472	-0.75077 62933 965817	0.522	-0.65008 76910 994983	0.572	-0.55861 62876 023392
0.473	-0.74865 98904 902041	0.523	-0.64817 38149 172142	0.573	-0.55686 95622 673975
0.474	-0.74654 79572 870606	0.524	-0.64626 35946 610949	0.574	-0.55512 58826 625706
0.475	-0.74444 04749 474958	0.525	-0.64435 70163 905133	0.575	-0.55338 52381 847866
0.476	-0.74233 74247 507170	0.526	-0.64245 40662 444272	0.576	-0.55164 76182 862458
0.477	-0.74023 87880 937958	0.527	-0.64055 47304 407747	0.577	-0.54991 30124 740375
0.478	-0.73814 45464 906811	0.528	-0.63865 89952 758756	0.578	-0.54818 14103 097596
0.479	-0.73605 46815 712218	0.529	-0.63676 68471 238377	0.579	-0.54645 28014 091418
0.480	-0.73396 91750 802004	0.530	-0.63487 82724 359695	0.580	-0.54472 71754 416720
0.481	-0.73188 80088 763759	0.531	-0.63299 32577 401982	0.581	-0.54300 45221 302258
0.482	-0.72981 11649 315367	0.532	-0.63111 17896 404927	0.582	-0.54128 48312 506992
0.483	-0.72773 86253 295644	0.533	-0.62923 38548 162925	0.583	-0.53956 80926 316447
0.484	-0.72567 03722 655053	0.534	-0.62735 94400 219422	0.584	-0.53785 42961 539100
0.485	-0.72360 63880 446539	0.535	-0.62548 85320 861305	0.585	-0.53614 34317 502806
0.486	-0.72154 66550 816433	0.536	-0.62362 11179 113351	0.586	-0.53443 54894 051244
0.487	-0.71949 11558 995473	0.537	-0.62175 71844 732724	0.587	-0.53273 04591 540406
0.488	-0.71743 98731 289899	0.538	-0.61989 67188 203526	0.588	-0.53102 83310 835101
0.489	-0.71539 27895 072650	0.539	-0.61803 97080 731399	0.589	-0.52932 90953 305503
0.490	-0.71334 98878 774648	0.540	-0.61618 61394 238170	0.590	-0.52763 27420 823719
0.491	-0.71131 11511 876165	0.541	-0.61433 60001 356555	0.591	-0.52593 92615 760389
0.492	-0.70927 65624 898289	0.542	-0.61248 92775 424908	0.592	-0.52424 86440 981314
0.493	-0.70724 61049 394469	0.543	-0.61064 59590 482016	0.593	-0.52256 08799 844116
0.494	-0.70521 97617 942145	0.544	-0.60880 60321 261944	0.594	-0.52087 59596 194921
0.495	-0.70319 75164 134468	0.545	-0.60696 94843 188930	0.595	-0.51919 38734 365073
0.496	-0.70117 93522 572096	0.546	-0.60513 63032 372320	0.596	-0.51751 46119 167873
0.497	-0.69916 52528 855083	0.547	-0.60330 64765 601558	0.597	-0.51583 81655 895350
0.498	-0.69715 52019 574841	0.548	-0.60147 99920 341215	0.598	-0.51416 45250 315053
0.499	-0.69514 91832 306184	0.549	-0.59965 68374 726064	0.599	-0.51249 36808 666877
0.500	-0.69314 71805 599453	0.550	-0.59783 70007 556204	0.600	-0.51082 56237 659907

TABLE VI.

Table   VI

NATURAL LOGARITHMS

x	$\ln x$	x	$\ln x$	x	$\ln x$
0.600	−0.51082 56237 659907	0.650	−0.43078 29160 924543	0.700	−0.35667 49439 387324
0.601	−0.50916 03444 469295	0.651	−0.42924 56367 735678	0.701	−0.35524 73919 475470
0.602	−0.50749 78336 733160	0.652	−0.42771 07170 554841	0.702	−0.35382 18749 563259
0.603	−0.50583 80822 549516	0.653	−0.42617 81497 057060	0.703	−0.35239 83871 714721
0.604	−0.50418 10810 473221	0.654	−0.42464 79275 249384	0.704	−0.35097 69228 240947
0.605	−0.50252 68209 512956	0.655	−0.42312 00433 468851	0.705	−0.34955 74761 698684
0.606	−0.50087 52929 128226	0.656	−0.42159 44900 380480	0.706	−0.34814 00414 888950
0.607	−0.49922 64879 226388	0.657	−0.42007 12604 975265	0.707	−0.34672 46130 855643
0.608	−0.49758 03970 159700	0.658	−0.41855 03476 568199	0.708	−0.34531 11852 884173
0.609	−0.49593 70112 722400	0.659	−0.41703 17444 796298	0.709	−0.34389 97524 500096
0.610	−0.49429 63218 147801	0.660	−0.41551 54439 616658	0.710	−0.34249 03089 467759
0.611	−0.49265 83198 105417	0.661	−0.41400 14391 304508	0.711	−0.34108 28491 788962
0.612	−0.49102 29964 698110	0.662	−0.41248 97230 451288	0.712	−0.33967 73675 701613
0.613	−0.48939 03430 459257	0.663	−0.41098 02887 962745	0.713	−0.33827 38585 678411
0.614	−0.48776 03508 349946	0.664	−0.40947 31295 057032	0.714	−0.33687 23166 425527
0.615	−0.48613 30111 756192	0.665	−0.40796 82383 262829	0.715	−0.33547 27362 881294
0.616	−0.48450 83154 486173	0.666	−0.40646 56084 417479	0.716	−0.33407 51120 214914
0.617	−0.48288 62550 767492	0.667	−0.40496 52330 665133	0.717	−0.33267 94383 825167
0.618	−0.48126 68215 244463	0.668	−0.40346 71054 454913	0.718	−0.33128 57099 339129
0.619	−0.47965 00062 975409	0.669	−0.40197 12188 539086	0.719	−0.32989 39212 610904
0.620	−0.47803 58009 429998	0.670	−0.40047 75665 971253	0.720	−0.32850 40669 720361
0.621	−0.47642 41970 486583	0.671	−0.39898 61420 104553	0.721	−0.32711 61416 971880
0.622	−0.47481 51862 429576	0.672	−0.39749 69384 589875	0.722	−0.32573 01400 893108
0.623	−0.47320 87601 946839	0.673	−0.39600 99493 374092	0.723	−0.32434 60568 233724
0.624	−0.47160 49106 127094	0.674	−0.39452 51680 698300	0.724	−0.32296 38865 964207
0.625	−0.47000 36292 457356	0.675	−0.39304 25881 096072	0.725	−0.32158 36241 274623
0.626	−0.46840 49078 820385	0.676	−0.39156 22029 391730	0.726	−0.32020 52641 573410
0.627	−0.46680 87383 492164	0.677	−0.39008 40060 698621	0.727	−0.31882 88014 486177
0.628	−0.46521 51125 139384	0.678	−0.38860 79910 417415	0.728	−0.31745 42307 854511
0.629	−0.46362 40222 816965	0.679	−0.38713 41514 234409	0.729	−0.31608 15469 734789
0.630	−0.46203 54595 965587	0.680	−0.38566 24808 119847	0.730	−0.31471 07448 397002
0.631	−0.46044 94164 409239	0.681	−0.38419 29728 326247	0.731	−0.31334 18192 323585
0.632	−0.45886 58848 352796	0.682	−0.38272 56211 386750	0.732	−0.31197 47650 208255
0.633	−0.45728 48568 379609	0.683	−0.38126 04194 113470	0.733	−0.31060 95770 954856
0.634	−0.45570 63245 449111	0.684	−0.37979 73613 595866	0.734	−0.30924 62503 676215
0.635	−0.45413 02800 894454	0.685	−0.37833 64407 199118	0.735	−0.30788 47797 693004
0.636	−0.45255 67156 420149	0.686	−0.37687 76512 562518	0.736	−0.30652 51602 532608
0.637	−0.45098 56234 099737	0.687	−0.37542 09867 597877	0.737	−0.30516 73867 928004
0.638	−0.44941 69956 373472	0.688	−0.37396 64410 487934	0.738	−0.30381 14543 816646
0.639	−0.44785 08246 046022	0.689	−0.37251 40079 684785	0.739	−0.30245 73580 339353
0.640	−0.44628 71026 284195	0.690	−0.37106 36813 908320	0.740	−0.30110 50927 839216
0.641	−0.44472 58220 614670	0.691	−0.36961 54552 144672	0.741	−0.29975 46536 860502
0.642	−0.44316 69752 921759	0.692	−0.36816 93233 644675	0.742	−0.29840 60358 147566
0.643	−0.44161 05547 445177	0.693	−0.36672 52797 922338	0.743	−0.29705 92342 643779
0.644	−0.44005 65528 777834	0.694	−0.36528 33184 753326	0.744	−0.29571 42441 490452
0.645	−0.43850 49621 863646	0.695	−0.36384 34334 173449	0.745	−0.29437 10606 025775
0.646	−0.43695 57751 995352	0.696	−0.36240 56186 477174	0.746	−0.29302 96787 783762
0.647	−0.43540 89844 812365	0.697	−0.36096 98682 216132	0.747	−0.29169 00938 493197
0.648	−0.43386 45826 298624	0.698	−0.35953 61762 197646	0.748	−0.29035 23010 076598
0.649	−0.43232 25622 780471	0.699	−0.35810 45367 483268	0.749	−0.28901 62954 649176
0.650	−0.43078 29160 924543	0.700	−0.35667 49439 387324	0.750	−0.28768 20724 517809

TABLE VI.

Table VI   *223*

NATURAL LOGARITHMS

x	ln x	x	ln x	x	ln x
0.750	-0.28768 20724 517809	0.800	-0.22314 35513 142098	0.850	-0.16251 89294 977749
0.751	-0.28634 96272 180023	0.801	-0.22189 43319 137778	0.851	-0.16134 31504 087629
0.752	-0.28501 89550 322973	0.802	-0.22064 66711 156226	0.852	-0.16016 87521 528213
0.753	-0.28369 00511 822435	0.803	-0.21940 05650 353754	0.853	-0.15899 57314 904579
0.754	-0.28236 29109 741810	0.804	-0.21815 60098 031707	0.854	-0.15782 40851 935672
0.755	-0.28103 75297 331123	0.805	-0.21691 30015 635737	0.855	-0.15665 38100 453768
0.756	-0.27971 39028 026041	0.806	-0.21567 15364 755088	0.856	-0.15548 49028 403950
0.757	-0.27839 20255 446883	0.807	-0.21443 16107 121883	0.857	-0.15431 73603 843573
0.758	-0.27707 18933 397654	0.808	-0.21319 32204 610417	0.858	-0.15315 11794 941748
0.759	-0.27575 35015 865071	0.809	-0.21195 63619 236454	0.859	-0.15198 63569 978817
0.760	-0.27443 68457 017603	0.810	-0.21072 10313 156526	0.860	-0.15082 28897 345836
0.761	-0.27312 19211 204512	0.811	-0.20948 72248 667241	0.861	-0.14966 07745 544063
0.762	-0.27180 87232 954908	0.812	-0.20825 49388 204591	0.862	-0.14850 00083 184440
0.763	-0.27049 72476 976800	0.813	-0.20702 41694 343265	0.863	-0.14734 05878 987091
0.764	-0.26918 74898 156166	0.814	-0.20579 49129 795968	0.864	-0.14618 25101 780814
0.765	-0.26787 94451 556012	0.815	-0.20456 71657 412743	0.865	-0.14502 57720 502577
0.766	-0.26657 31092 415458	0.816	-0.20334 09240 180300	0.866	-0.14387 03704 197019
0.767	-0.26526 84776 148809	0.817	-0.20211 61841 221342	0.867	-0.14271 63022 015952
0.768	-0.26396 55458 344649	0.818	-0.20089 29423 793900	0.868	-0.14156 35643 217869
0.769	-0.26266 43094 764931	0.819	-0.19967 11951 290676	0.869	-0.14041 21537 167450
0.770	-0.26136 47641 344075	0.820	-0.19845 09387 238383	0.870	-0.13926 20673 335076
0.771	-0.26006 69054 188076	0.821	-0.19723 21695 297088	0.871	-0.13811 33021 296343
0.772	-0.25877 07289 573609	0.822	-0.19601 48839 259571	0.872	-0.13696 58550 731574
0.773	-0.25747 62303 947151	0.823	-0.19479 90783 050672	0.873	-0.13581 97231 425348
0.774	-0.25618 34053 924099	0.824	-0.19358 47490 726654	0.874	-0.13467 49033 266016
0.775	-0.25489 22496 287901	0.825	-0.19237 18926 474561	0.875	-0.13353 13926 245226
0.776	-0.25360 27587 989183	0.826	-0.19116 05054 611590	0.876	-0.13238 91880 457456
0.777	-0.25231 49286 144896	0.827	-0.18995 05839 584457	0.877	-0.13124 82866 099540
0.778	-0.25102 87548 037454	0.828	-0.18874 21245 968774	0.878	-0.13010 86853 470204
0.779	-0.24974 42331 113888	0.829	-0.18753 51238 468421	0.879	-0.12897 03812 969601
0.780	-0.24846 13592 984996	0.830	-0.18632 95781 914934	0.880	-0.12783 33715 098849
0.781	-0.24718 01291 424511	0.831	-0.18512 54841 266889	0.881	-0.12669 76530 459575
0.782	-0.24590 05384 368260	0.832	-0.18392 28381 609285	0.882	-0.12556 32229 753457
0.783	-0.24462 25829 913340	0.833	-0.18272 16368 152944	0.883	-0.12443 00783 781770
0.784	-0.24334 62586 317292	0.834	-0.18152 18766 233903	0.884	-0.12329 82163 444936
0.785	-0.24207 15611 997286	0.835	-0.18032 35541 312816	0.885	-0.12216 76339 742075
0.786	-0.24079 84865 529305	0.836	-0.17912 66658 974354	0.886	-0.12103 83283 770561
0.787	-0.23952 70305 647338	0.837	-0.17793 12084 926617	0.887	-0.11991 02966 725576
0.788	-0.23825 71891 242579	0.838	-0.17673 71785 000540	0.888	-0.11878 35359 899670
0.789	-0.23698 89581 362628	0.839	-0.17554 45725 149309	0.889	-0.11765 80434 682325
0.790	-0.23572 23335 210699	0.840	-0.17435 33871 447778	0.890	-0.11653 38162 559515
0.791	-0.23445 73112 144832	0.841	-0.17316 36190 091890	0.891	-0.11541 08515 113277
0.792	-0.23319 38871 677112	0.842	-0.17197 52647 398103	0.892	-0.11428 91464 021277
0.793	-0.23193 20573 472891	0.843	-0.17078 83209 802816	0.893	-0.11316 86981 056380
0.794	-0.23067 18177 350013	0.844	-0.16960 27843 861799	0.894	-0.11204 95038 086229
0.795	-0.22941 31643 278052	0.845	-0.16841 86516 249632	0.895	-0.11093 15607 072817
0.796	-0.22815 60931 377540	0.846	-0.16723 59193 759138	0.896	-0.10981 48660 072066
0.797	-0.22690 06001 919220	0.847	-0.16605 45843 300827	0.897	-0.10869 94169 233409
0.798	-0.22564 66815 323283	0.848	-0.16487 46431 902340	0.898	-0.10758 52106 799374
0.799	-0.22439 43332 158624	0.849	-0.16369 60926 707897	0.899	-0.10647 22445 105168
0.800	-0.22314 35513 142098	0.850	-0.16251 89294 977749	0.900	-0.10536 05156 578263

TABLE VI.

Table VI

NATURAL LOGARITHMS

x	$\ln x$	x	$\ln x$	x	$\ln x$
0.900	−0.10536 05156 578263	0.950	−0.05129 32943 875505	1.000	0.00000 00000 000000
0.901	−0.10425 00213 737991	0.951	−0.05024 12164 367467	1.001	0.00099 95003 330835
0.902	−0.10314 07589 195134	0.952	−0.04919 02441 907717	1.002	0.00199 80026 626731
0.903	−0.10203 27255 651516	0.953	−0.04814 03753 279349	1.003	0.00299 55089 797985
0.904	−0.10092 59185 899606	0.954	−0.04709 16075 338505	1.004	0.00399 20212 695375
0.905	−0.09982 03352 822109	0.955	−0.04604 39385 014068	1.005	0.00498 75415 110391
0.906	−0.09871 59729 391577	0.956	−0.04499 73659 307358	1.006	0.00598 20716 775475
0.907	−0.09761 28288 670004	0.957	−0.04395 18875 291828	1.007	0.00697 56137 364252
0.908	−0.09651 09003 808438	0.958	−0.04290 75010 112765	1.008	0.00796 81696 491769
0.909	−0.09541 01848 046582	0.959	−0.04186 42040 986988	1.009	0.00895 97413 714719
0.910	−0.09431 06794 712413	0.960	−0.04082 19945 202551	1.010	0.00995 03308 531681
0.911	−0.09321 23817 221787	0.961	−0.03978 08700 118446	1.011	0.01093 99400 383344
0.912	−0.09211 52889 078057	0.962	−0.03874 08283 164306	1.012	0.01192 85708 652738
0.913	−0.09101 93983 871686	0.963	−0.03770 18671 840115	1.013	0.01291 62252 665463
0.914	−0.08992 47075 279870	0.964	−0.03666 39843 715914	1.014	0.01390 29051 689914
0.915	−0.08883 12137 066157	0.965	−0.03562 71776 431511	1.015	0.01488 86124 937507
0.916	−0.08773 89143 080068	0.966	−0.03459 14447 696191	1.016	0.01587 33491 562901
0.917	−0.08664 78067 256722	0.967	−0.03355 67835 288427	1.017	0.01685 71170 664229
0.918	−0.08555 78883 616466	0.968	−0.03252 31917 055600	1.018	0.01783 99181 283310
0.919	−0.08446 91566 264500	0.969	−0.03149 06670 913708	1.019	0.01882 17542 405878
0.920	−0.08338 16089 390511	0.970	−0.03045 92074 847085	1.020	0.01980 26272 961797
0.921	−0.08229 52427 268302	0.971	−0.02942 88106 908121	1.021	0.02078 25391 825285
0.922	−0.08121 00554 255432	0.972	−0.02839 94745 216980	1.022	0.02176 14917 815127
0.923	−0.08012 60444 792849	0.973	−0.02737 11967 961320	1.023	0.02273 94869 694894
0.924	−0.07904 32073 404529	0.974	−0.02634 39753 396020	1.024	0.02371 65266 173160
0.925	−0.07796 15414 697119	0.975	−0.02531 78079 842899	1.025	0.02469 26125 903715
0.926	−0.07688 10443 359577	0.976	−0.02429 26925 690446	1.026	0.02566 77467 485778
0.927	−0.07580 17134 162819	0.977	−0.02326 86269 393543	1.027	0.02664 19309 464212
0.928	−0.07472 35461 959365	0.978	−0.02224 56089 473197	1.028	0.02761 51670 329734
0.929	−0.07364 65401 682985	0.979	−0.02122 36364 516267	1.029	0.02858 74568 519126
0.930	−0.07257 06928 348354	0.980	−0.02020 27073 175194	1.030	0.02955 88022 415444
0.931	−0.07149 60017 050700	0.981	−0.01918 28194 167740	1.031	0.03052 92050 348229
0.932	−0.07042 24642 965459	0.982	−0.01816 39706 276712	1.032	0.03149 86670 593710
0.933	−0.06935 00781 347932	0.983	−0.01714 61588 349705	1.033	0.03246 71901 375015
0.934	−0.06827 88407 532944	0.984	−0.01612 93819 298836	1.034	0.03343 47760 862374
0.935	−0.06720 87496 934501	0.985	−0.01511 36378 100482	1.035	0.03440 14267 173324
0.936	−0.06613 98025 045450	0.986	−0.01409 89243 795016	1.036	0.03536 71438 372913
0.937	−0.06507 19967 437149	0.987	−0.01308 52395 486555	1.037	0.03633 19292 473903
0.938	−0.06400 53299 759124	0.988	−0.01207 25812 342692	1.038	0.03729 57847 436969
0.939	−0.06293 97997 738741	0.989	−0.01106 09473 594249	1.039	0.03825 87121 170903
0.940	−0.06187 54037 180875	0.990	−0.01005 03358 535014	1.040	0.03922 07131 532813
0.941	−0.06081 21393 967574	0.991	−0.00904 07446 521491	1.041	0.04018 17896 328318
0.942	−0.05975 00044 057740	0.992	−0.00803 21716 972643	1.042	0.04114 19433 311752
0.943	−0.05868 89963 486796	0.993	−0.00702 46149 369645	1.043	0.04210 11760 186354
0.944	−0.05762 91128 366364	0.994	−0.00601 80723 255630	1.044	0.04305 94894 604470
0.945	−0.05657 03514 883943	0.995	−0.00501 25418 235443	1.045	0.04401 68854 167743
0.946	−0.05551 27099 302588	0.996	−0.00400 80213 975388	1.046	0.04497 33656 427312
0.947	−0.05445 61857 960588	0.997	−0.00300 45090 202987	1.047	0.04592 89318 883998
0.948	−0.05340 07767 271152	0.998	−0.00200 20026 706731	1.048	0.04688 35858 988504
0.949	−0.05234 64803 722092	0.999	−0.00100 05003 335835	1.049	0.04783 73294 141601
0.950	−0.05129 32943 875505	1.000	0.00000 00000 000000	1.050	0.04879 01641 694320

TABLE VI.

x	$\ln x$	x	$\ln x$	x	$\ln x$
1.050	0.04879 01641 694320	1.100	0.09531 01798 043249	1.150	0.13976 19423 751587
1.051	0.04974 20918 948141	1.101	0.09621 88577 405429	1.151	0.14063 11297 397456
1.052	0.05069 31143 155181	1.102	0.09712 67107 307227	1.152	0.14149 95622 736995
1.053	0.05164 32331 518384	1.103	0.09803 37402 713654	1.153	0.14236 72412 869220
1.054	0.05259 24501 191706	1.104	0.09893 99478 549036	1.154	0.14323 41680 859078
1.055	0.05354 07669 280298	1.105	0.09984 53349 697161	1.155	0.14410 03439 737569
1.056	0.05448 81852 840697	1.106	0.10074 99031 001431	1.156	0.14496 57702 501857
1.057	0.05543 47068 881006	1.107	0.10165 36537 264998	1.157	0.14583 04482 115395
1.058	0.05638 03334 361076	1.108	0.10255 65883 250921	1.158	0.14669 43791 508035
1.059	0.05732 50666 192694	1.109	0.10345 87083 682300	1.159	0.14755 75643 576147
1.060	0.05826 89081 239758	1.110	0.10436 00153 242428	1.160	0.14842 00051 182733
1.061	0.05921 18596 318461	1.111	0.10526 05106 574929	1.161	0.14928 17027 157544
1.062	0.06015 39228 197471	1.112	0.10616 01958 283906	1.162	0.15014 26584 297195
1.063	0.06109 50993 598109	1.113	0.10705 90722 934078	1.163	0.15100 28735 365274
1.064	0.06203 53909 194526	1.114	0.10795 71415 050923	1.164	0.15186 23493 092461
1.065	0.06297 47991 613884	1.115	0.10885 44049 120821	1.165	0.15272 10870 176639
1.066	0.06391 33257 436528	1.116	0.10975 08639 591192	1.166	0.15357 90879 283006
1.067	0.06485 09723 196163	1.117	0.11064 65200 870637	1.167	0.15443 63533 044189
1.068	0.06578 77405 380031	1.118	0.11154 13747 329074	1.168	0.15529 28844 060353
1.069	0.06672 36320 429082	1.119	0.11243 54293 297882	1.169	0.15614 86824 899314
1.070	0.06765 86484 738148	1.120	0.11332 86853 070032	1.170	0.15700 37488 096648
1.071	0.06859 27914 656117	1.121	0.11422 11440 900229	1.171	0.15785 80846 155803
1.072	0.06952 60626 486102	1.122	0.11511 28071 005046	1.172	0.15871 16911 548209
1.073	0.07045 84636 485614	1.123	0.11600 36757 563061	1.173	0.15956 45696 713384
1.074	0.07138 99960 866729	1.124	0.11689 37514 714993	1.174	0.16041 67214 059047
1.075	0.07232 06615 796261	1.125	0.11778 30356 563835	1.175	0.16126 81475 961223
1.076	0.07325 04617 395927	1.126	0.11867 15297 174986	1.176	0.16211 88494 764352
1.077	0.07417 93981 742515	1.127	0.11955 92350 576392	1.177	0.16296 88282 781397
1.078	0.07510 74724 868054	1.128	0.12044 61530 758672	1.178	0.16381 80852 293950
1.079	0.07603 46862 759976	1.129	0.12133 22851 675250	1.179	0.16466 66215 552339
1.080	0.07696 10411 361283	1.130	0.12221 76327 242492	1.180	0.16551 44384 775734
1.081	0.07788 65386 570712	1.131	0.12310 21971 339834	1.181	0.16636 15372 152253
1.082	0.07881 11804 242898	1.132	0.12398 59797 809912	1.182	0.16720 79189 839065
1.083	0.07973 49680 188536	1.133	0.12486 89820 458693	1.183	0.16805 35849 962497
1.084	0.08065 79030 174545	1.134	0.12575 12053 055603	1.184	0.16889 85364 618139
1.085	0.08157 99869 924229	1.135	0.12663 26509 333660	1.185	0.16974 27745 870945
1.086	0.08250 12215 117437	1.136	0.12751 33202 989596	1.186	0.17058 63005 755337
1.087	0.08342 16081 390724	1.137	0.12839 32147 683990	1.187	0.17142 91156 275310
1.088	0.08434 11484 337509	1.138	0.12927 23357 041392	1.188	0.17227 12209 404532
1.089	0.08525 98439 508234	1.139	0.13015 06844 650451	1.189	0.17311 26177 086448
1.090	0.08617 76962 410523	1.140	0.13102 82624 064041	1.190	0.17395 33071 234380
1.091	0.08709 47068 509338	1.141	0.13190 50708 799386	1.191	0.17479 32903 731631
1.092	0.08801 08773 227133	1.142	0.13278 11112 338185	1.192	0.17563 25686 431580
1.093	0.08892 62091 944015	1.143	0.13365 63848 126736	1.193	0.17647 11431 157791
1.094	0.08984 07039 997895	1.144	0.13453 08929 576062	1.194	0.17730 90149 704103
1.095	0.09075 43632 684641	1.145	0.13540 46370 062030	1.195	0.17814 61853 834740
1.096	0.09166 71885 258238	1.146	0.13627 76182 925478	1.196	0.17898 26555 284400
1.097	0.09257 91812 930932	1.147	0.13714 98381 472336	1.197	0.17981 84265 758361
1.098	0.09349 03430 873389	1.148	0.13802 12978 973747	1.198	0.18065 34996 932576
1.099	0.09440 06754 214843	1.149	0.13889 19988 666186	1.199	0.18148 78760 453772
1.100	0.09531 01798 043249	1.150	0.13976 19423 751587	1.200	0.18232 15567 939546

TABLE VI.

Table   VI

NATURAL LOGARITHMS

x	$\ln x$	x	$\ln x$	x	$\ln x$
1.200	0.18232 15567 939546	1.250	0.22314 35513 142098	1.300	0.26236 42644 674911
1.201	0.18315 45430 978465	1.251	0.22394 32314 847741	1.301	0.26313 31995 303682
1.202	0.18398 68361 130158	1.252	0.22474 22726 779068	1.302	0.26390 15437 863775
1.203	0.18481 84369 925418	1.253	0.22554 06759 139312	1.303	0.26466 92981 427081
1.204	0.18564 93468 866293	1.254	0.22633 84422 107290	1.304	0.26543 64635 044612
1.205	0.18647 95669 426183	1.255	0.22713 55725 837472	1.305	0.26620 30407 746567
1.206	0.18730 90983 049937	1.256	0.22793 20680 460069	1.306	0.26696 90308 542393
1.207	0.18813 79421 153944	1.257	0.22872 79296 081104	1.307	0.26773 44346 420849
1.208	0.18896 60995 126232	1.258	0.22952 31582 782488	1.308	0.26849 92530 350070
1.209	0.18979 35716 326556	1.259	0.23031 77550 622101	1.309	0.26926 34869 277629
1.210	0.19062 03596 086497	1.260	0.23111 17209 633866	1.310	0.27002 71372 130602
1.211	0.19144 64645 709552	1.261	0.23190 50569 827825	1.311	0.27079 02047 815628
1.212	0.19227 18876 471227	1.262	0.23269 77641 190214	1.312	0.27155 26905 218973
1.213	0.19309 66299 619131	1.263	0.23348 98433 683541	1.313	0.27231 45953 206591
1.214	0.19392 06926 373065	1.264	0.23428 12957 246657	1.314	0.27307 59200 624188
1.215	0.19474 40767 925118	1.265	0.23507 21221 794836	1.315	0.27383 66656 297279
1.216	0.19556 67835 439753	1.266	0.23586 23237 219844	1.316	0.27459 68329 031255
1.217	0.19638 88140 053901	1.267	0.23665 19013 390020	1.317	0.27535 64227 611440
1.218	0.19721 01692 877053	1.268	0.23744 08560 150342	1.318	0.27611 54360 803155
1.219	0.19803 08504 991345	1.269	0.23822 91887 322506	1.319	0.27687 38737 351775
1.220	0.19885 08587 451652	1.270	0.23901 69004 704999	1.320	0.27763 17365 982795
1.221	0.19967 01951 285676	1.271	0.23980 39922 073170	1.321	0.27838 90255 401883
1.222	0.20048 88607 494036	1.272	0.24059 04649 179304	1.322	0.27914 57414 294945
1.223	0.20130 68567 050353	1.273	0.24137 63195 752695	1.323	0.27990 18851 328186
1.224	0.20212 41840 901343	1.274	0.24216 15571 499716	1.324	0.28065 74575 148165
1.225	0.20294 08439 966903	1.275	0.24294 61786 103895	1.325	0.28141 24594 381855
1.226	0.20375 68375 140197	1.276	0.24373 01849 225981	1.326	0.28216 68917 636708
1.227	0.20457 21657 287744	1.277	0.24451 35770 504022	1.327	0.28292 07553 500705
1.228	0.20538 68297 249507	1.278	0.24529 63559 553431	1.328	0.28367 40510 542421
1.229	0.20620 08305 838978	1.279	0.24607 85225 967056	1.329	0.28442 67797 311083
1.230	0.20701 41693 843261	1.280	0.24686 00779 315258	1.330	0.28517 89422 336624
1.231	0.20782 68472 023165	1.281	0.24764 10229 145972	1.331	0.28593 05394 129746
1.232	0.20863 88651 113280	1.282	0.24842 13584 984783	1.332	0.28668 15721 181974
1.233	0.20945 02241 822072	1.283	0.24920 10856 334994	1.333	0.28743 20411 965716
1.234	0.21026 09254 831961	1.284	0.24998 02052 677694	1.334	0.28818 19474 934320
1.235	0.21107 09700 799405	1.285	0.25075 87183 471831	1.335	0.28893 12918 522129
1.236	0.21188 03590 354990	1.286	0.25153 66258 154276	1.336	0.28968 00751 144540
1.237	0.21268 90934 103508	1.287	0.25231 39286 139896	1.337	0.29042 82981 198061
1.238	0.21349 71742 624044	1.288	0.25309 06276 821619	1.338	0.29117 59617 060367
1.239	0.21430 46026 470054	1.289	0.25386 67239 570503	1.339	0.29192 30667 090355
1.240	0.21511 13796 169455	1.290	0.25464 22183 735807	1.340	0.29266 96139 628200
1.241	0.21591 75062 224702	1.291	0.25541 71118 645054	1.341	0.29341 56042 995415
1.242	0.21672 29835 112870	1.292	0.25619 14053 604101	1.342	0.29416 10385 494901
1.243	0.21752 78125 285741	1.293	0.25696 50997 897204	1.343	0.29490 59175 411005
1.244	0.21833 19943 169877	1.294	0.25773 81960 787088	1.344	0.29565 02421 009578
1.245	0.21913 55299 166709	1.295	0.25851 06951 515011	1.345	0.29639 40130 538024
1.246	0.21993 84203 652614	1.296	0.25928 25979 300830	1.346	0.29713 72312 225361
1.247	0.22074 06666 978994	1.297	0.26005 39053 343068	1.347	0.29787 98974 282269
1.248	0.22154 22699 472359	1.298	0.26082 46182 818983	1.348	0.29862 20124 901153
1.249	0.22234 32311 434406	1.299	0.26159 47376 884625	1.349	0.29936 35772 256188
1.250	0.22314 35513 142098	1.300	0.26236 42644 674911	1.350	0.30010 45924 503381

TABLE VI.

Table VI 227

NATURAL LOGARITHMS

x	$\ln x$	x	$\ln x$	x	$\ln x$
1.350	0.30010 45924 503381	1.400	0.33647 22366 212129	1.450	0.37156 35564 324830
1.351	0.30084 50589 780618	1.401	0.33718 62673 548700	1.451	0.37225 29739 020508
1.352	0.30158 49776 207723	1.402	0.33789 97886 123983	1.452	0.37294 19164 026043
1.353	0.30232 43491 886510	1.403	0.33861 28011 203239	1.453	0.37363 03845 881459
1.354	0.30306 31744 900833	1.404	0.33932 53056 036194	1.454	0.37431 83791 113276
1.355	0.30380 14543 316642	1.405	0.34003 73027 857091	1.455	0.37500 59006 234558
1.356	0.30453 91895 182038	1.406	0.34074 87933 884732	1.456	0.37569 29497 744942
1.357	0.30527 63808 527321	1.407	0.34145 97781 322520	1.457	0.37637 95272 130678
1.358	0.30601 30291 365044	1.408	0.34217 02577 358507	1.458	0.37706 56335 864664
1.359	0.30674 91351 690067	1.409	0.34288 02329 165432	1.459	0.37775 12695 406486
1.360	0.30748 46997 479606	1.410	0.34358 97043 900769	1.460	0.37843 64357 202451
1.361	0.30821 97236 693290	1.411	0.34429 86728 706770	1.461	0.37912 11327 685624
1.362	0.30895 42077 273206	1.412	0.34500 71390 710503	1.462	0.37980 53613 275868
1.363	0.30968 81527 143956	1.413	0.34571 51037 023904	1.463	0.38048 91220 379873
1.364	0.31042 15594 212704	1.414	0.34642 25674 743810	1.464	0.38117 24155 391198
1.365	0.31115 44286 369231	1.415	0.34712 95310 952009	1.465	0.38185 52424 690306
1.366	0.31188 67611 485983	1.416	0.34783 59952 715280	1.466	0.38253 76034 644597
1.367	0.31261 85577 418125	1.417	0.34854 19607 085434	1.467	0.38321 94991 608447
1.368	0.31334 98192 003587	1.418	0.34924 74281 099358	1.468	0.38390 09301 923238
1.369	0.31408 05463 063118	1.419	0.34995 23981 779056	1.469	0.38458 18971 917403
1.370	0.31481 07398 400335	1.420	0.35065 68716 131694	1.470	0.38526 24007 906449
1.371	0.31554 04005 801773	1.421	0.35136 08491 149636	1.471	0.38594 24416 193005
1.372	0.31626 95293 036935	1.422	0.35206 43313 810491	1.472	0.38662 20203 066845
1.373	0.31699 81267 858340	1.423	0.35276 73191 077153	1.473	0.38730 11374 804932
1.374	0.31772 61938 001576	1.424	0.35346 98129 897840	1.474	0.38797 97937 671449
1.375	0.31845 37311 185346	1.425	0.35417 18137 206138	1.475	0.38865 79897 917831
1.376	0.31918 07395 111519	1.426	0.35487 33219 921042	1.476	0.38933 57261 782808
1.377	0.31990 72197 465178	1.427	0.35557 43384 946994	1.477	0.39001 30035 492427
1.378	0.32063 31725 914668	1.428	0.35627 48639 173926	1.478	0.39068 98225 260100
1.379	0.32135 85988 111648	1.429	0.35697 48989 477304	1.479	0.39136 61837 286627
1.380	0.32208 34991 691133	1.430	0.35767 44442 718159	1.480	0.39204 20877 760237
1.381	0.32280 78744 271551	1.431	0.35837 35005 743139	1.481	0.39271 75352 856617
1.382	0.32353 17253 454782	1.432	0.35907 20685 384539	1.482	0.39339 25268 738951
1.383	0.32425 50526 826212	1.433	0.35977 01488 460348	1.483	0.39406 70631 557950
1.384	0.32497 78571 954778	1.434	0.36046 77421 774286	1.484	0.39474 11447 451887
1.385	0.32570 01396 393018	1.435	0.36116 48492 115844	1.485	0.39541 47722 546629
1.386	0.32642 19007 677115	1.436	0.36186 14706 260324	1.486	0.39608 79462 955674
1.387	0.32714 31413 326945	1.437	0.36255 76070 968879	1.487	0.39676 06674 780180
1.388	0.32786 38620 846128	1.438	0.36325 32592 988549	1.488	0.39743 29364 109001
1.389	0.32858 40637 722067	1.439	0.36394 84279 052308	1.489	0.39810 47537 018719
1.390	0.32930 37471 426004	1.440	0.36464 31135 879093	1.490	0.39877 61199 573678
1.391	0.33002 29129 413059	1.441	0.36533 73170 173850	1.491	0.39944 70357 826014
1.392	0.33074 15619 122279	1.442	0.36603 10388 627573	1.492	0.40011 75017 815691
1.393	0.33145 96947 976686	1.443	0.36672 42797 917338	1.493	0.40078 75185 570533
1.394	0.33217 73123 383321	1.444	0.36741 70404 706345	1.494	0.40145 70867 106256
1.395	0.33289 44152 733290	1.445	0.36810 93215 643955	1.495	0.40212 62068 426497
1.396	0.33361 10043 401807	1.446	0.36880 11237 365729	1.496	0.40279 48795 522855
1.397	0.33432 70802 748248	1.447	0.36949 24476 493468	1.497	0.40346 31054 374913
1.398	0.33504 26438 116185	1.448	0.37018 32939 635246	1.498	0.40413 08850 950277
1.399	0.33575 76956 833441	1.449	0.37087 36633 385453	1.499	0.40479 82191 204607
1.400	0.33647 22366 212129	1.450	0.37156 35564 324830	1.500	0.40546 51081 081644

TABLE VI.

NATURAL LOGARITHMS

Table VI

x	ln x	x	ln x	x	ln x
1.500	0.40546 51081 081644	1.550	0.43825 49309 311553	1.600	0.47000 36292 457356
1.501	0.40613 15526 513249	1.551	0.43889 98841 944018	1.601	0.47062 84340 145776
1.502	0.40679 75533 419430	1.552	0.43954 44217 610270	1.602	0.47125 28486 461675
1.503	0.40746 31107 708374	1.553	0.44018 85441 665500	1.603	0.47187 68736 274159
1.504	0.40812 82255 276481	1.554	0.44083 22519 454557	1.604	0.47250 05094 443228
1.505	0.40879 28982 008391	1.555	0.44147 55456 311975	1.605	0.47312 37565 819792
1.506	0.40945 71293 777018	1.556	0.44211 84257 561999	1.606	0.47374 66155 245699
1.507	0.41012 09196 443584	1.557	0.44276 08928 518613	1.607	0.47436 90867 553755
1.508	0.41078 42695 857643	1.558	0.44340 29474 485565	1.608	0.47499 11707 567746
1.509	0.41144 71797 857118	1.559	0.44404 45900 756395	1.609	0.47561 28680 102462
1.510	0.41210 96508 268330	1.560	0.44468 58212 614457	1.610	0.47623 41789 963716
1.511	0.41277 16832 906025	1.561	0.44532 66415 332950	1.611	0.47685 51041 948373
1.512	0.41343 32777 573413	1.562	0.44596 70514 174942	1.612	0.47747 56440 844365
1.513	0.41409 44348 062189	1.563	0.44660 70514 393396	1.613	0.47809 57991 430718
1.514	0.41475 51550 152570	1.564	0.44724 66421 231193	1.614	0.47871 55698 477571
1.515	0.41541 54389 613325	1.565	0.44788 58239 921165	1.615	0.47933 49566 746199
1.516	0.41607 52872 201799	1.566	0.44852 45975 686114	1.616	0.47995 39600 989036
1.517	0.41673 47003 663952	1.567	0.44916 29633 738838	1.617	0.48057 25805 949698
1.518	0.41739 36789 734382	1.568	0.44980 09219 282161	1.618	0.48119 08186 362999
1.519	0.41805 22236 136358	1.569	0.45043 84737 508955	1.619	0.48180 86746 954981
1.520	0.41871 03348 581850	1.570	0.45107 56193 602167	1.620	0.48242 61492 442927
1.521	0.41936 80132 771558	1.571	0.45171 23592 734841	1.621	0.48304 32427 535391
1.522	0.42002 52594 394941	1.572	0.45234 86940 070148	1.622	0.48365 99556 932212
1.523	0.42068 20739 130248	1.573	0.45298 46240 761408	1.623	0.48427 62885 324542
1.524	0.42133 84572 644545	1.574	0.45362 01499 952115	1.624	0.48489 22417 394862
1.525	0.42199 44100 593749	1.575	0.45425 52722 775964	1.625	0.48550 78157 817008
1.526	0.42264 99328 622653	1.576	0.45488 99914 356874	1.626	0.48612 30111 256188
1.527	0.42330 50262 364954	1.577	0.45552 43079 809013	1.627	0.48673 78282 369007
1.528	0.42395 96907 443287	1.578	0.45615 82224 236825	1.628	0.48735 22675 803486
1.529	0.42461 39269 469252	1.579	0.45679 17352 735050	1.629	0.48796 63296 199081
1.530	0.42526 77354 043441	1.580	0.45742 48470 388754	1.630	0.48858 00148 186710
1.531	0.42592 11166 755467	1.581	0.45805 75582 273350	1.631	0.48919 33236 388768
1.532	0.42657 40713 183996	1.582	0.45868 98693 454621	1.632	0.48980 62565 419153
1.533	0.42722 65998 896771	1.583	0.45932 17808 988751	1.633	0.49041 88139 883281
1.534	0.42787 87029 450644	1.584	0.45995 32933 922341	1.634	0.49103 09964 378111
1.535	0.42853 03810 391605	1.585	0.46058 44073 292439	1.635	0.49164 28043 492167
1.536	0.42918 16347 254804	1.586	0.46121 51232 126562	1.636	0.49225 42381 805553
1.537	0.42983 24645 564588	1.587	0.46184 54415 442720	1.637	0.49286 52983 889979
1.538	0.43048 28710 834522	1.588	0.46247 53628 249440	1.638	0.49347 59854 308777
1.539	0.43113 28548 567422	1.589	0.46310 48875 545789	1.639	0.49408 62997 616926
1.540	0.43178 24164 255378	1.590	0.46373 40162 321402	1.640	0.49469 62418 361071
1.541	0.43243 15563 379787	1.591	0.46436 27493 556498	1.641	0.49530 58121 079538
1.542	0.43308 02751 411377	1.592	0.46499 10874 221913	1.642	0.49591 50110 302365
1.543	0.43372 85733 810238	1.593	0.46561 90309 279115	1.643	0.49652 38390 551310
1.544	0.43437 64516 025844	1.594	0.46624 65803 680233	1.644	0.49713 22966 339882
1.545	0.43502 39103 497088	1.595	0.46687 37362 368079	1.645	0.49774 03842 173352
1.546	0.43567 09501 652302	1.596	0.46750 04990 276170	1.646	0.49834 81022 548781
1.547	0.43631 75715 909291	1.597	0.46812 68692 328754	1.647	0.49895 54511 955033
1.548	0.43696 37751 675354	1.598	0.46875 28473 440829	1.648	0.49956 24314 872800
1.549	0.43760 95614 347316	1.599	0.46937 84338 518172	1.649	0.50016 90435 774619
1.550	0.43825 49309 311553	1.600	0.47000 36292 457356	1.650	0.50077 52879 124892

TABLE VI.

x	$\ln x$	x	$\ln x$	x	$\ln x$
1.650	0.50077 52879 124892	1.700	0.53062 82510 621704	1.750	0.55961 57879 354227
1.651	0.50138 11649 379910	1.701	0.53121 63134 137247	1.751	0.56018 70533 037148
1.652	0.50198 66750 987863	1.702	0.53180 40301 511824	1.752	0.56075 79925 141997
1.653	0.50259 18188 388871	1.703	0.53239 14016 805512	1.753	0.56132 86059 390974
1.654	0.50319 65966 014996	1.704	0.53297 84284 071240	1.754	0.56189 88939 499913
1.655	0.50380 10088 290262	1.705	0.53356 51107 354801	1.755	0.56246 88569 178291
1.656	0.50440 50559 630679	1.706	0.53415 14490 694874	1.756	0.56303 84952 129249
1.657	0.50500 87384 444259	1.707	0.53473 74438 123036	1.757	0.56360 78092 049601
1.658	0.50561 20567 131032	1.708	0.53532 30953 663781	1.758	0.56417 67992 629853
1.659	0.50621 50112 083074	1.709	0.53590 84041 334538	1.759	0.56474 54657 554211
1.660	0.50681 76023 684519	1.710	0.53649 33705 145685	1.760	0.56531 38090 500604
1.661	0.50741 98306 311578	1.711	0.53707 79949 100564	1.761	0.56588 18295 140691
1.662	0.50802 16964 332564	1.712	0.53766 22777 195504	1.762	0.56644 95275 139878
1.663	0.50862 32002 107906	1.713	0.53824 62193 419829	1.763	0.56701 69034 157332
1.664	0.50922 43423 990168	1.714	0.53882 98201 755880	1.764	0.56758 39575 845996
1.665	0.50982 51234 324071	1.715	0.53941 30806 179032	1.765	0.56815 06903 852601
1.666	0.51042 55437 446509	1.716	0.53999 60010 657705	1.766	0.56871 71021 817683
1.667	0.51102 56037 686569	1.717	0.54057 85819 153385	1.767	0.56928 31933 375593
1.668	0.51162 53039 365550	1.718	0.54116 08235 620636	1.768	0.56984 89642 154517
1.669	0.51222 46446 796980	1.719	0.54174 27264 007122	1.769	0.57041 44151 776482
1.670	0.51282 36264 286637	1.720	0.54232 42908 253617	1.770	0.57097 95465 857378
1.671	0.51342 22496 132567	1.721	0.54290 55172 294024	1.771	0.57154 43588 006965
1.672	0.51402 05146 625099	1.722	0.54348 64060 055391	1.772	0.57210 88521 828892
1.673	0.51461 84220 046869	1.723	0.54406 69575 457926	1.773	0.57267 30270 920708
1.674	0.51521 59720 672836	1.724	0.54464 71722 415014	1.774	0.57323 68838 873877
1.675	0.51581 31652 770298	1.725	0.54522 70504 833231	1.775	0.57380 04229 273791
1.676	0.51641 00020 598913	1.726	0.54580 65926 612362	1.776	0.57436 36445 699783
1.677	0.51700 64828 410718	1.727	0.54638 57991 645415	1.777	0.57492 65491 725143
1.678	0.51760 26080 450144	1.728	0.54696 46703 818639	1.778	0.57548 91370 917128
1.679	0.51819 83780 954038	1.729	0.54754 32067 011534	1.779	0.57605 14086 836981
1.680	0.51879 37934 151676	1.730	0.54812 14085 096876	1.780	0.57661 33643 039938
1.681	0.51938 88544 264786	1.731	0.54869 92761 940722	1.781	0.57717 50043 075246
1.682	0.51998 35615 507563	1.732	0.54927 68101 402434	1.782	0.57773 63290 486176
1.683	0.52057 79152 086690	1.733	0.54985 40107 334690	1.783	0.57829 73388 810034
1.684	0.52117 19158 201350	1.734	0.55043 08783 583501	1.784	0.57885 80341 578176
1.685	0.52176 55638 043250	1.735	0.55100 74133 988225	1.785	0.57941 84152 316024
1.686	0.52235 88595 796637	1.736	0.55158 36162 381584	1.786	0.57997 84824 543073
1.687	0.52295 18035 638312	1.737	0.55215 94872 589679	1.787	0.58053 82361 772910
1.688	0.52354 43961 737654	1.738	0.55273 50268 432003	1.788	0.58109 76767 513224
1.689	0.52413 66378 256630	1.739	0.55331 02353 721460	1.789	0.58165 68045 265821
1.690	0.52472 85289 349821	1.740	0.55388 51132 264377	1.790	0.58221 56198 526636
1.691	0.52532 00699 164432	1.741	0.55445 96607 860520	1.791	0.58277 41230 785747
1.692	0.52591 12611 840315	1.742	0.55503 38784 303111	1.792	0.58333 23145 527387
1.693	0.52650 21031 509983	1.743	0.55560 77665 378839	1.793	0.58389 01946 229958
1.694	0.52709 25962 298627	1.744	0.55618 13254 867879	1.794	0.58444 77636 366044
1.695	0.52768 27408 324136	1.745	0.55675 45556 543905	1.795	0.58500 50219 402422
1.696	0.52827 25373 697113	1.746	0.55732 74574 174105	1.796	0.58556 19698 800079
1.697	0.52886 19862 520893	1.747	0.55790 00311 519195	1.797	0.58611 86078 014220
1.698	0.52945 10878 891556	1.748	0.55847 22772 333437	1.798	0.58667 49360 494285
1.699	0.53003 98426 897950	1.749	0.55904 41960 364650	1.799	0.58723 09549 683961
1.700	0.53062 82510 621704	1.750	0.55961 57879 354227	1.800	0.58778 66649 021190

TABLE VI.

Table   VI

NATURAL LOGARITHMS

x	$\ln x$	x	$\ln x$	x	$\ln x$
1.800	0.58778 66649 021190	1.850	0.61518 56390 902335	1.900	0.64185 38861 723948
1.801	0.58834 20661 938190	1.851	0.61572 60335 913605	1.901	0.64238 00635 062921
1.802	0.58889 71591 861462	1.852	0.61626 61362 239876	1.902	0.64290 59641 231986
1.803	0.58945 19442 211802	1.853	0.61680 59473 032227	1.903	0.64343 15883 140124
1.804	0.59000 64216 404319	1.854	0.61734 54671 436634	1.904	0.64395 69363 691736
1.805	0.59056 05917 848442	1.855	0.61788 46960 593985	1.905	0.64448 20085 786643
1.806	0.59111 44549 947937	1.856	0.61842 36343 640088	1.906	0.64500 68052 320104
1.807	0.59166 80116 100914	1.857	0.61896 22823 705687	1.907	0.64553 13266 182820
1.808	0.59222 12619 699848	1.858	0.61950 06403 916468	1.908	0.64605 55730 260948
1.809	0.59277 42064 131581	1.859	0.62003 87087 393070	1.909	0.64657 95447 436106
1.810	0.59332 68452 777344	1.860	0.62057 64877 251099	1.910	0.64710 32420 585385
1.811	0.59387 91789 012763	1.861	0.62111 39776 601137	1.911	0.64762 66652 581360
1.812	0.59443 12076 207876	1.862	0.62165 11788 548753	1.912	0.64814 98146 292095
1.813	0.59498 29317 727140	1.863	0.62218 80916 194514	1.913	0.64867 26904 581158
1.814	0.59553 43516 929449	1.864	0.62272 47162 633994	1.914	0.64919 52930 307625
1.815	0.59608 54677 168141	1.865	0.62326 10530 957789	1.915	0.64971 76226 326093
1.816	0.59663 62801 791016	1.866	0.62379 71024 251521	1.916	0.65023 96795 486688
1.817	0.59718 67894 140341	1.867	0.62433 28645 595856	1.917	0.65076 14640 635074
1.818	0.59773 69957 552871	1.868	0.62486 83398 066509	1.918	0.65128 29764 612465
1.819	0.59828 68995 359852	1.869	0.62540 35284 734258	1.919	0.65180 42170 255629
1.820	0.59883 65010 887040	1.870	0.62593 84308 664953	1.920	0.65232 51860 396902
1.821	0.59938 58007 454709	1.871	0.62647 30472 919526	1.921	0.65284 58837 864196
1.822	0.59993 47988 377666	1.872	0.62700 73780 554003	1.922	0.65336 63105 481007
1.823	0.60048 34956 965260	1.873	0.62754 14234 619515	1.923	0.65388 64666 066427
1.824	0.60103 18916 521396	1.874	0.62807 51838 162304	1.924	0.65440 63522 435147
1.825	0.60157 99870 344548	1.875	0.62860 86594 223741	1.925	0.65492 59677 397475
1.826	0.60212 77821 727767	1.876	0.62914 18505 840329	1.926	0.65544 53133 759338
1.827	0.60267 52773 958697	1.877	0.62967 47576 043718	1.927	0.65596 43894 322293
1.828	0.60322 24730 319583	1.878	0.63020 73807 860712	1.928	0.65648 31961 883539
1.829	0.60376 93694 087286	1.879	0.63073 97204 313283	1.929	0.65700 17339 235920
1.830	0.60431 59668 533296	1.880	0.63127 17768 418578	1.930	0.65752 00029 167942
1.831	0.60486 22656 923737	1.881	0.63180 35503 188933	1.931	0.65803 80034 463774
1.832	0.60540 82662 519385	1.882	0.63233 50411 631879	1.932	0.65855 57357 903263
1.833	0.60595 39688 575680	1.883	0.63286 62496 750154	1.933	0.65907 32002 261938
1.834	0.60649 93738 342731	1.884	0.63339 71761 541713	1.934	0.65959 03970 311026
1.835	0.60704 44815 065336	1.885	0.63392 78208 999741	1.935	0.66010 73264 817451
1.836	0.60758 92921 982987	1.886	0.63445 81842 112658	1.936	0.66062 39888 543853
1.837	0.60813 38062 329886	1.887	0.63498 82663 864132	1.937	0.66114 03844 248588
1.838	0.60867 80239 334953	1.888	0.63551 80677 233089	1.938	0.66165 65134 685745
1.839	0.60922 19456 221840	1.889	0.63604 75885 193725	1.939	0.66217 23762 605148
1.840	0.60976 55716 208943	1.890	0.63657 68290 715510	1.940	0.66268 79730 752368
1.841	0.61030 89022 509408	1.891	0.63710 57896 763204	1.941	0.66320 33041 868732
1.842	0.61085 19378 331151	1.892	0.63763 44706 296865	1.942	0.66371 83698 691332
1.843	0.61139 46786 876862	1.893	0.63816 28722 271858	1.943	0.66423 31703 953030
1.844	0.61193 71251 344021	1.894	0.63869 09947 638865	1.944	0.66474 77060 382473
1.845	0.61247 92774 924905	1.895	0.63921 88385 343897	1.945	0.66526 19770 704096
1.846	0.61302 11360 806604	1.896	0.63974 64038 328301	1.946	0.66577 59837 638133
1.847	0.61356 27012 171029	1.897	0.64027 36909 528772	1.947	0.66628 97263 900626
1.848	0.61410 39732 194924	1.898	0.64080 07001 877361	1.948	0.66680 32052 203434
1.849	0.61464 49524 049878	1.899	0.64132 74318 301488	1.949	0.66731 64205 254238
1.850	0.61518 56390 902335	1.900	0.64185 38861 723948	1.950	0.66782 93725 756554

TABLE VI.

Table VI 231

NATURAL LOGARITHMS

x	$\ln x$	x	$\ln x$	x	$\ln x$
1.950	0.66782 93725 756554	2.000	0.69314 71805 599453	2.050	0.71783 97931 503168
1.951	0.66834 20616 409742	2.001	0.69364 70556 015964	2.051	0.71832 74790 902436
1.952	0.66885 44879 909007	2.002	0.69414 66808 930288	2.052	0.71881 49273 085231
1.953	0.66936 66518 945419	2.003	0.69464 60566 836812	2.053	0.71930 21380 367965
1.954	0.66987 85536 205910	2.004	0.69514 51832 226184	2.054	0.71978 91115 063665
1.955	0.67039 01934 373291	2.005	0.69564 40607 585325	2.055	0.72027 58479 481979
1.956	0.67090 15716 126256	2.006	0.69614 26895 397438	2.056	0.72076 23475 929187
1.957	0.67141 26884 139392	2.007	0.69664 10698 142011	2.057	0.72124 86106 708201
1.958	0.67192 35441 083186	2.008	0.69713 92018 294828	2.058	0.72173 46374 118579
1.959	0.67243 41389 624037	2.009	0.69763 70858 327974	2.059	0.72222 04280 456524
1.960	0.67294 44732 424259	2.010	0.69813 47220 709844	2.060	0.72270 59828 014897
1.961	0.67345 45472 142092	2.011	0.69863 21107 905150	2.061	0.72319 13019 083220
1.962	0.67396 43611 431713	2.012	0.69912 92522 374928	2.062	0.72367 63855 947682
1.963	0.67447 39152 943240	2.013	0.69962 61466 576544	2.063	0.72416 12340 891148
1.964	0.67498 32099 322741	2.014	0.70012 27942 963706	2.064	0.72464 58476 193163
1.965	0.67549 22453 212246	2.015	0.70061 91953 986463	2.065	0.72513 02264 129961
1.966	0.67600 10217 249748	2.016	0.70111 53502 091222	2.066	0.72561 43706 974468
1.967	0.67650 95394 069220	2.017	0.70161 12589 720747	2.067	0.72609 82806 996312
1.968	0.67701 77986 300617	2.018	0.70210 69219 314172	2.068	0.72658 19566 461827
1.969	0.67752 57996 569885	2.019	0.70260 23393 307004	2.069	0.72706 53987 634060
1.970	0.67803 35427 498971	2.020	0.70309 75114 131134	2.070	0.72754 86072 772777
1.971	0.67854 10281 705832	2.021	0.70359 24384 214840	2.071	0.72803 15824 134471
1.972	0.67904 82561 804437	2.022	0.70408 71205 982797	2.072	0.72851 43243 972366
1.973	0.67955 52270 404783	2.023	0.70458 15581 856084	2.073	0.72899 68334 536425
1.974	0.68006 19410 112898	2.024	0.70507 57514 252191	2.074	0.72947 91098 073356
1.975	0.68056 83983 530852	2.025	0.70556 97005 585025	2.075	0.72996 11536 826616
1.976	0.68107 45993 256761	2.026	0.70606 34058 264916	2.076	0.73044 29653 036422
1.977	0.68158 05441 884799	2.027	0.70655 68674 698630	2.077	0.73092 45448 939753
1.978	0.68208 62332 005204	2.028	0.70705 00857 289367	2.078	0.73140 58926 770357
1.979	0.68259 16666 204287	2.029	0.70754 30608 436777	2.079	0.73188 70088 758759
1.980	0.68309 68447 064439	2.030	0.70803 57930 536960	2.080	0.73236 78937 132266
1.981	0.68360 17677 164139	2.031	0.70852 82825 982476	2.081	0.73284 85474 114974
1.982	0.68410 64359 077962	2.032	0.70902 05297 162355	2.082	0.73332 89701 927771
1.983	0.68461 08495 376589	2.033	0.70951 25346 462096	2.083	0.73380 91622 788349
1.984	0.68511 50088 626811	2.034	0.71000 42976 263682	2.084	0.73428 91238 911205
1.985	0.68561 89141 391537	2.035	0.71049 58188 945583	2.085	0.73476 88552 507648
1.986	0.68612 25656 229808	2.036	0.71098 70986 882763	2.086	0.73524 83565 785807
1.987	0.68662 59635 696798	2.037	0.71147 81372 446688	2.087	0.73572 76280 950637
1.988	0.68712 91082 343823	2.038	0.71196 89348 005331	2.088	0.73620 66700 203923
1.989	0.68763 19998 718351	2.039	0.71245 94915 923181	2.089	0.73668 54825 744287
1.990	0.68813 46387 364010	2.040	0.71294 98078 561250	2.090	0.73716 40659 767196
1.991	0.68863 70250 820592	2.041	0.71343 98838 277077	2.091	0.73764 24204 464965
1.992	0.68913 91591 624065	2.042	0.71392 97197 424738	2.092	0.73812 05462 026765
1.993	0.68964 10412 306577	2.043	0.71441 93158 354850	2.093	0.73859 84434 638627
1.994	0.69014 26715 396466	2.044	0.71490 86723 414580	2.094	0.73907 61124 483451
1.995	0.69064 40503 418268	2.045	0.71539 77894 947651	2.095	0.73955 35533 741011
1.996	0.69114 51778 892722	2.046	0.71588 66675 294347	2.096	0.74003 07664 587957
1.997	0.69164 60544 336782	2.047	0.71637 53066 791525	2.097	0.74050 77519 197829
1.998	0.69214 66802 263618	2.048	0.71686 37071 772614	2.098	0.74098 45099 741054
1.999	0.69264 70555 182630	2.049	0.71735 18692 567627	2.099	0.74146 10408 384959
2.000	0.69314 71805 599453	2.050	0.71783 97931 503168	2.100	0.74193 73447 293773

TABLE VI.

2% INTEREST FACTORS FOR ANNUAL COMPOUNDING INTEREST

	Single Payment		Equal-Payment Series			
n	Compound-amount factor	Present-worth factor	Compound-amount factor	Sinking-fund factor	Present-worth factor	Capital-recovery factor
	SP *i-n* ( )	PS *i-n* ( )	SR *i-n* ( )	RS *i-n* ( )	PR *i-n* ( )	RP *i-n* ( )
1	1.020	0.98039	1.000	1.00000	0.98039	1.02000
2	1.040	0.96117	2.020	0.49505	1.94156	0.51505
3	1.061	0.94232	3.060	0.32675	2.88388	0.34675
4	1.082	0.92385	4.122	0.24262	3.80773	0.26262
5	1.104	0.90573	5.204	0.19216	4.71346	0.21216
6	1.126	0.88797	6.308	0.15853	5.60143	0.17853
7	1.149	0.87056	7.434	0.13451	6.47199	0.15451
8	1.172	0.85349	8.583	0.11651	7.32548	0.13651
9	1.195	0.83676	9.755	0.10252	8.16224	0.12252
10	1.219	0.82035	10.950	0.09133	8.98258	0.11133
11	1.243	0.80426	12.169	0.08218	9.78685	0.10218
12	1.268	0.78849	13.412	0.07456	10.57534	0.09456
13	1.294	0.77303	14.680	0.06812	11.34837	0.08812
14	1.319	0.75788	15.974	0.06260	12.10625	0.08260
15	1.346	0.74301	17.293	0.05783	12.84926	0.07783
16	1.373	0.72845	18.639	0.05365	13.57771	0.07365
17	1.400	0.71416	20.012	0.04997	14.29187	0.06997
18	1.428	0.70016	21.412	0.04670	14.99203	0.06670
19	1.457	0.68643	22.841	0.04378	15.67846	0.06378
20	1.486	0.67297	24.297	0.04116	16.35143	0.06116
21	1.516	0.65978	25.783	0.03878	17.01121	0.05878
22	1.546	0.64684	27.299	0.03663	17.65805	0.05663
23	1.577	0.63416	28.845	0.03467	18.29220	0.05467
24	1.608	0.62172	30.422	0.03287	18.91392	0.05287
25	1.641	0.60953	32.030	0.03122	19.52346	0.05122
26	1.673	0.59758	33.671	0.02970	20.12104	0.04970
27	1.707	0.58586	35.344	0.02829	20.70690	0.04829
28	1.741	0.57437	37.051	0.02699	21.28127	0.04699
29	1.776	0.56311	38.792	0.02578	21.84438	0.04578
30	1.811	0.55207	40.568	0.02465	22.39646	0.04465
35	2.000	0.50003	49.994	0.02000	24.99862	0.04000
40	2.208	0.45289	60.402	0.01656	27.35548	0.03656
45	2.438	0.41020	71.893	0.01391	29.49016	0.03391
50	2.692	0.37153	84.579	0.01182	31.42361	0.03182
55	2.972	0.33650	98.587	0.01014	33.17479	0.03014
60	3.281	0.30478	114.052	0.00877	34.76088	0.02877
70	4.000	0.25003	149.978	0.00667	37.49862	0.02667
80	4.875	0.20511	193.772	0.00516	39.74451	0.02516
90	5.943	0.16826	247.157	0.00405	41.58693	0.02405
100	7.245	0.13803	312.232	0.00320	43.09835	0.02320

TABLE VII.

TABLE VII. INTEREST FACTORS FOR ANNUAL COMPOUNDING INTEREST
H. G. Thuesen and W. J. Fabrycky, *ENGINEERING ECONOMY*, Third Edition, © 1964. Reprinted by permission of Prentice-Hall, Inc., Englewood Cliffs, N. J., pp. 490–498.

Table VII 233

3% INTEREST FACTORS FOR ANNUAL COMPOUNDING INTEREST

n	Single Payment		Equal-Payment Series			
	Compound-amount factor	Present-worth factor	Compound-amount factor	Sinking-fund factor	Present-worth factor	Capital-recovery factor
	SP *i-n* ( )	PS *i-n* ( )	SR *i-n* ( )	RS *i-n* ( )	PR *i-n* ( )	RP *i-n* ( )
1	1.030	0.97087	1.000	1.00000	0.97087	1.03000
2	1.061	0.94260	2.030	0.49261	1.91347	0.52261
3	1.093	0.91514	3.091	0.32353	2.82861	0.35353
4	1.126	0.88849	4.184	0.23903	3.71710	0.26903
5	1.159	0.86261	5.309	0.18835	4.57971	0.21835
6	1.194	0.83748	6.468	0.15460	5.41719	0.18460
7	1.230	0.81309	7.662	0.13051	6.23028	0.16051
8	1.267	0.78941	8.892	0.11246	7.01969	0.14246
9	1.305	0.76642	10.159	0.09843	7.78611	0.12843
10	1.344	0.74409	11.464	0.08723	8.53020	0.11723
11	1.384	0.72242	12.808	0.07808	9.25263	0.10808
12	1.426	0.70138	14.192	0.07046	9.95401	0.10046
13	1.469	0.68095	15.618	0.06403	10.63496	0.09403
14	1.513	0.66112	17.086	0.05853	11.29608	0.08853
15	1.558	0.64186	18.599	0.05377	11.93794	0.08377
16	1.605	0.62317	20.157	0.04961	12.56111	0.07961
17	1.653	0.60502	21.762	0.04595	13.16612	0.07595
18	1.702	0.58739	23.414	0.04271	13.75352	0.07271
19	1.754	0.57029	25.117	0.03981	14.32380	0.06981
20	1.806	0.55368	26.870	0.03722	14.87748	0.06722
21	1.860	0.53755	28.677	0.03487	15.41503	0.06487
22	1.916	0.52189	30.537	0.03275	15.93692	0.06275
23	1.974	0.50669	32.453	0.03081	16.44361	0.06081
24	2.033	0.49193	34.426	0.02905	16.93555	0.05905
25	2.094	0.47761	36.459	0.02743	17.41315	0.05743
26	2.157	0.46369	38.553	0.02594	17.87685	0.05594
27	2.221	0.45019	40.710	0.02456	18.32704	0.05456
28	2.288	0.43708	42.931	0.02329	18.76411	0.05329
29	2.357	0.42435	45.219	0.02211	19.18846	0.05211
30	2.427	0.41199	47.575	0.02102	19.60045	0.05102
35	2.814	0.35538	60.462	0.01654	21.48722	0.04654
40	3.262	0.30656	75.401	0.01326	23.11478	0.04326
45	3.782	0.26444	92.720	0.01079	24.51872	0.04079
50	4.384	0.22811	112.797	0.00887	25.72977	0.03887
55	5.082	0.19677	136.072	0.00735	26.77443	0.03735
60	5.892	0.16973	163.054	0.00613	27.67557	0.03613
70	7.918	0.12630	230.594	0.00434	29.12342	0.03434
80	10.641	0.09398	321.363	0.00311	30.20077	0.03311
90	14.300	0.06993	443.340	0.00226	31.00241	0.03226
100	19.219	0.05203	607.288	0.00165	31.59891	0.03165

TABLE VII.

Table VII

4% INTEREST FACTORS FOR ANNUAL COMPOUNDING INTEREST

	Single Payment		Equal-Payment Series			
n	Compound-amount factor	Present-worth factor	Compound-amount factor	Sinking-fund factor	Present-worth factor	Capital-recovery factor
	SP i-n ( )	PS i-n ( )	SR i-n ( )	RS i-n ( )	PR i-n ( )	RP i-n ( )
1	1.040	0.96154	1.000	1.00000	0.96154	1.04000
2	1.082	0.92456	2.040	0.49020	1.88609	0.53020
3	1.125	0.88900	3.122	0.32035	2.77509	0.36035
4	1.170	0.85480	4.246	0.23549	3.62990	0.27549
5	1.217	0.82193	5.416	0.18463	4.45182	0.22463
6	1.265	0.79031	6.633	0.15076	5.24214	0.19076
7	1.316	0.75992	7.898	0.12661	6.00205	0.16661
8	1.369	0.73069	9.214	0.10853	6.73275	0.14853
9	1.423	0.70259	10.583	0.09449	7.43533	0.13449
10	1.480	0.67556	12.006	0.08329	8.11090	0.12329
11	1.539	0.64958	13.486	0.07415	8.76048	0.11415
12	1.601	0.62460	15.026	0.06655	9.38507	0.10655
13	1.665	0.60057	16.627	0.06014	9.98565	0.10014
14	1.732	0.57748	18.292	0.05467	10.56312	0.09467
15	1.801	0.55526	20.024	0.04994	11.11839	0.08994
16	1.873	0.53391	21.825	0.04582	11.65230	0.08582
17	1.948	0.51337	23.698	0.04220	12.16567	0.08220
18	2.026	0.49363	25.645	0.03899	12.65930	0.07899
19	2.107	0.47464	27.671	0.03614	13.13394	0.07614
20	2.191	0.45639	29.778	0.03358	13.59033	0.07358
21	2.279	0.43883	31.969	0.03128	14.02916	0.07128
22	2.370	0.42196	34.248	0.02920	14.45112	0.06920
23	2.465	0.40573	36.618	0.02731	14.85684	0.06731
24	2.563	0.39012	39.083	0.02559	15.24696	0.06559
25	2.666	0.37512	41.646	0.02401	15.62208	0.06401
26	2.772	0.36069	44.312	0.02257	15.98277	0.06257
27	2.883	0.34682	47.084	0.02124	16.32959	0.06124
28	2.999	0.33348	49.968	0.02001	16.66306	0.06001
29	3.119	0.32065	52.966	0.01888	16.98372	0.05888
30	3.243	0.30832	56.085	0.01783	17.29203	0.05783
35	3.946	0.25342	73.652	0.01358	18.66461	0.05358
40	4.801	0.20829	95.026	0.01052	19.79277	0.05052
45	5.841	0.17120	121.029	0.00826	20.72004	0.04826
50	7.107	0.14071	152.667	0.00655	21.48218	0.04655
55	8.646	0.11566	191.159	0.00523	22.10861	0.04523
60	10.520	0.09506	237.991	0.00420	22.62349	0.04420
70	15.572	0.06422	364.291	0.00275	23.39452	0.04275
80	23.050	0.04338	551.245	0.00181	23.91539	0.04181
90	34.119	0.02931	827.984	0.00121	24.26728	0.04121
100	50.505	0.01980	1237.624	0.00081	24.50500	0.04081

TABLE VII.

Table VII 235

5% INTEREST FACTORS FOR ANNUAL COMPOUNDING INTEREST

n	Single Payment		Equal-Payment Series			
	Compound-amount factor	Present-worth factor	Compound-amount factor	Sinking-fund factor	Present-worth factor	Capital-recovery factor
	SP *i-n* ( )	PS *i-n* ( )	SR *i-n* ( )	RS *i-n* ( )	PR *i-n* ( )	RP *i-n* ( )
1	1.050	0.95238	1.000	1.00000	0.95238	1.05000
2	1.103	0.90703	2.050	0.48780	1.85941	0.53780
3	1.158	0.86384	3.153	0.31721	2.72325	0.36721
4	1.216	0.82270	4.310	0.23201	3.54595	0.28201
5	1.276	0.78353	5.526	0.18097	4.32948	0.23097
6	1.340	0.74622	6.802	0.14702	5.07569	0.19702
7	1.407	0.71068	8.142	0.12282	5.78637	0.17282
8	1.477	0.67684	9.549	0.10742	6.46321	0.15472
9	1.551	0.64461	11.027	0.09069	7.10782	0.14069
10	1.629	0.61391	12.578	0.07950	7.72174	0.12950
11	1.710	0.58468	14.207	0.07039	8.30641	0.12039
12	1.796	0.55684	15.917	0.06283	8.86325	0.11283
13	1.886	0.53032	17.713	0.05646	9.39357	0.10646
14	1.980	0.50507	19.599	0.05102	9.89864	0.10102
15	2.079	0.48102	21.579	0.04634	10.37966	0.09634
16	2.183	0.45811	23.657	0.04227	10.83777	0.09227
17	2.292	0.43630	25.840	0.03870	11.27407	0.08870
18	2.407	0.41552	28.132	0.03555	11.68959	0.08555
19	2.527	0.39573	30.539	0.03275	12.08532	0.08275
20	2.653	0.37689	33.066	0.03024	12.46221	0.08024
21	2.786	0.35894	35.719	0.02800	12.82115	0.07800
22	2.925	0.34185	38.505	0.02597	13.16300	0.07597
23	3.072	0.32557	41.430	0.02414	13.48857	0.07414
24	3.225	0.31007	44.502	0.02247	13.79864	0.07247
25	3.386	0.29530	47.727	0.02095	14.09394	0.07095
26	3.556	0.28124	51.113	0.01956	14.37519	0.06956
27	3.733	0.26785	54.669	0.01829	14.64303	0.06829
28	3.920	0.25509	58.403	0.01712	14.89813	0.06712
29	4.116	0.24295	62.323	0.01605	15.14107	0.06605
30	4.322	0.23138	66.439	0.01505	15.37245	0.06505
35	5.516	0.18129	90.320	0.01107	16.37419	0.06107
40	7.040	0.14205	120.800	0.00828	17.15909	0.05828
45	8.985	0.11130	159.700	0.00626	17.77407	0.05626
50	11.467	0.08720	209.348	0.00478	18.25593	0.05478
55	14.636	0.06833	272.713	0.00367	18.63347	0.05367
60	18.679	0.05354	353.584	0.00283	18.92929	0.05283
70	30.426	0.03287	588.529	0.00170	19.34268	0.05170
80	49.561	0.02018	971.229	0.00103	19.59646	0.05103
90	80.730	0.01239	1594.608	0.00063	19.75226	0.05063
100	131.501	0.00760	2610.026	0.00038	19.84791	0.05038

TABLE VII.

Table VII

6% INTEREST FACTORS FOR ANNUAL COMPOUNDING INTEREST

	Single Payment		Equal-Payment Series			
	Compound-amount factor	Present-worth factor	Compound-amount factor	Sinking-fund factor	Present-worth factor	Capital-recovery factor
n	SP i-n ( )	PS i-n ( )	SR i-n ( )	RS i-n ( )	PR i-n ( )	RP i-n ( )
1	1.060	0.94340	1.000	1.00000	0.94340	1.06000
2	1.124	0.89000	2.060	0.48544	1.83339	0.54544
3	1.191	0.83962	3.184	0.31411	2.67301	0.37411
4	1.262	0.79209	4.375	0.22859	3.46510	0.28859
5	1.338	0.74726	5.637	0.17740	4.21236	0.23740
6	1.419	0.70496	6.975	0.14336	4.91732	0.20336
7	1.504	0.66506	8.394	0.11914	5.58238	0.17914
8	1.594	0.62741	9.897	0.10104	6.20979	0.16104
9	1.689	0.59190	11.491	0.08702	6.80169	0.14702
10	1.791	0.55839	13.181	0.07587	7.36009	0.13587
11	1.898	0.52679	14.972	0.06679	7.88687	0.12679
12	2.012	0.49697	16.870	0.05928	8.38384	0.11928
13	2.133	0.46884	18.882	0.05296	8.85268	0.11296
14	2.261	0.44230	21.015	0.04758	9.29498	0.10758
15	2.397	0.41727	23.276	0.04296	9.71225	0.10296
16	2.540	0.39365	25.673	0.03895	10.10590	0.09895
17	2.693	0.37136	28.213	0.03544	10.47726	0.09544
18	2.854	0.35034	30.906	0.03236	10.82760	0.09236
19	3.026	0.33051	33.760	0.02962	11.15812	0.08962
20	3.207	0.31180	36.786	0.02718	11.46992	0.08718
21	3.400	0.29416	39.993	0.02500	11.76408	0.08500
22	3.604	0.27751	43.392	0.02305	12.04158	0.08305
23	3.820	0.26180	46.996	0.02128	12.30338	0.08128
24	4.049	0.24698	50.816	0.01968	12.55036	0.07968
25	4.292	0.23300	54.865	0.01823	12.78336	0.07823
26	4.549	0.21981	59.156	0.01690	13.00317	0.07690
27	4.822	0.20737	63.706	0.01570	13.21053	0.07570
28	5.112	0.19563	68.528	0.01459	13.40616	0.07459
29	5.418	0.18456	73.640	0.01358	13.59072	0.07358
30	5.743	0.17411	79.058	0.01265	13.76483	0.07265
35	7.686	0.13011	111.435	0.00897	14.49825	0.06897
40	10.286	0.09722	154.762	0.00646	15.04630	0.06646
45	13.765	0.07265	212.744	0.00470	15.45583	0.06470
50	18.420	0.05429	290.336	0.00344	15.76186	0.06344
55	24.650	0.04057	394.172	0.00254	15.99054	0.06254
60	32.988	0.03031	533.128	0.00188	16.16143	0.06188
70	59.076	0.01693	967.932	0.00103	16.38454	0.06103
80	105.796	0.00945	1746.600	0.00057	16.50913	0.06057
90	189.465	0.00528	3141.075	0.00032	16.57870	0.06032
100	339.302	0.00295	5638.369	0.00018	16.61755	0.06018

TABLE VII.

Table VII 237

7% INTEREST FACTORS FOR ANNUAL COMPOUNDING INTEREST

n	Single Payment		Equal-Payment Series			
	Compound-amount factor	Present-worth factor	Compound-amount factor	Sinking-fund factor	Present-worth factor	Capital-recovery factor
	SP *i-n* ( )	PS *i-n* ( )	SR *i-n* ( )	RS *i-n* ( )	PR *i-n* ( )	RP *i-n* ( )
1	1.070	0.93458	1.000	1.00000	0.93458	1.07000
2	1.145	0.87344	2.070	0.48309	1.80802	0.55309
3	1.225	0.81630	3.215	0.31105	2.62432	0.38105
4	1.311	0.76290	4.440	0.22523	3.38721	0.29523
5	1.403	0.71299	5.751	0.17389	4.10020	0.24389
6	1.501	0.66634	7.153	0.13980	4.76654	0.20980
7	1.606	0.62275	8.654	0.11555	5.38929	0.18555
8	1.718	0.58201	10.260	0.09747	5.97130	0.16747
9	1.838	0.54393	11.978	0.08349	6.51523	0.15349
10	1.967	0.50835	13.816	0.07238	7.02358	0.14238
11	2.105	0.47509	15.784	0.06336	7.49867	0.13336
12	2.252	0.44401	17.888	0.05590	7.94269	0.12590
13	2.410	0.41496	20.141	0.04965	8.35765	0.11965
14	2.579	0.38782	22.550	0.04434	8.74547	0.11434
15	2.759	0.36245	25.129	0.03979	9.10791	0.10979
16	2.952	0.33873	27.888	0.03586	9.44665	0.10586
17	3.159	0.31657	30.840	0.03243	9.76322	0.10243
18	3.380	0.29586	33.999	0.02941	10.05909	0.09941
19	3.617	0.27651	37.379	0.02675	10.33559	0.09675
20	3.870	0.25842	40.995	0.02439	10.59401	0.09439
21	4.141	0.24151	44.865	0.02229	10.83553	0.09229
22	4.430	0.22571	49.006	0.02041	11.06124	0.09041
23	4.741	0.21095	53.436	0.01871	11.27219	0.08871
24	5.072	0.19715	58.177	0.01719	11.46933	0.08719
25	5.427	0.18425	63.249	0.01581	11.65358	0.08581
26	5.807	0.17220	68.676	0.01456	11.82578	0.08456
27	6.214	0.16093	74.484	0.01343	11.98671	0.08343
28	6.649	0.15040	80.698	0.01239	12.13711	0.08239
29	7.114	0.14056	87.347	0.01145	12.27767	0.08145
30	7.612	0.13137	94.461	0.01059	12.40904	0.08059
35	10.677	0.09366	138.237	0.00723	12.94767	0.07723
40	14.974	0.06678	199.635	0.00501	13.33171	0.07501
45	21.002	0.04761	285.749	0.00350	13.60552	0.07350
50	29.457	0.03395	406.529	0.00246	13.80075	0.07246
55	41.315	0.02420	575.929	0.00174	13.93994	0.07174
60	57.946	0.01726	813.520	0.00123	14.03918	0.07123
70	113.989	0.00877	1614.134	0.00062	14.16039	0.07062
80	224.234	0.00446	3189.062	0.00031	14.22201	0.07031
90	441.103	0.00227	6287.185	0.00016	14.25333	0.07016
100	867.716	0.00115	12381.661	0.00008	14.26925	0.07008

TABLE VII.

Table VII

8% INTEREST FACTORS FOR ANNUAL COMPOUNDING INTEREST

n	Single Payment		Equal-Payment Series			
	Compound-amount factor	Present-worth factor	Compound-amount factor	Sinking-fund factor	Present-worth factor	Capital-recovery factor
	SP *i-n* ( )	PS *i-n* ( )	SR *i-n* ( )	RS *i-n* ( )	PR *i-n* ( )	RP *i-n* ( )
1	1.080	0.92593	1.000	1.00000	0.92593	1.08000
2	1.166	0.85734	2.080	0.48077	1.78326	0.56077
3	1.260	0.79383	3.246	0.30803	2.57710	0.38803
4	1.360	0.73503	4.506	0.22192	3.31213	0.30192
5	1.469	0.68058	5.867	0.17046	3.99271	0.25046
6	1.587	0.63017	7.336	0.13632	4.62288	0.21632
7	1.714	0.58349	8.923	0.11207	5.20637	0.19207
8	1.851	0.54027	10.637	0.09401	5.74664	0.17401
9	1.999	0.50025	12.488	0.08008	6.24689	0.16008
10	2.159	0.46319	14.487	0.06903	6.71008	0.14903
11	2.332	0.42888	16.645	0.06008	7.13896	0.14008
12	2.518	0.39711	18.977	0.05270	7.53608	0.13270
13	2.720	0.36770	21.495	0.04652	7.90378	0.12652
14	2.937	0.34046	24.215	0.04130	8.24424	0.12130
15	3.172	0.31524	27.152	0.03683	8.55948	0.11683
16	3.426	0.29189	30.324	0.03298	8.85137	0.11298
17	3.700	0.27027	33.750	0.02963	9.12164	0.10963
18	3.996	0.25025	37.450	0.02670	9.37189	0.10670
19	4.316	0.23171	41.446	0.02413	9.60360	0.10413
20	4.661	0.21455	45.762	0.02185	9.81815	0.10185
21	5.034	0.19866	50.423	0.01983	10.01680	0.09983
22	5.437	0.18394	55.457	0.01803	10.20074	0.09803
23	5.871	0.17032	60.893	0.01642	10.37106	0.09642
24	6.341	0.15770	66.765	0.01498	10.52876	0.09498
25	6.848	0.14602	73.106	0.01368	10.67478	0.09368
26	7.396	0.13520	79.954	0.01251	10.80998	0.09251
27	7.988	0.12519	87.351	0.01145	10.93516	0.09145
28	8.627	0.11591	95.339	0.01049	11.05108	0.09049
29	9.317	0.10733	103.966	0.00962	11.15841	0.08962
30	10.063	0.09938	113.283	0.00883	11.25778	0.08883
35	14.785	0.06763	172.317	0.00580	11.65457	0.08580
40	21.725	0.04603	259.057	0.00386	11.92461	0.08386
45	31.920	0.03133	386.506	0.00259	12.10840	0.08259
50	46.902	0.02132	573.770	0.00174	12.23348	0.08174
55	68.914	0.01451	848.923	0.00118	12.31861	0.08118
60	101.257	0.00988	1253.213	0.00080	12.37655	0.08080
70	218.606	0.00457	2720.080	0.00037	12.44282	0.08037
80	471.955	0.00212	5886.935	0.00017	12.47351	0.08017
90	1018.915	0.00098	12723.937	0.00008	12.48773	0.08008
100	2199.761	0.00045	27484.515	0.00004	12.49432	0.08004

TABLE VII.

Table   VII   *239*

9% INTEREST FACTORS FOR ANNUAL COMPOUNDING INTEREST

n	Single Payment		Equal-Payment Series			
	Compound-amount factor	Present-worth factor	Compound-amount factor	Sinking-fund factor	Present-worth factor	Capital-recovery factor
	SP *i-n* ( )	PS *i-n* ( )	SR *i-n* ( )	RS *i-n* ( )	PR *i-n* ( )	RP *i-n* ( )⁻
1	1.090	0.91743	1.000	1.00000	0.91743	1.09000
2	1.188	0.84168	2.090	0.47847	1.75911	0.56847
3	1.295	0.77218	3.278	0.30505	2.53129	0.39505
4	1.412	0.70843	4.573	0.21867	3.23972	0.30867
5	1.539	0.64993	5.985	0.16709	3.88965	0.25709
6	1.677	0.59627	7.523	0.13292	4.48592	0.22292
7	1.828	0.54703	9.200	0.10869	5.03295	0.19869
8	1.993	0.50187	11.028	0.09067	5.53482	0.18067
9	2.172	0.46043	13.021	0.07680	5.99525	0.16680
10	2.367	0.42241	15.193	0.06582	6.41766	0.15582
11	2.580	0.38753	17.560	0.05695	6.80519	0.14695
12	2.813	0.35553	20.141	0.04965	7.16072	0.13965
13	3.066	0.32618	22.953	0.04357	7.48690	0.13357
14	3.342	0.29925	26.019	0.03843	7.78615	0.12843
15	3.642	0.27454	29.361	0.03406	8.06069	0.12406
16	3.970	0.25187	33.003	0.03030	8.31256	0.12030
17	4.328	0.23107	36.974	0.02705	8.54363	0.11705
18	4.717	0.21199	41.301	0.02421	8.75562	0.11421
19	5.142	0.19449	46.018	0.02173	8.95011	0.11173
20	5.604	0.17843	51.160	0.01955	9.12855	0.10955
21	6.109	0.16370	56.765	0.01762	9.29224	0.10762
22	6.659	0.15018	62.873	0.01590	9.44243	0.10590
23	7.258	0.13778	69.532	0.01438	9.58021	0.10438
24	7.911	0.12640	76.790	0.01302	9.70661	0.10302
25	8.623	0.11597	84.701	0.01181	9.82258	0.10181
26	9.399	0.10639	93.324	0.01072	9.92897	0.10072
27	10.245	0.09761	102.723	0.00973	10.02658	0.09973
28	11.167	0.08955	112.968	0.00885	10.11613	0.09885
29	12.172	0.08215	124.135	0.00806	10.19828	0.09806
30	13.268	0.07537	136.308	0.00734	10.27365	0.09734
35	20.414	0.04899	215.711	0.00464	10.56682	0.09464
40	31.409	0.03184	337.882	0.00296	10.75736	0.09296
45	48.327	0.02069	525.859	0.00190	10.88120	0.09190
50	74.358	0.01345	815.084	0.00123	10.96168	0.09123
55	114.408	0.00874	1260.092	0.00079	11.01399	0.09079
60	176.031	0.00568	1944.792	0.00051	11.04799	0.09051
70	416.730	0.00240	4619.223	0.00022	11.08445	0.09022
80	986.552	0.00101	10950.573	0.00009	11.09985	0.09009
90	2335.526	0.00043	25939.182	0.00004	11.10635	0.09004
100	5529.041	0.00018	61422.674	0.00002	11.10910	0.09002

TABLE VII.

Table VII

10% INTEREST FACTORS FOR ANNUAL COMPOUNDING INTEREST

n	Single Payment		Equal-Payment Series			
	Compound-amount factor	Present-worth factor	Compound-amount factor	Sinking-fund factor	Present-worth factor	Capital-recovery factor
	SP i-n ( )	PS i-n ( )	SR i-n ( )	RS i-n ( )	PR i-n ( )	RP i-n ( )
1	1.100	0.90909	1.000	1.00000	0.90909	1.10000
2	1.210	0.82645	2.100	0.47619	1.73554	0.57619
3	1.331	0.75131	3.310	0.30211	2.48685	0.40211
4	1.464	0.68301	4.641	0.21547	3.16987	0.31547
5	1.611	0.62092	6.105	0.16380	3.79079	0.26380
6	1.772	0.56447	7.716	0.12961	4.35526	0.22961
7	1.949	0.51316	9.487	0.10541	4.86842	0.20541
8	2.144	0.46651	11.436	0.08744	5.33493	0.18744
9	2.358	0.42410	13.579	0.07364	5.75902	0.17364
10	2.594	0.38554	15.937	0.06275	6.14457	0.16275
11	2.853	0.35049	18.531	0.05396	6.49506	0.15396
12	3.138	0.31863	21.384	0.04676	6.81369	0.14676
13	3.452	0.28966	24.523	0.04078	7.10336	0.14078
14	3.797	0.26333	27.975	0.03575	7.36669	0.13575
15	4.177	0.23939	31.772	0.03147	7.60608	0.13147
16	4.595	0.21763	35.950	0.02782	7.82371	0.12782
17	5.054	0.19784	40.545	0.02466	8.02155	0.12466
18	5.560	0.17986	45.599	0.02193	8.20141	0.12193
19	6.116	0.16351	51.159	0.01955	8.36492	0.11955
20	6.727	0.14864	57.275	0.01746	8.51356	0.11746
21	7.400	0.13513	64.002	0.01562	8.64869	0.11562
22	8.140	0.12285	71.403	0.01401	8.77154	0.11401
23	8.954	0.11168	79.543	0.01257	8.88322	0.11257
24	9.850	0.10153	88.497	0.01130	8.98474	0.11130
25	10.835	0.09230	98.347	0.01017	9.07704	0.11017
26	11.918	0.08391	109.182	0.00916	9.16095	0.10916
27	13.110	0.07628	121.100	0.00826	9.23722	0.10826
28	14.421	0.06934	134.210	0.00745	9.30657	0.10745
29	15.863	0.06304	148.631	0.00673	9.36961	0.10673
30	17.449	0.05731	164.494	0.00608	9.42691	0.10608
35	28.102	0.03558	271.024	0.00369	9.64416	0.10369
40	45.259	0.02209	442.593	0.00226	9.77905	0.10226
45	72.890	0.01372	718.905	0.00139	9.86281	0.10139
50	117.391	0.00852	1163.908	0.00086	9.91481	0.10086
55	189.059	0.00529	1880.591	0.00053	9.94711	0.10053
60	304.482	0.00328	3034.816	0.00033	9.96716	0.10033
70	789.747	0.00127	7887.469	0.00013	9.98734	0.10013
80	2048.400	0.00049	20474.000	0.00005	9.99512	0.10005
90	5313.022	0.00019	53120.222	0.00002	9.99812	0.10002
100	13780.611	0.00007	137796.110	0.00001	9.99927	0.10001

TABLE VII.

Table   VIII

241

$$\sum_{x=c}^{\infty} \frac{e^{-a}a^x}{x!}$$

a

c	0.1	0.2	0.3	0.4	0.5	0.6	0.7	0.8	0.9	1.0
0	1.0000	1.0000	1.0000	1.0000	1.0000	1.0000	1.0000	1.0000	1.0000	1.0000
1	.0952	.1813	.2592	.3297	.3935	.4512	.5034	.5507	.5934	.6321
2	.0047	.0175	.0369	.0616	.0902	.1219	.1558	.1912	.2275	.2642
3	.0002	.0011	.0036	.0079	.0144	.0231	.0341	.0474	.0629	.0803
4	.0000	.0001	.0003	.0008	.0018	.0034	.0058	.0091	.0135	.0190
5	.0000	.0000	.0000	.0001	.0002	.0004	.0008	.0014	.0023	.0037
6	.0000	.0000	.0000	.0000	.0000	.0000	.0001	.0002	.0003	.0006
7	.0000	.0000	.0000	.0000	.0000	.0000	.0000	.0000	.0000	.0001

a

c	1.1	1.2	1.3	1.4	1.5	1.6	1.7	1.8	1.9	2.0
0	1.0000	1.0000	1.0000	1.0000	1.0000	1.0000	1.0000	1.0000	1.0000	1.0000
1	.6671	.6988	.7275	.7534	.7769	.7981	.8173	.8347	.8504	.8647
2	.3010	.3374	.3732	.4082	.4422	.4751	.5068	.5372	.5663	.5940
3	.0996	.1205	.1429	.1665	.1912	.2166	.2428	.2694	.2963	.3233
4	.0257	.0338	.0431	.0537	.0656	.0788	.0932	.1087	.1253	.1429
5	.0054	.0077	.0107	.0143	.0186	.0237	.0296	.0364	.0441	.0527
6	.0010	.0015	.0022	.0032	.0045	.0060	.0080	.0104	.0132	.0166
7	.0001	.0003	.0004	.0006	.0009	.0013	.0019	.0026	.0034	.0045
8	.0000	.0000	.0001	.0001	.0002	.0003	.0004	.0006	.0008	.0011
9	.0000	.0000	.0000	.0000	.0000	.0000	.0001	.0001	.0002	.0002

TABLE VIII.   CUMULATIVE POISSON DISTRIBUTION

Table VIII

a

c	2.1	2.2	2.3	2.4	2.5	2.6	2.7	2.8	2.9	3.0
0	1.0000	1.0000	1.0000	1.0000	1.0000	1.0000	1.0000	1.0000	1.0000	1.0000
1	.8775	.8892	.8997	.9093	.9179	.9257	.9328	.9392	.9450	.9502
2	.6204	.6454	.6691	.6916	.7127	.7326	.7513	.7689	.7854	.8009
3	.3504	.3773	.4040	.4303	.4562	.4816	.5064	.5305	.5540	.5768
4	.1614	.1806	.2007	.2213	.2424	.2640	.2859	.3081	.3304	.3528
5	.0621	.0725	.0838	.0959	.1088	.1226	.1371	.1523	.1682	.1847
6	.0204	.0249	.0300	.0357	.0420	.0490	.0567	.0651	.0742	.0839
7	.0059	.0075	.0094	.0116	.0142	.0172	.0206	.0244	.0287	.0335
8	.0015	.0020	.0026	.0033	.0042	.0053	.0066	.0081	.0099	.0119
9	.0003	.0005	.0006	.0009	.0011	.0015	.0019	.0024	.0031	.0038
10	.0001	.0001	.0001	.0002	.0003	.0004	.0005	.0007	.0009	.0011
11	.0000	.0000	.0000	.0000	.0001	.0001	.0001	.0002	.0002	.0003
12	.0000	.0000	.0000	.0000	.0000	.0000	.0000	.0000	.0001	.0001

a

c	3.1	3.2	3.3	3.4	3.5	3.6	3.7	3.8	3.9	4.0
0	1.0000	1.0000	1.0000	1.0000	1.0000	1.0000	1.0000	1.0000	1.0000	1.0000
1	.9550	.9592	.9631	.9666	.9698	.9727	.9753	.9776	.9798	.9817
2	.8153	.8288	.8414	.8532	.8641	.8743	.8838	.8926	.9008	.9084
3	.5988	.6201	.6406	.6603	.6792	.6973	.7146	.7311	.7469	.7619
4	.3752	.3975	.4197	.4416	.4634	.4848	.5058	.5265	.5468	.5665
5	.2018	.2194	.2374	.2558	.2746	.2936	.3128	.3322	.3516	.3712
6	.0943	.1054	.1171	.1295	.1424	.1559	.1699	.1844	.1994	.2149
7	.0388	.0446	.0510	.0579	.0653	.0733	.0818	.0909	.1005	.1107
8	.0142	.0168	.0198	.0231	.0267	.0308	.0352	.0401	.0454	.0511
9	.0047	.0057	.0069	.0083	.0099	.0117	.0137	.0160	.0185	.0214
10	.0014	.0018	.0022	.0027	.0033	.0040	.0048	.0058	.0069	.0081
11	.0004	.0005	.0006	.0008	.0010	.0013	.0016	.0019	.0023	.0028
12	.0001	.0001	.0002	.0002	.0003	.0004	.0005	.0006	.0007	.0009
13	.0000	.0000	.0000	.0001	.0001	.0001	.0001	.0002	.0002	.0003
14	.0000	.0000	.0000	.0000	.0000	.0000	.0000	.0000	.0001	.0001

TABLE VIII.   CUMULATIVE POISSON DISTRIBUTION

Table VIII 243

a

c	4.1	4.2	4.3	4.4	4.5	4.6	4.7	4.8	4.9	5.0
0	1.0000	1.0000	1.0000	1.0000	1.0000	1.0000	1.0000	1.0000	1.0000	1.0000
1	.9834	.9850	.9864	.9877	.9889	.9899	.9909	.9918	.9926	.9933
2	.9155	.9220	.9281	.9337	.9389	.9437	.9482	.9523	.9561	.9596
3	.7762	.7898	.8026	.8149	.8264	.8374	.8477	.8575	.8667	.8753
4	.5858	.6046	.6228	.6406	.6577	.6743	.6903	.7058	.7207	.7350
5	.3907	.4102	.4296	.4488	.4679	.4868	.5054	.5237	.5418	.5595
6	.2307	.2469	.2633	.2801	.2971	.3142	.3316	.3490	.3665	.3840
7	.1214	.1325	.1442	.1564	.1689	.1820	.1954	.2092	.2233	.2378
8	.0573	.0639	.0710	.0786	.0866	.0951	.1040	.1133	.1231	.1334
9	.0245	.0279	.0317	.0358	.0403	.0451	.0503	.0558	.0618	.0681
10	.0095	.0111	.0129	.0149	.0171	.0195	.0222	.0251	.0283	.0318
11	.0034	.0041	.0048	.0057	.0067	.0078	.0090	.0104	.0120	.0137
12	.0011	.0014	.0017	.0020	.0024	.0029	.0034	.0040	.0047	.0055
13	.0003	.0004	.0005	.0007	.0008	.0010	.0012	.0014	.0017	.0020
14	.0001	.0001	.0002	.0002	.0003	.0003	.0004	.0005	.0006	.0007
15	.0000	.0000	.0000	.0001	.0001	.0001	.0001	.0001	.0002	.0002
16	.0000	.0000	.0000	.0000	.0000	.0000	.0000	.0000	.0001	.0001

a

c	5.1	5.2	5.3	5.4	5.5	5.6	5.7	5.8	5.9	6.0
0	1.0000	1.0000	1.0000	1.0000	1.0000	1.0000	1.0000	1.0000	1.0000	1.0000
1	.9939	.9945	.9950	.9955	.9959	.9963	.9967	.9970	.9973	.9975
2	.9628	.9658	.9686	.9711	.9734	.9756	.9776	.9794	.9811	.9826
3	.8835	.8912	.8984	.9052	.9116	.9176	.9232	.9285	.9334	.9380
4	.7487	.7619	.7746	.7867	.7983	.8094	.8200	.8300	.8396	.8488
5	.5769	.5939	.6105	.6267	.6425	.6579	.6728	.6873	.7013	.7149
6	.4016	.4191	.4365	.4539	.4711	.4881	.5050	.5217	.5381	.5543
7	.2526	.2676	.2829	.2983	.3140	.3297	.3456	.3616	.3776	.3937
8	.1440	.1551	.1665	.1783	.1905	.2030	.2159	.2290	.2424	.2560
9	.0748	.0819	.0894	.0974	.1056	.1143	.1234	.1328	.1426	.1528

TABLE VIII.   CUMULATIVE POISSON DISTRIBUTION

Table VIII

a

c	5.1	5.2	5.3	5.4	5.5	5.6	5.7	5.8	5.9	6.0
10	.0356	.0397	.0441	.0488	.0538	.0591	.0648	.0708	.0772	.0839
11	.0156	.0177	.0200	.0225	.0253	.0282	.0314	.0349	.0386	.0426
12	.0063	.0073	.0084	.0096	.0110	.0125	.0141	.0160	.0179	.0201
13	.0024	.0028	.0033	.0038	.0045	.0051	.0059	.0068	.0078	.0088
14	.0008	.0010	.0012	.0014	.0017	.0020	.0023	.0027	.0031	.0036
15	.0003	.0003	.0004	.0005	.0006	.0007	.0009	.0010	.0012	.0014
16	.0001	.0001	.0001	.0002	.0002	.0002	.0003	.0004	.0004	.0005
17	.0000	.0000	.0000	.0001	.0001	.0001	.0001	.0001	.0001	.0002
18	.0000	.0000	.0000	.0000	.0000	.0000	.0000	.0000	.0000	.0001

a

c	6.1	6.2	6.3	6.4	6.5	6.6	6.7	6.8	6.9	7.0
0	1.0000	1.0000	1.0000	1.0000	1.0000	1.0000	1.0000	1.0000	1.0000	1.0000
1	.9978	.9980	.9982	.9983	.9985	.9986	.9988	.9989	.9990	.9991
2	.9841	.9854	.9866	.9877	.9887	.9897	.9905	.9913	.9920	.9927
3	.9423	.9464	.9502	.9537	.9570	.9600	.9629	.9656	.9680	.9704
4	.8575	.8658	.8736	.8811	.8882	.8948	.9012	.9072	.9129	.9182
5	.7281	.7408	.7531	.7649	.7763	.7873	.7978	.8080	.8177	.8270
6	.5702	.5859	.6012	.6163	.6310	.6453	.6594	.6730	.6863	.6993
7	.4098	.4258	.4418	.4577	.4735	.4892	.5047	.5201	.5353	.5503
8	.2699	.2840	.2983	.3127	.3272	.3419	.3567	.3715	.3864	.4013
9	.1633	.1741	.1852	.1967	.2084	.2204	.2327	.2452	.2580	.2709
10	.0910	.0984	.1061	.1142	.1226	.1314	.1404	.1498	.1505	.1695
11	.0469	.0514	.0563	.0614	.0668	.0726	.0786	.0849	.0916	.0985
12	.0224	.0250	.0277	.0307	.0339	.0373	.0409	.0448	.0490	.0534
13	.0100	.0113	.0127	.0143	.0160	.0179	.0199	.0221	.0245	.0270
14	.0042	.0048	.0055	.0063	.0071	.0080	.0091	.0102	.0115	.0128
15	.0016	.0019	.0022	.0026	.0030	.0034	.0039	.0044	.0050	.0057
16	.0006	.0007	.0008	.0010	.0012	.0014	.0016	.0018	.0021	.0024
17	.0002	.0003	.0003	.0004	.0004	.0005	.0006	.0007	.0008	.0010
18	.0001	.0001	.0001	.0001	.0002	.0002	.0002	.0003	.0003	.0004
19	.0000	.0000	.0000	.0000	.0001	.0001	.0001	.0001	.0001	.0001

TABLE VIII. CUMULATIVE POISSON DISTRIBUTION

Table   VIII   245

c	7.1	7.2	7.3	7.4	7.5	7.6	7.7	7.8	7.9	8.0
0	1.0000	1.0000	1.0000	1.0000	1.0000	1.0000	1.0000	1.0000	1.0000	1.0000
1	.9992	.9993	.9993	.9994	.9994	.9995	.9995	.9996	.9996	.9997
2	.9933	.9939	.9944	.9949	.9953	.9957	.9961	.9964	.9967	.9970
3	.9725	.9745	.9764	.9781	.9797	.9812	.9826	.9839	.9851	.9862
4	.9233	.9281	.9326	.9368	.9409	.9446	.9482	.9515	.9547	.9576
5	.8359	.8445	.8527	.8605	.8679	.8751	.8819	.8883	.8945	.9004
6	.7119	.7241	.7360	.7474	.7586	.7693	.7797	.7897	.7994	.8088
7	.5651	.5796	.5940	.6080	.6218	.6354	.6486	.6616	.6743	.6866
8	.4162	.4311	.4459	.4607	.4754	.4900	.5044	.5188	.5330	.5470
9	.2840	.2973	.3108	.3243	.3380	.3518	.3657	.3796	.3935	.4075
10	.1798	.1904	.2012	.2123	.2236	.2351	.2469	.2589	.2710	.2834
11	.1058	.1133	.1212	.1293	.1378	.1465	.1555	.1648	.1743	.1841
12	.0580	.0629	.0681	.0735	.0792	.0852	.0915	.0980	.1048	.1119
13	.0297	.0327	.0358	.0391	.0427	.0464	.0504	.0546	.0591	.0638
14	.0143	.0159	.0176	.0195	.0216	.0238	.0261	.0286	.0313	.0342
15	.0065	.0073	.0082	.0092	.0103	.0114	.0127	.0141	.0156	.0173
16	.0028	.0031	.0036	.0041	.0046	.0052	.0059	.0066	.0074	.0082
17	.0011	.0013	.0015	.0017	.0020	.0022	.0026	.0029	.0033	.0037
18	.0004	.0005	.0006	.0007	.0008	.0009	.0011	.0012	.0014	.0016
19	.0002	.0002	.0002	.0003	.0003	.0004	.0004	.0005	.0006	.0006
20	.0001	.0001	.0001	.0001	.0001	.0001	.0002	.0002	.0002	.0003
21	.0000	.0000	.0000	.0000	.0000	.0000	.0001	.0001	.0001	.0001

a

c	8.1	8.2	8.3	8.4	8.5	8.6	8.7	8.8	8.9	9.0
0	1.0000	1.0000	1.0000	1.0000	1.0000	1.0000	1.0000	1.0000	1.0000	1.0000
1	.9997	.9997	.9998	.9998	.9998	.9998	.9998	.9998	.9999	.9999
2	.9972	.9975	.9977	.9979	.9981	.9982	.9984	.9985	.9987	.9988
3	.9873	.9882	.9891	.9900	.9907	.9914	.9921	.9927	.9932	.9938
4	.9604	.9630	.9654	.9677	.9699	.9719	.9738	.9756	.9772	.9788

TABLE VIII.   CUMULATIVE POISSON DISTRIBUTION

Table VIII

a

c	8.1	8.2	8.3	8.4	8.5	8.6	8.7	8.8	8.9	9.0
5	.9060	.9113	.9163	.9211	.9256	.9299	.9340	.9379	.9416	.9450
6	.8178	.8264	.8347	.8427	.8504	.8578	.8648	.8716	.8781	.8843
7	.6987	.7104	.7219	.7330	.7438	.7543	.7645	.7744	.7840	.7932
8	.5609	.5746	.5881	.6013	.6144	.6272	.6398	.6522	.6643	.6761
9	.4214	.4353	.4493	.4631	.4769	.4906	.5042	.5177	.5311	.5443
10	.2959	.3085	.3212	.3341	.3470	.3600	.3731	.3863	.3994	.4126
11	.1942	.2045	.2150	.2257	.2366	.2478	.2591	.2706	.2822	.2940
12	.1193	.1269	.1348	.1429	.1513	.1600	.1689	.1780	.1874	.1970
13	.0687	.0739	.0793	.0850	.0909	.0971	.1035	.1102	.1171	.1242
14	.0372	.0405	.0439	.0476	.0514	.0555	.0597	.0642	.0689	.0739
15	.0190	.0209	.0229	.0251	.0274	.0299	.0325	.0353	.0383	.0415
16	.0092	.0102	.0113	.0125	.0138	.0152	.0168	.0184	.0202	.0220
17	.0042	.0047	.0053	.0059	.0066	.0074	.0082	.0091	.0101	.0111
18	.0018	.0021	.0023	.0027	.0030	.0034	.0038	.0043	.0048	.0053
19	.0008	.0009	.0010	.0011	.0013	.0015	.0017	.0019	.0022	.0024
20	.0003	.0003	.0004	.0005	.0005	.0006	.0007	.0008	.0009	.0011
21	.0001	.0001	.0002	.0002	.0002	.0002	.0003	.0003	.0004	.0004
22	.0000	.0000	.0001	.0001	.0001	.0001	.0001	.0001	.0002	.0002
23	.0000	.0000	.0000	.0000	.0000	.0000	.0000	.0000	.0001	.0001

a

c	9.1	9.2	9.3	9.4	9.5	9.6	9.7	9.8	9.9	10
0	1.0000	1.0000	1.0000	1.0000	1.0000	1.0000	1.0000	1.0000	1.0000	1.0000
1	.9999	.9999	.9999	.9999	.9999	.9999	.9999	.9999	1.0000	1.0000
2	.9989	.9990	.9991	.9991	.9992	.9993	.9993	.9994	.9995	.9995
3	.9942	.9947	.9951	.9955	.9958	.9962	.9965	.9967	.9970	.9972
4	.9802	.9816	.9828	.9840	.9851	.9862	.9871	.9880	.9889	.9897
5	.9483	.9514	.9544	.9571	.9597	.9622	.9645	.9667	.9688	.9707
6	.8902	.8959	.9014	.9065	.9115	.9162	.9207	.9250	.9290	.9329
7	.8022	.8108	.8192	.8273	.8351	.8426	.8498	.8567	.8634	.8699
8	.6877	.6990	.7101	.7208	.7313	.7416	.7515	.7612	.7706	.7798
9	.5574	.5704	.5832	.5958	.6082	.6204	.6324	.6442	.6558	.6672

TABLE VIII. CUMULATIVE POISSON DISTRIBUTION

Table   VIII   247

a

c	9.1	9.2	9.3	9.4	9.5	9.6	9.7	9.8	9.9	10
10	.4258	.4389	.4521	.4651	.4782	.4911	.5040	.5168	.5295	.5421
11	.3059	.3180	.3301	.3424	.3547	.3671	.3795	.3920	.4045	.4170
12	.2068	.2168	.2270	.2374	.2480	.2588	.2697	.2807	.2919	.3032
13	.1316	.1393	.1471	.1552	.1636	.1721	.1809	.1899	.1991	.2084
14	.0790	.0844	.0900	.0958	.1019	.1081	.1147	.1214	.1284	.1355
15	.0448	.0483	.0520	.0559	.0600	.0643	.0688	.0735	.0784	.0835
16	.0240	.0262	.0285	.0309	.0335	.0362	.0391	.0421	.0454	.0487
17	.0122	.0135	.0148	.0162	.0177	.0194	.0211	.0230	.0249	.0270
18	.0059	.0066	.0073	.0081	.0089	.0098	.0108	.0119	.0130	.0143
19	.0027	.0031	.0034	.0038	.0043	.0048	.0053	.0059	.0065	.0072
20	.0012	.0014	.0015	.0017	.0020	.0022	.0025	.0028	.0031	.0035
21	.0005	.0006	.0007	.0008	.0009	.0010	.0011	.0013	.0014	.0016
22	.0002	.0002	.0003	.0003	.0004	.0004	.0005	.0005	.0006	.0007
23	.0001	.0001	.0001	.0001	.0001	.0002	.0002	.0002	.0003	.0003
24	.0000	.0000	.0000	.0000	.0001	.0001	.0001	.0001	.0001	.0001

a

c	11	12	13	14	15	16	17	18	19	20
0	1.0000	1.0000	1.0000	1.0000	1.0000	1.0000	1.0000	1.0000	1.0000	1.0000
1	1.0000	1.0000	1.0000	1.0000	1.0000	1.0000	1.0000	1.0000	1.0000	1.0000
2	.9998	.9999	1.0000	1.0000	1.0000	1.0000	1.0000	1.0000	1.0000	1.0000
3	.9988	.9995	.9998	.9999	1.0000	1.0000	1.0000	1.0000	1.0000	1.0000
4	.9951	.9977	.9990	.9995	.9998	.9999	1.0000	1.0000	1.0000	1.0000
5	.9849	.9924	.9963	.9982	.9991	.9996	.9998	.9999	1.0000	1.0000
6	.9625	.9797	.9893	.9945	.9972	.9986	.9993	.9997	.9998	.9999
7	.9214	.9542	.9741	.9858	.9924	.9960	.9979	.9990	.9995	.9997
8	.8568	.9105	.9460	.9684	.9820	.9900	.9946	.9971	.9985	.9992
9	.7680	.8450	.9002	.9379	.9626	.9780	.9874	.9929	.9961	.9979
10	.6595	.7576	.8342	.8906	.9301	.9567	.9739	.9846	.9911	.9950
11	.5401	.6528	.7483	.8243	.8815	.9226	.9509	.9696	.9817	.9892
12	.4207	.5384	.6468	.7400	.8152	.8730	.9153	.9451	.9653	.9786
13	.3113	.4240	.5369	.6415	.7324	.8069	.8650	.9083	.9394	.9610
14	.2187	.3185	.4270	.5356	.6368	.7255	.7991	.8574	.9016	.9339

TABLE VIII.   CUMULATIVE POISSON DISTRIBUTION

Table VIII

a

c	11	12	13	14	15	16	17	18	19	20
15	.1460	.2280	.3249	.4296	.5343	.6325	.7192	.7919	.8503	.8951
16	.0926	.1556	.2364	.3306	.4319	.5333	.6285	.7133	.7852	.8435
17	.0559	.1013	.1645	.2441	.3359	.4340	.5323	.6250	.7080	.7789
18	.0322	.0630	.1095	.1728	.2511	.3407	.4360	.5314	.6216	.7030
19	.0177	.0374	.0698	.1174	.1805	.2577	.3450	.4378	.5305	.6186
20	.0093	.0213	.0427	.0765	.1248	.1878	.2637	.3491	.4394	.5297
21	.0047	.0116	.0250	.0479	.0830	.1318	.1945	.2693	.3528	.4409
22	.0023	.0061	.0141	.0288	.0531	.0892	.1385	.2009	.2745	.3563
23	.0010	.0030	.0076	.0167	.0327	.0582	.0953	.1449	.2069	.2794
24	.0005	.0015	.0040	.0093	.0195	.0367	.0633	.1011	.1510	.2125
25	.0002	.0007	.0020	.0050	.0112	.0223	.0406	.0683	.1067	.1568
26	.0001	.0003	.0010	.0026	.0062	.0131	.0252	.0446	.0731	.1122
27	.0000	.0001	.0005	.0013	.0033	.0075	.0152	.0282	.0486	.0779
28	.0000	.0001	.0002	.0006	.0017	.0041	.0088	.0173	.0313	.0525
29	.0000	.0000	.0001	.0003	.0009	.0022	.0050	.0103	.0195	.0343
30	.0000	.0000	.0000	.0001	.0004	.0011	.0027	.0059	.0118	.0218
31	.0000	.0000	.0000	.0001	.0002	.0006	.0014	.0033	.0070	.0135
32	.0000	.0000	.0000	.0000	.0001	.0003	.0007	.0018	.0040	.0081
33	.0000	.0000	.0000	.0000	.0000	.0001	.0004	.0010	.0022	.0047
34	.0000	.0000	.0000	.0000	.0000	.0001	.0002	.0005	.0012	.0027
35	.0000	.0000	.0000	.0000	.0000	.0000	.0001	.0002	.0006	.0015
36	.0000	.0000	.0000	.0000	.0000	.0000	.0000	.0001	.0003	.0008
37	.0000	.0000	.0000	.0000	.0000	.0000	.0000	.0001	.0002	.0004
38	.0000	.0000	.0000	.0000	.0000	.0000	.0000	.0000	.0001	.0002
39	.0000	.0000	.0000	.0000	.0000	.0000	.0000	.0000	.0000	.0001
40	.0000	.0000	.0000	.0000	.0000	.0000	.0000	.0000	.0000	.0001

TABLE VIII. CUMULATIVE POISSON DISTRIBUTION

Table IX 249

x	e^x	e^{-x}	Sinh x	Cosh x	L e^x	L Sinh x	L Cosh x
0.00	1.0000	1.0000	0.0000	1.0000	.00000	$-\infty$	.00000
.01	.0101	.99005	.0100	.0001	.00434	.00001	.00002
.02	.0202	.98020	.0200	.0002	.00869	.30106	.00009
.03	.0305	.97045	.0300	.0005	.01303	.47719	.00020
.04	.0408	.96079	.0400	.0008	.01737	.60218	.00035
0.05	1.0513	.95123	0.0500	1.0013	.02171	.69915	.00054
.06	.0618	.94176	.0600	.0018	.02606	.77841	.00078
.07	.0725	.93239	.0701	.0025	.03040	.84545	.00106
.08	.0833	.92312	.0801	.0032	.03474	.90355	.00139
.09	.0942	.91393	.0901	.0041	.03909	.95483	.00176
0.10	1.1052	.90484	0.1002	1.0050	.04343	.00072	.00217
.11	.1163	.89583	.1102	.0061	.04777	.04227	.00262
.12	.1275	.88692	.1203	.0072	.05212	.08022	.00312
.13	.1388	.87810	.1304	.0085	.05646	.11517	.00366
.14	.1503	.86936	.1405	.0098	.06080	.14755	.00424
0.15	1.1618	.86071	0.1506	1.0113	.06514	.17772	.00487
.16	.1735	.85214	.1607	.0128	.06949	.20597	.00554
.17	.1853	.84366	.1708	.0145	.07383	.23254	.00625
.18	.1972	.83527	.1810	.0162	.07817	.25762	.00700
.19	.2092	.82696	.1911	.0181	.08252	.28136	.00779
0.20	1.2214	.81873	0.2013	1.0201	.08686	.30392	.00863
.21	.2337	.81058	.2115	.0221	.09120	.32541	.00951
.22	.2461	.80252	.2218	.0243	.09554	.34592	.01043
.23	.2586	.79453	.2320	.0266	.09989	.36555	.01139
.24	.2712	.78663	.2423	.0289	.10423	.38437	.01239
0.25	1.2840	.77880	0.2526	1.0314	.10857	.40245	.01343
.26	.2969	.77105	.2629	.0340	.11292	.41986	.01452
.27	.3100	.76338	.2733	.0367	.11726	.43663	.01564
.28	.3231	.75578	.2837	.0395	.12160	.45282	.01681
.29	.3364	.74826	.2941	.0423	.12595	.46847	.01801
0.30	1.3499	.74082	0.3045	1.0453	.13029	.48362	.01926
.31	.3634	.73345	.3150	.0484	.13463	.49830	.02054
.32	.3771	.72615	.3255	.0516	.13897	.51254	.02187
.33	.3910	.71892	.3360	.0549	.14332	.52637	.02323
.34	.4049	.71177	.3466	.0584	.14766	.53981	.02463
0.35	1.4191	.70469	0.3572	1.0619	.15200	.55290	.02607
.36	.4333	.69768	.3678	.0655	.15635	.56564	.02755
.37	.4477	.69073	.3785	.0692	.16069	.57807	.02907
.38	.4623	.68386	.3892	.0731	.16503	.59019	.03063
.39	.4770	.67706	.4000	.0770	.16937	,60202	.03222
0.40	1.4918	.67032	0.4108	1.0811	.17372	.61358	.03385
.41	.5068	.66365	.4216	.0852	.17806	.62488	.03552
.42	.5220	.65705	.4325	.0895	.18240	.63594	.03723
.43	.5373	.65051	.4434	.0939	.18675	.64677	.03897
.44	.5527	.64404	.4543	.0984	.19109	.65738	.04075
0.45	1.5683	.63763	0.4653	1.1030	.19543	.66777	.04256
.46	.5841	.63128	.4764	.1077	.19978	.67797	.04441
.47	.6000	.62500	.4875	.1125	.20412	.68797	.04630
.48	.6161	.61878	.4986	.1174	.20846	.69779	.04822
.49	.6323	.61263	.5098	.1225	.21280	.70744	.05018
0.50	1.6487	.60653	0.5211	1.1276	.21715	.71692	.05217

TABLE IX. EXPONENTIAL AND HYPERBOLIC FUNCTIONS

TABLE IX.   EXPONENTIAL AND HYPERBOLIC FUNCTIONS
From *Mathematical Tables and Formulas*, by Robert D. Carmichael and Edwin R. Smith, © 1931, Dover Publications, Inc., New York, pp. 202–208.

Table IX

x	e^x	e^{-x}	Sinh x	Cosh x	L e^x	L Sinh x	L Cosh x
0.50	1.6487	.60653	0.5211	1.1276	.21715	.71692	.05217
.51	.6653	.60050	.5324	.1329	.22149	.72624	.05419
.52	.6820	.59452	.5438	.1383	.22583	.73540	.05625
.53	.6989	.58860	.5552	.1438	.23018	.74442	.05834
.54	.7160	.58275	.5666	.1494	.23452	.75330	.06046
0.55	1.7333	.57695	0.5782	1.1551	.23886	.76204	.06262
.56	.7507	.57121	.5897	.1609	.24320	.77065	.06481
.57	.7683	.56553	.6014	.1669	.24755	.77914	.06703
.58	.7860	.55990	.6131	.1730	.25189	.78751	.06929
.59	.8040	.55433	.6248	.1792	.25623	.79576	.07157
0.60	1.8221	.54881	0.6367	1.1855	.26058	.80390	.07389
.61	.8404	.54335	.6485	.1919	.26492	.81194	.07624
.62	.8589	.53794	.6605	.1984	.26926	.81987	.07861
.63	.8776	.53259	.6725	.2051	.27361	.82770	.08102
.64	.8965	.52729	.6846	.2119	.27795	.83543	.08346
0.65	1.9155	.52205	0.6967	1.2188	.28229	.84308	.08593
.66	.9348	.51685	.7090	.2258	.28663	.85063	.08843
.67	.9542	.51171	.7213	.2330	.29098	.85809	.09095
.68	.9739	.50662	.7336	.2402	.29532	.86548	.09351
.69	.9937	.50158	.7461	.2476	.29966	.87278	.09609
0.70	2.0138	.49659	0.7586	1.2552	.30401	.88000	.09870
.71	.0340	.49164	.7712	.2628	.30835	.88715	.10134
.72	.0544	.48675	.7838	.2706	.31269	.89423	.10401
.73	.0751	.48191	.7966	.2785	.31703	.90123	.10670
.74	.0959	.47711	.8094	.2865	.32138	.90817	.10942
0.75	2.1170	.47237	0.8223	1.2947	.32572	.91504	.11216
.76	.1383	.46767	.8353	.3030	.33006	.92185	.11493
.77	.1598	.46301	.8484	.3114	.33441	.92859	.11773
.78	.1815	.45841	.8615	.3199	.33875	.93527	.12055
.79	.2034	.45384	.8748	.3286	.34309	.94190	.12340
0.80	2.2255	.44933	0.8881	1.3374	.34744	.94846	.12627
.81	.2479	.44486	.9015	.3464	.35178	.95498	.12917
.82	.2705	.44043	.9150	.3555	.35612	.96144	.13209
.83	.2933	.43605	.9286	.3647	.36046	.96784	.13503
.84	.3164	.43171	.9423	.3740	.36481	.97420	.13800
0.85	2.3396	.42741	0.9561	1.3835	.36915	.98051	.14099
.86	.3632	.42316	.9700	.3932	.37349	.98677	.14400
.87	.3869	.41895	.9840	.4029	.37784	.99299	.14704
.88	.4109	.41478	.9981	.4128	.38218	.99916	.15009
.89	.4351	.41066	1.0122	.4229	.38652	.00528	.15317
0.90	2.4596	.40657	1.0265	1.4331	.39087	.01137	.15627
.91	.4843	.40252	.0409	.4434	.39521	.01741	.15939
.92	.5093	.39852	.0554	.4539	.39955	.02341	.16254
.93	.5345	.39455	.0700	.4645	.40389	.02937	.16570
.94	.5600	.39063	.0847	.4753	.40824	.03530	.16888
0.95	2.5857	.38674	1.0995	1.4862	.41258	.04119	.17208
.96	.6117	.38289	.1144	.4973	.41692	.04704	.17531
.97	.6379	.37908	.1294	.5085	.42127	.05286	.17855
.98	.6645	.37531	.1446	.5199	.42561	.05864	.18181
.99	.6912	.37158	.1598	.5314	.42995	.06439	.18509
1.00	2.7183	.36788	1.1752	1.5431	.43429	.07011	.18839

TABLE IX.   EXPONENTIAL AND HYPERBOLIC FUNCTIONS

Table IX 251

x	eˣ	e⁻ˣ	Sinh x	Cosh x	L eˣ	L Sinh x	L Cosh x
1.00	2.7183	.36788	1.1752	1.5431	.43429	.07011	.18839
.01	.7456	.36422	.1907	.5549	.43864	.07580	.19171
.02	.7732	.36059	.2063	.5669	.44298	.08146	.19504
.03	.8011	.35701	.2220	.5790	.44732	.08708	.19839
.04	.8292	.35345	.2379	.5913	.45167	.09268	.20176
1.05	2.8577	.34994	1.2539	1.6038	.45601	.09825	.20515
.06	.8864	.34646	.2700	.6164	.46035	.10379	.20855
.07	.9154	.34301	.2862	.6292	.46470	.10930	.21197
.08	.9447	.33960	.3025	.6421	.46904	.11479	.21541
.09	.9743	.33622	.3190	.6552	.47338	.12025	.21886
1.10	3.0042	.33287	1.3356	1.6685	.47772	.12569	.22233
.11	.0344	.32956	.3524	.6820	.48207	.13111	.22582
.12	.0649	.32628	.3693	.6956	.48641	.13649	.22931
.13	.0957	.32303	.3863	.7093	.49075	.14186	.23283
.14	.1268	.31982	.4035	.7233	.49510	.14720	.23636
1.15	3.1582	.31664	1.4208	1.7374	.49944	.15253	.23990
.16	.1899	.31349	.4382	.7517	.50378	.15783	.24346
.17	.2220	.31037	.4558	.7662	.50812	.16311	.24703
.18	.2544	.30728	.4735	.7808	.51247	.16836	.25062
.19	.2871	.30422	.4914	.7957	.51681	.17360	.25422
1.20	3.3201	.30119	1.5095	1.8107	.52115	.17882	.25784
.21	.3535	.29820	.5276	.8258	.52550	.18402	.26146
.22	.3872	.29523	.5460	.8412	.52984	.18920	.26510
.23	.4212	.29229	.5645	.8568	.53418	.19437	.26876
.24	.4556	.28938	.5831	.8725	.53853	.19951	.27242
1.25	3.4903	.28650	1.6019	1.8884	.54287	.20464	.27610
.26	.5254	.28365	.6209	.9045	.54721	.20975	.27979
.27	.5609	.28083	.6400	.9208	.55155	.21485	.28349
.28	.5966	.27804	.6593	.9373	.55590	.21993	.28721
.29	.6328	.27527	.6788	.9540	.56024	.22499	.29093
1.30	3.6693	.27253	1.6984	1.9709	.56458	.23004	.29467
.31	.7062	.26982	.7182	.9880	.56893	.23507	.29842
.32	.7434	.26714	.7381	2.0053	.57327	.24009	.30217
.33	.7810	.26448	.7583	.0228	.57761	.24509	.30594
.34	.8190	.26185	.7786	.0404	.58195	.25008	.30972
1.35	3.8574	.25924	1.7991	2.0583	.58630	.25505	.31352
.36	.8962	.25666	.8198	.0764	.59064	.26002	.31732
.37	.9354	.25411	.8406	.0947	.59498	.26496	.32113
.38	.9749	.25158	.8617	.1132	.59933	.26990	.32495
.39	4.0149	.24908	.8829	.1320	.60367	.27482	.32878
1.40	4.0552	.24660	1.9043	2.1509	.60801	.27974	.33262
.41	.0960	.24414	.9259	.1700	.61236	.28464	.33647
.42	.1371	.24171	.9477	.1894	.61670	.28952	.34033
.43	.1787	.23931	.9697	.2090	.62104	.29440	.34420
.44	.2207	.23693	.9919	.2288	.62538	.29926	.34807
1.45	4.2631	.23457	2.0143	2.2488	.62973	.30412	.35196
.46	.3060	.23224	.0369	.2691	.63407	.30896	.35585
.47	.3492	.22993	.0597	.2896	.63841	.31379	.35976
.48	.3929	.22764	.0827	.3103	.64276	.31862	.36367
.49	.4371	.22537	.1059	.3312	.64710	.32343	.36759
1.50	4.4817	.22313	2.1293	2.3524	.65144	.32823	.37151

TABLE IX. EXPONENTIAL AND HYPERBOLIC FUNCTIONS

Table IX

x	e^x	e^{-x}	Sinh x	Cosh x	L e^x	L Sinh x	L Cosh x
1.50	4.4817	.22313	2.1293	2.3524	.65144	.32823	.37151
.51	.5267	.22091	.1529	.3738	.65578	.33303	.37545
.52	.5722	.21871	.1768	.3955	.66013	.33781	.37939
.53	.6182	.21654	.2008	.4174	.66447	.34258	.38334
.54	.6646	.21438	.2251	.4395	.66881	.34735	.38730
1.55	4.7115	.21225	2.2496	2.4619	.67316	.35211	.39126
.56	.7588	.21014	.2743	.4845	.67750	.35686	.39524
.57	.8066	.20805	.2993	.5073	.68184	.36160	.39921
.58	.8550	.20598	.3245	.5305	.68619	.36633	.40320
.59	.9037	.20393	.3499	.5538	.69053	.37105	.40719
1.60	4.9530	.20190	2.3756	2.5775	.69487	.37577	.41119
.61	5.0028	.19989	.4015	.6013	.69921	.38048	.41520
.62	.0531	.19790	.4276	.6255	.70356	.38518	.41921
.63	.1039	.19593	.4540	.6499	.70790	.38987	.42323
.64	.1552	.19398	.4806	.6746	.71224	.39456	.42725
1.65	5.2070	.19205	2.5075	2.6995	.71659	.39923	.43129
.66	.2593	.19014	.5346	.7247	.72093	.40391	.43532
.67	.3122	.18825	.5620	.7502	.72527	.40857	.43937
.68	.3656	.18637	.5896	.7760	.72961	.41323	.44341
.69	.4195	.18452	.6175	.8020	.73396	.41788	.44747
1.70	5.4739	.18268	2.6456	2.8283	.73830	.42253	.45153
.71	.5290	.18087	.6740	.8549	.74264	.42717	.45559
.72	.5845	.17907	.7027	.8818	.74699	.43180	.45966
.73	.6407	.17728	.7317	.9090	.75133	.43643	.46374
.74	.6973	.17552	.7609	.9364	.75567	.44105	.46782
1.75	5.7546	.17377	2.7904	2.9642	.76002	.44567	.47191
.76	.8124	.17204	.8202	.9922	.76436	.45028	.47600
.77	.8709	.17033	.8503	3.0206	.76870	.45488	.48009
.78	.9299	.16864	.8806	.0492	.77304	.45948	.48419
.79	.9895	.16696	.9112	.0782	.77739	.46408	.48830
1.80	6.0496	.16530	2.9422	3.1075	.78173	.46867	.49241
.81	.1104	.16365	.9734	.1371	.78607	.47325	.49652
.82	.1719	.16203	3.0049	.1669	.79042	.47783	.50064
.83	.2339	.16041	.0367	.1972	.79476	.48241	.50476
.84	.2965	.15882	.0689	.2277	.79910	.48698	.50889
1.85	6.3598	.15724	3.1013	3.2585	.80344	.49154	.51302
.86	.4237	.15567	.1340	.2897	.80779	.49610	.51716
.87	.4883	.15412	.1671	.3212	.81213	.50066	.52130
.88	.5535	.15259	.2005	.3530	.81647	.50521	.52544
.89	.6194	.15107	.2341	.3852	.82082	.50976	.52959
1.90	6.6859	.14957	3.2682	3.4177	.82516	.51430	.53374
.91	.7531	.14808	.3025	.4506	.82950	.51884	.53789
.92	.8210	.14661	.3372	.4838	.83385	.52338	.54205
.93	.8895	.14515	.3722	.5173	.83819	.52791	.54621
.94	.9588	.14370	.4075	.5512	.84253	.53244	.55038
1.95	7.0287	.14227	3.4432	3.5855	.84687	.53696	.55455
.96	.0993	.14086	.4792	.6201	.85122	.54148	.55872
.97	.1707	.13946	.5156	.6551	.85556	.54600	.56290
.98	.2427	.13807	.5523	.6904	.85990	.55051	.56707
.99	.3155	.13670	.5894	.7261	.86425	.55502	.57126
2.00	7.3891	.13534	3.6269	3.7622	.86859	.55953	.57544

TABLE IX.   EXPONENTIAL AND HYPERBOLIC FUNCTIONS

Table IX 253

x	e^x	e^{-x}	Sinh x	Cosh x	L e^x	L Sinh x	L Cosh x
2.00	7.3891	.13534	3.6269	3.7622	.86859	.55953	.57544
.01	.4633	.13399	.6647	.7987	.87293	.56403	.57963
.02	.5383	.13266	.7028	.8355	.87727	.56853	.58382
.03	.6141	.13134	.7414	.8727	.88162	.57303	.58802
.04	.6906	.13003	.7803	.9103	.88596	.57753	.59221
2.05	7.7679	.12873	3.8196	3.9483	.89030	.58202	.59641
.06	.8460	.12745	.8593	.9867	.89465	.58650	.60061
.07	.9248	.12619	.8993	4.0255	.89899	.59099	.60482
.08	8.0045	.12493	.9398	.0647	.90333	.59547	.60903
.09	.0849	.12369	.9806	.1043	.90768	.59995	.61324
2.10	8.1662	.12246	4.0219	4.1443	.91202	.60443	.61745
.11	.2482	.12124	.0635	.1847	.91636	.60890	.62167
.12	.3311	.12003	.1056	.2256	.92070	.61337	.62589
.13	.4149	.11884	.1480	.2669	.92505	.61784	.63011
.14	.4994	.11765	.1909	.3085	.92939	.62231	.63433
2.15	8.5849	.11648	4.2342	4.3507	.93373	.62677	.63856
.16	.6711	.11533	.2779	.3932	.93808	.63123	.64278
.17	.7583	.11418	.3221	.4362	.94242	.63569	.64701
.18	.8463	.11304	.3666	.4797	.94676	.64015	.65125
.19	.9352	.11192	.4116	.5236	.95110	.64460	.65548
2.20	9.0250	.11080	4.4571	4.5679	.95545	.64905	.65972
.21	.1157	.10970	.5030	.6127	.95979	.65350	.66396
.22	.2073	.10861	.5494	.6580	.96413	.65795	.66820
.23	.2999	.10753	.5962	.7037	.96848	.66240	.67244
.24	.3933	.10646	.6434	.7499	.97282	.66684	.67668
2.25	9.4877	.10540	4.6912	4.7966	.97716	.67128	.68093
.26	.5831	.10435	.7394	.8437	.98151	.67572	.68518
.27	.6794	.10331	.7880	.8914	.98585	.68016	.68943
.28	.7767	.10228	.8372	.9395	.99019	.68459	.69368
.29	.8749	.10127	.8868	.9881	.99453	.68903	.69794
2.30	9.9742	.10026	4.9370	5.0372	.99888	.69346	.70219
.31	10.074	.09926	.9876	.0868	.00322	.69789	.70645
.32	.176	.09827	5.0387	.1370	.00756	.70232	.71071
.33	.278	.09730	.0903	.1876	.01191	.70675	.71497
.34	.381	.09633	.1425	.2388	.01625	.71117	.71923
2.35	10.486	.09537	5.1951	5.2905	.02059	.71559	.72349
.36	.591	.09442	.2483	.3427	.02493	.72002	.72776
.37	.697	.09348	.3020	.3954	.02928	.72444	.73203
.38	.805	.09255	.3562	.4487	.03362	.72885	.73630
.39	.913	.09163	.4109	.5026	.03796	.73327	.74056
2.40	11.023	.09072	5.4662	5.5569	.04231	.73769	.74484
.41	.134	.08982	.5221	.6119	.04665	.74210	.74911
.42	.246	.08892	.5785	.6674	.05099	.74652	.75338
.43	.359	.08804	.6354	.7235	.05534	.75093	.75766
.44	.473	.08716	.6929	.7801	.05968	.75534	.76194
2.45	11.588	.08629	5.7510	5.8373	.06402	.75975	.76621
.46	.705	.08543	.8097	.8951	.06836	.76415	.77049
.47	.822	.08458	.8689	.9535	.07271	.76856	.77477
.48	.941	.08374	.9288	6.0125	.07705	.77296	.77906
.49	12.061	.08291	.9892	.0721	.08139	.77737	.78334
2.50	12.182	.08208	6.0502	6.1323	.08574	.78177	.78762

TABLE IX. EXPONENTIAL AND HYPERBOLIC FUNCTIONS

Table IX

x	e^x	e^{-x}	Sinh x	Cosh x	L e^x	L Sinh x	L Cosh x
2.50	12.182	.08208	6.0502	6.1323	.08574	.78177	.78762
.51	.305	.08127	.1118	.1931	.09008	.78617	.79191
.52	.429	.08046	.1741	.2545	.09442	.79057	.79619
.53	.554	.07966	.2369	.3166	.09877	.79497	.80048
.54	.680	.07887	.3004	.3793	.10311	.79937	.80477
2.55	12.807	.07808	6.3645	6.4426	.10745	.80377	.80906
.56	.936	.07730	.4293	.5066	.11179	.80816	.81335
.57	13.066	.07654	.4946	.5712	.11614	.81256	.81764
.58	.197	.07577	.5607	.6365	.12048	.81695	.82194
.59	.330	.07502	.6274	.7024	.12482	.82134	.82623
2.60	13.464	.07427	6.6947	6.7690	.12917	.82573	.83052
.61	.599	.07353	.7628	.8363	.13351	.83012	.83482
.62	.736	.07280	.8315	.9043	.13785	.83451	.83912
.63	.874	.07208	.9008	.9729	.14219	.83890	.84341
.64	14.013	.07136	.9709	7.0423	.14654	.84329	.84771
2.65	14.154	.07065	7.0417	7.1123	.15088	.84768	.85201
.66	.296	.06995	.1132	.1831	.15522	.85206	.85631
.67	.440	.06925	.1854	.2546	.15957	.85645	.86061
.68	.585	.06856	.2583	.3268	.16391	.86083	.86492
.69	.732	.06788	.3319	.3998	.16825	.86522	.86922
2.70	14.880	.06721	7.4063	7.4735	.17260	.86960	.87352
.71	15.029	.06654	.4814	.5479	.17694	.87398	.87783
.72	.180	.06587	.5572	.6231	.18128	.87836	.88213
.73	.333	.06522	.6338	.6991	.18562	.88274	.88644
.74	.487	.06457	.7112	.7758	.18997	.88712	.89074
2.75	15.643	.06393	7.7894	7.8533	.19431	.89150	.89505
.76	.800	.06329	.8683	.9316	.19865	.89588	.89936
.77	.959	.06266	.9480	8.0106	.20300	.90026	.90367
.78	16.119	.06204	8.0285	.0905	.20734	.90463	.90798
.79	.281	.06142	.1098	.1712	.21168	.90901	.91229
2.80	16.445	.06081	8.1919	8.2527	.21602	.91339	.91660
.81	.610	.06020	.2749	.3351	.22037	.91776	.92091
.82	.777	.05961	.3586	.4182	.22471	.92213	.92522
.83	.945	.05901	.4432	.5022	.22905	.92651	.92953
.84	17.116	.05843	.5287	.5871	.23340	.93088	.93385
2.85	17.288	.05784	8.6150	8.6728	.23774	.93525	.93816
.86	.462	.05727	.7021	.7594	.24208	.93963	.94247
.87	.637	.05670	.7902	.8469	.24643	.94400	.94679
.88	.814	.05613	.8791	.9352	.25077	.94837	.95110
.89	.993	.05558	.9689	9.0244	.25511	.95274	.95542
2.90	18.174	.05502	9.0596	9.1146	.25945	.95711	.95974
.91	.357	.05448	.1512	.2056	.26380	.96148	.96405
.92	.541	.05393	.2437	.2976	.26814	.96584	.96837
.93	.728	.05340	.3371	.3905	.27248	.97021	.97269
.94	.916	.05287	.4315	.4844	.27683	.97458	.97701
2.95	19.106	.05234	9.5268	9.5791	.28117	.97895	.98133
.96	.298	.05182	.6231	.6749	.28551	.98331	.98565
.97	.492	.05130	.7203	.7716	.28985	.98768	.98997
.98	.688	.05079	.8185	.8693	.29420	.99205	.99429
.99	.886	.05029	.9177	.9680	.29854	.99641	.99861
3.00	20.086	.04979	10.018	10.068	.30288	.00078	.00293

TABLE IX.   EXPONENTIAL AND HYPERBOLIC FUNCTIONS

Table IX 255

x	e^x	e^{-x}	Sinh x	Cosh x	L e^x	L Sinh x	L Cosh x
3.00	20.086	.04979	10.018	10.068	.30288	.00078	.00293
.05	21.115	.04736	10.534	10.581	.32460	.02259	.02454
.10	22.198	.04505	11.076	11.122	.34631	.04440	.04616
.15	23.336	.04285	11.647	11.689	.36803	.06620	.06779
.20	24.533	.04076	12.246	12.287	.38974	.08799	.08943
3.25	25.790	.03877	12.876	12.915	.41146	.10977	.11108
.30	27.113	.03688	13.538	13.575	.43317	.13155	.13273
.35	28.503	.03508	14.234	14.269	.45489	.15332	.15439
.40	29.964	.03337	14.965	14.999	.47660	.17509	.17605
.45	31.500	.03175	15.734	15.766	.49832	.19685	.19772
3.50	33.115	.03020	16.543	16.573	.52003	.21860	.21940
.55	34.813	.02872	17.392	17.421	.54175	.24036	.24107
.60	36.598	.02732	18.285	18.313	.56346	.26211	.26275
.65	38.475	.02599	19.224	19.250	.58517	.28385	.28444
.70	40.447	.02472	20.211	20.236	.60689	.30559	.30612
3.75	42.521	.02352	21.249	21.272	.62860	.32733	.32781
.80	44.701	.02237	22.339	22.362	.65032	.34907	.34951
.85	46.993	.02128	23.486	23.507	.67203	.37081	.37120
.90	49.402	.02024	24.691	24.711	.69375	.39254	.39290
.95	51.935	.01925	25.958	25.977	.71546	.41427	.41459
4.00	54.598	.01832	27.290	27.308	.73718	.43600	.43629
.10	60.340	.01657	30.162	30.178	.78061	.47946	.47970
.20	66.686	.01500	33.336	33.351	.82404	.52291	.52310
.30	73.700	.01357	36.843	36.857	.86747	.56636	.56652
.40	81.451	.01228	40.719	40.732	.91090	.60980	.60993
4.50	90.017	.01111	45.003	45.014	.95433	.65324	.65335
.60	99.484	.01005	49.737	49.747	.99775	.69668	.69677
.70	109.95	.00910	54.969	54.978	.04118	.74012	.74019
.80	121.51	.00823	60.751	60.759	.08461	.78355	.78361
.90	134.29	.00745	67.141	67.149	.12804	.82699	.82704
5.00	148.41	.00674	74.203	74.210	.17147	.87042	.87046
.10	164.02	.00610	82.008	82.014	.21490	.91386	.91389
.20	181.27	.00552	90.633	90.639	.25833	.95729	.95731
.30	200.34	.00499	100.17	100.17	.30176	.00072	.00074
.40	221.41	.00452	110.70	110.71	.34519	.04415	.04417
5.50	244.69	.00409	122.34	122.35	.38862	.08758	.08760
.60	270.43	.00370	135.21	135.22	.43205	.13101	.13103
.70	298.87	.00335	149.43	149.44	.47548	.17444	.17445
.80	330.30	.00303	165.15	165.15	.51891	.21787	.21788
.90	365.04	.00274	182.52	182.52	.56234	.26130	.26131
6.00	403.43	.00248	201.71	201.72	.60577	.30473	.30474
6.25	518.01	.00193	259.01	259.01	.71434	.41331	.41331
6.50	665.14	.00150	332.57	332.57	.82291	.52188	.52189
6.75	854.06	.00117	427.03	427.03	.93149	.63046	.63046
7.00	1096.6	.00091	548.32	548.32	.04006	.73903	.73903
7.50	1808.0	.00055	904.02	904.02	.25721	.95618	.95618
8.00	2981.0	.00034	1490.5	1490.5	.47436	.17333	.17333
8.50	4914.8	.00020	2457.4	2457.4	.69150	.39047	.39047
9.00	8103.1	.00012	4051.5	4051.5	.90865	.60762	.60762
9.50	13360.	.00007	6679.9	6679.9	.12580	.82477	.82477
10.00	22026.	.00005	11013.	11013.	.34294	.04191	.04191

TABLE IX.   EXPONENTIAL AND HYPERBOLIC FUNCTIONS

Line \ Col.	(1)	(2)	(3)	(4)	(5)	(6)
1	10480	15011	01536	02011	81647	91646
2	22368	46573	25595	85393	30995	89198
3	24130	48360	22527	97265	76393	64809
4	42167	93093	06243	61680	07856	16376
5	37570	39975	81837	16656	06121	91782
6	77921	06907	11008	42751	27756	53498
7	99562	72905	56420	69994	98872	31016
8	96301	91977	05463	07972	18876	20922
9	89579	14342	63661	10281	17453	18103
10	85475	36857	53342	53988	53060	59533
11	28918	69578	88231	33276	70997	79936
12	63553	40961	48235	03427	49626	69445
13	09429	93969	52636	92737	88974	33488
14	10365	61129	87529	85689	48237	52267
15	07119	97336	71048	08178	77233	13916
16	51085	12765	51821	51259	77452	16308
17	02368	21382	52404	60268	89368	19885
18	01011	54092	33362	94904	31273	04146
19	52162	53916	46369	58586	23216	14513
20	07056	97628	33787	09998	42698	06691
21	48663	91245	85828	14346	09172	30168
22	54164	58492	22421	74103	47070	25306
23	32639	32363	05597	24200	13363	38005
24	29334	27001	87637	87308	58731	00256
25	02488	33062	28834	07351	19731	92420
26	81525	72295	04839	96423	24878	82651
27	29676	20591	68086	26432	46901	20849
28	00742	57392	39064	66432	84673	40027
29	05366	04213	25669	26422	44407	44048
30	91921	26418	64117	94305	26766	25940
31	00582	04711	87917	77341	42206	35126
32	00725	69884	62797	56170	86324	88072
33	69011	65795	95876	55293	18988	27354
34	25976	57948	29888	88604	67917	48708
35	09763	83473	73577	12908	30883	18317
36	91567	42595	27958	30134	04024	86385
37	17955	56349	90999	49127	20044	59931
38	46503	18584	18845	49618	02304	51038
39	92157	89634	94824	78171	84610	82834
40	14577	62765	35605	48263	39667	47358
41	98427	07523	33362	64270	01638	92477
42	34914	63976	88720	82765	34476	17032
43	70060	28277	39475	46473	23219	53416
44	53976	54914	06990	67245	68350	82948
45	76072	29515	40980	07391	58745	25774
46	90725	52210	83974	29992	65831	38857
47	64364	67412	33339	31926	14883	24413
48	08962	00358	31662	25388	61642	34072
49	95012	68379	93526	70765	10592	04542
50	15664	10493	20492	38391	91132	21999

TABLE X.   TABLE OF RANDOM NUMBERS

TABLE X.   TABLE OF RANDOM NUMBERS
"Table of 105,000 Random Decimal Digits," Interstate Commerce Commission, Bureau of Transport Economics and Statistics, Washington, D.C., 1949.

Table X 257

(7)	(8)	(9)	(10)	(11)	(12)	(13)	(14)
69179	14194	62590	36207	20969	99570	91291	90700
27982	53402	93965	34095	52666	19174	39615	99505
15179	24830	49340	32081	30680	19655	63348	58629
39440	53537	71341	57004	00849	74917	97758	16379
60468	81305	49684	60672	14110	06927	01263	54613
18602	70659	90655	15053	21916	81825	44394	42880
71194	18738	44013	48840	63213	21069	10634	12952
94595	56869	69014	60045	18425	84903	42508	32307
57740	84378	25331	12566	58678	44947	05585	56941
38867	62300	08158	17983	16439	11458	18593	64952
56865	05859	90106	31595	01547	85590	91610	78188
18663	72695	52180	20847	12234	90511	33703	90322
36320	17617	30015	08272	84115	27156	30613	74952
67689	93394	01511	26358	85104	20285	29975	89868
47564	81056	97735	85977	29372	74461	28551	90707
60756	92144	49442	53900	70960	63990	75601	40719
55322	44819	01188	65255	64835	44919	05944	55157
18594	29852	71585	85030	51132	01915	92747	64951
83149	98736	23495	64350	94738	17752	35156	35749
76988	13602	51851	46104	88916	19509	25625	58104
90229	04734	59193	22178	30421	61666	99904	32812
76468	26384	58151	06646	21524	15227	96909	44592
94342	28728	35806	06912	17012	64161	18296	22851
45834	15398	46557	41135	10367	07684	36188	18510
60952	61280	50001	67658	32586	86679	50720	94953
66566	14778	76797	14780	13300	87074	79666	95725
89768	81536	86645	12659	92259	57102	80428	25280
32832	61362	98947	96067	64760	64584	96096	98253
37937	63904	45766	66134	75470	66520	34693	90449
39972	22209	71500	64568	91402	42416	07844	69618
74087	99547	81817	42607	43808	76655	62028	76630
76222	36086	84637	93161	76038	65855	77919	88006
26575	08625	40801	59920	29841	80150	12777	48501
18912	82271	65424	69774	33611	54262	85963	03547
28290	35797	05998	41688	34952	37888	38917	88050
29880	99730	55536	84855	29080	09250	79656	73211
06115	20542	18059	02008	73708	83517	36103	42791
20655	58727	28168	15475	56942	53389	20562	87338
09922	25417	44137	48413	25555	21246	35509	20468
56873	56307	61607	49518	89636	20103	77490	18062
66969	98420	04880	45585	46565	04102	46880	45709
87589	40836	32427	70002	70663	88863	77775	69348
94970	25832	69975	94884	19661	72828	00102	66794
11398	42878	80287	88267	47363	46634	06541	97809
22987	80059	39911	96189	41151	14222	60697	59583
50490	83765	55657	14361	31720	57375	56228	41546
59744	92351	97473	89286	35931	04110	23726	51900
81249	35648	56891	69352	48373	45578	78547	81788
76463	54328	02349	17247	28865	14777	62730	92277
59516	81652	27195	48223	46751	22923	32261	85653

TABLE X.   TABLE OF RANDOM NUMBERS

Table X

Col. Line	(1)	(2)	(3)	(4)	(5)	(6)
51	16408	81899	04153	53381	79401	21438
52	18629	81953	05520	91962	04739	13092
53	73115	35101	47498	87637	99016	71060
54	57491	16703	23167	49323	45021	33132
55	30405	83946	23792	14422	15059	45799
56	16631	35006	85900	98275	32388	52390
57	96773	20206	42559	78985	05300	22164
58	38935	64202	14349	82674	66523	44133
59	31624	76384	17403	53363	44167	64486
60	78919	19474	23632	27889	47914	02584
61	03931	33309	57047	74211	63445	17361
62	74426	33278	43972	10119	89917	15665
63	09066	00903	20795	95452	92648	45454
64	42238	12426	87025	14267	20979	04508
65	16153	08002	26504	41744	81959	65642
66	21457	40742	29820	96783	29400	21840
67	21581	57802	02050	89728	17937	37621
68	55612	78095	83197	33732	05810	24813
69	44657	66999	99324	51281	84463	60563
70	91340	84979	46949	81973	37949	61023
71	91227	21199	31935	27022	84067	05462
72	50001	38140	66321	19924	72163	09538
73	65390	05224	72958	28609	81406	39147
74	27504	96131	83944	41575	10573	08619
75	37169	94851	39117	89632	00959	16487
76	11508	70225	51111	38351	19444	66499
77	37449	30362	06694	54690	04052	53115
78	46515	70331	85922	38329	57015	15765
79	30986	81223	42416	58353	21532	30502
80	63798	64995	46583	09785	44160	78128
81	82486	84846	99254	67632	43218	50076
82	21885	32906	92431	09060	64297	51674
83	60336	98782	07408	53458	13564	59089
84	43937	46891	24010	25560	86355	33941
85	97656	63175	89303	16275	07100	92063
86	03299	01221	05418	38982	55758	92237
87	79626	06486	03574	17668	07785	76020
88	85636	68335	47539	03129	65651	11977
89	18039	14367	61337	06177	12143	46609
90	08362	15656	60627	36478	65648	16764
91	79556	29068	04142	16268	15387	12856
92	92608	82674	27072	32534	17075	27698
93	23982	25835	40055	67006	12293	02753
94	09915	96306	05908	97901	28395	14186
95	59037	33300	26695	62247	69927	76123
96	42488	78077	69882	61657	34136	79180
97	46764	86273	63003	93017	31204	36692
98	03237	45430	55417	63282	90816	17349
99	86591	81482	52667	61582	14972	90053
100	38534	01715	94964	87288	65680	43772

TABLE X.   TABLE OF RANDOM NUMBERS

Table X   *259*

(7)	(8)	(9)	(10)	(11)	(12)	(13)	(14)
83035	92350	36693	31238	59649	91754	72772	02338
97662	24822	94730	06496	35090	04822	86774	98289
88824	71013	18735	20286	23153	72924	35165	43040
12544	41035	80780	45393	44812	12515	98931	91202
22716	19792	09983	74353	68668	30429	70735	25499
16815	69298	82732	38480	73817	32523	41961	44437
24369	54224	35083	19687	11052	91491	60383	19746
00697	35552	35970	19124	63318	29686	03387	59846
64758	75366	76554	31601	12614	33072	60332	92325
37680	20801	72152	39339	34806	08930	85001	87820
62825	39908	05607	91284	68833	25570	38818	46920
52872	73823	73144	88662	88970	74492	51805	99378
09552	88815	16553	51125	79375	97596	16296	66092
64535	31355	86064	29472	47689	05974	52468	16834
74240	56302	00033	67107	77510	70625	28725	34191
15035	34537	33310	06116	95240	15957	16572	06004
47075	42080	97403	48626	68995	43805	33386	21597
86902	60397	16489	03264	88525	42786	05269	92532
79312	93454	68876	25471	93911	25650	12682	73572
43997	15263	80644	43942	89203	71795	99533	50501
35216	14486	29891	68607	41867	14951	91696	85065
12151	06878	91903	18749	34405	56087	82790	70925
25549	48542	42627	45233	57202	94617	23772	07896
64482	73923	36152	05184	94142	25299	84387	34925
65536	49071	39782	17095	02330	74301	00275	48280
71945	05422	13442	78675	84081	66938	93654	59894
62757	95348	78662	11163	18651	50245	34971	52924
97161	17869	45349	61796	66345	81073	49106	79860
32305	86482	05174	07901	54339	58861	74818	46942
83991	42865	92520	83531	80377	35909	81250	54238
21361	64816	51202	88124	41870	52689	51275	83556
64126	62570	26123	05155	59194	52799	28225	85762
26445	29789	85205	41001	12535	12133	14645	23541
25786	54990	71899	15475	95434	98227	21824	19585
21942	18611	47348	20203	18534	03862	78095	50136
26759	86367	21216	98442	08303	56613	91511	75928
79924	25651	83325	88428	85076	72811	22717	50585
02510	26113	99447	68645	34327	15152	55230	93448
32989	74014	64708	00533	35398	58408	13261	47908
53412	09013	07832	41574	17639	82163	60859	75567
66227	38358	22478	73373	88732	09443	82558	05250
98204	63863	11951	34648	88022	56148	34925	57031
14827	23235	35071	99704	37543	11601	35503	85171
00821	80703	70426	75647	76310	88717	37890	40129
50842	43834	86654	70959	79725	93872	28117	19233
97526	43092	04098	73571	80799	76536	71255	64239
40202	35275	87306	55543	53203	18098	47625	88684
88298	90183	36600	78406	06216	95787	42579	90730
89534	76036	49199	43716	97548	04379	46370	28672
39560	12918	86537	62738	19636	51132	25739	56947

TABLE X.   TABLE OF RANDOM NUMBERS

Table X

Col. Line	(1)	(2)	(3)	(4)	(5)	(6)
101	13284	16834	74151	92027	24670	36665
102	21224	00370	30420	03883	94648	89428
103	99052	47887	81085	64933	66279	80432
104	00199	50993	98603	38452	87890	94624
105	60578	06483	28733	37867	07936	98710
106	91240	18312	17441	01929	18163	69201
107	97458	14229	12063	59611	32249	90466
108	35249	38646	34475	72417	60514	69257
109	38980	46600	11759	11900	46743	27860
110	10750	52745	38749	87365	58959	53731
111	36247	27850	73958	20673	37800	63835
112	70994	66986	99744	72438	01174	42159
113	99638	94702	11463	18148	81386	80431
114	72055	15774	43857	99805	10419	76939
115	24038	65541	85788	55835	38835	59399
116	74976	14631	35908	28221	39470	91548
117	35553	71628	70189	26436	63407	91178
118	35676	12797	51434	82976	42010	26344
119	74815	67523	72985	23183	02446	63594
120	45246	88048	65173	50989	91060	89894
121	76509	47069	86378	41797	11910	49672
122	19689	90332	04315	21358	97248	11188
123	42751	35318	97513	61537	54955	08159
124	11946	22681	45045	13964	57517	59419
125	96518	48688	20996	11090	48396	57177
126	35726	58643	76869	84622	39098	36083
127	39737	42750	48968	70536	84864	64952
128	97025	66492	56177	04049	80312	48028
129	62814	08075	09788	56350	76787	51591
130	25578	22950	15227	83291	41737	79599
131	68763	69576	88991	49662	46704	63362
132	17900	00813	64361	60725	88974	61005
133	71944	60227	63551	71109	05624	43836
134	54684	93691	85132	64399	29182	44324
135	25946	27623	11258	65204	52832	50880
136	01353	39318	44961	44972	91766	90262
137	99083	88191	27662	99113	57174	35571
138	52021	45406	37945	75234	24327	86978
139	78755	47744	43776	83098	03225	14281
140	25282	69106	59180	16257	22810	43609
141	11959	94202	02743	86847	79725	51811
142	11644	13792	98190	01424	30078	28197
143	06307	97912	68110	59812	95448	43244
144	76285	75714	89585	99296	52640	46518
145	55322	07598	39600	60866	63007	20007
146	78017	90928	90220	92503	83375	26986
147	44768	43342	20696	26331	43140	69744
148	25100	19336	14605	86603	51680	97678
149	83612	46623	62876	85197	07824	91392
150	41347	81666	82961	60413	71020	83658

TABLE X. TABLE OF RANDOM NUMBERS

Table X 261

(7)	(8)	(9)	(10)	(11)	(12)	(13)	(14)
00770	22878	02179	51602	07270	76517	97275	45960
41583	17564	27395	63904	41548	49197	82277	24120
65793	83287	34142	13241	30590	97760	35848	91983
69721	57484	67501	77638	44331	11257	71131	11059
98539	27186	31237	80612	44488	97819	70401	95419
31211	54288	39296	37318	65724	90401	79017	62077
33216	19358	02591	54263	88449	01912	07436	50813
12489	51924	86871	92446	36607	11458	30440	52639
77940	39298	97838	95145	32378	68038	89351	37005
89295	59062	39404	13198	59960	70408	29812	83126
71051	84724	52492	22342	78071	17456	96104	18327
11392	20724	54322	36923	70009	23233	65438	59685
90628	52506	02016	85151	88598	47821	00265	82525
25993	03544	21560	83471	43989	90770	22965	44247
13790	35112	01324	39520	76210	22467	83275	32286
12854	30166	09073	75887	36782	00268	97121	57676
90348	55359	80392	41012	36270	77786	89578	21059
92920	92155	58807	54644	58581	95331	78629	73344
98924	20633	58842	85961	07648	70164	34994	67662
36036	32819	68559	99221	49475	50558	34698	71800
88575	97966	32466	10083	54728	81972	58975	30761
39062	63312	52496	07349	79178	33692	57352	72862
00337	80778	27507	95478	21252	12746	37554	97775
58045	44067	58716	58840	45557	96345	33271	53464
83867	86464	14342	21545	46717	72364	86954	55580
72505	92265	23107	60278	05822	46760	44294	07672
38404	94317	65402	13589	01055	79044	19308	83623
26408	43591	75528	65341	49044	95495	81256	53214
54509	49295	85830	59860	30883	89660	96142	18354
96191	71845	86899	70694	24290	01551	80092	82118
56625	00481	73323	91427	15264	06969	57048	54149
99709	30666	26451	11528	44323	34778	60342	60388
58254	26160	32116	63403	35404	57146	10909	07346
14491	55226	78793	34107	30374	48429	51376	09559
22273	05554	99521	73791	85744	29276	70326	60251
56073	06606	51826	18893	83448	31915	97764	75091
99884	13951	71057	53961	61448	74909	07322	80960
22644	87779	23753	99926	63898	54886	18051	96314
83637	55984	13300	52212	58781	14905	46502	04472
12224	25643	89884	31149	85423	32581	34374	70873
12998	76844	05320	54236	53891	70226	38632	84776
55583	05197	47714	68440	22016	79204	06862	94451
31262	88880	13040	16458	43813	89416	42482	33939
55486	90754	88932	19937	57119	23251	55619	23679
66819	84164	61131	81429	60676	42807	78286	29015
74399	30885	88567	29169	72816	53357	15428	86932
82928	24988	94237	46138	77426	39039	55596	12655
24261	02464	86563	74812	60069	71674	15478	47642
58317	37726	84628	42221	10268	20692	15699	29167
0241-	33322	66036	98712	46795	16308	28413	05417

TABLE X.   TABLE OF RANDOM NUMBERS

Table X

Col. / Line	(1)	(2)	(3)	(4)	(5)	(6)
151	38128	51178	75096	13609	16110	73533
152	60950	00455	73254	96067	50717	13878
153	90524	17320	29832	96118	75792	25326
154	49897	18278	67160	39408	97056	43517
155	18494	99209	81060	19488	65596	59787
156	65373	72984	30171	37741	70203	94094
157	40653	12843	04213	70925	95360	55774
158	51638	22238	56344	44587	83231	50317
159	69742	99303	62578	83575	30337	07488
160	58012	74072	67488	74580	47992	69482
161	18348	19855	42887	08279	43206	47077
162	59614	09193	58064	29086	44385	45740
163	75688	28630	39210	52897	62748	72658
164	13941	77802	69101	70061	35460	34576
165	96656	86420	96475	86458	54463	96419
166	03363	82042	15942	14549	38324	87094
167	70366	08390	69155	25496	13240	57407
168	47870	36605	12927	16043	53257	93796
169	79504	77606	22761	30518	28373	73898
170	46967	74841	50923	15339	37755	98995
171	14558	50769	35444	59030	87516	48193
172	12440	25057	01132	38611	28135	68089
173	32293	29938	68653	10497	98919	46587
174	10640	21875	72462	77981	56550	55999
175	47615	23169	39571	56972	20628	21788
176	16948	11128	71624	72754	49084	96303
177	21258	61092	66634	70335	92448	17354
178	15072	48853	15178	30730	47481	48490
179	99154	57412	09858	65671	70655	71479
180	08759	61089	23706	32994	35426	36666
181	67323	57839	61114	62192	47547	58023
182	09255	13986	84834	20764	72206	89393
183	36304	74712	00374	10107	85061	69228
184	15884	67429	86612	47367	10242	44880
185	18745	32031	35303	08134	33925	03004
186	72934	40086	88292	65728	38300	42323
187	17626	02944	20910	57662	80181	38579
188	27117	61399	50967	41399	81636	16663
189	93995	18678	90012	63645	85701	85269
190	67392	89421	09623	80725	62620	84162
191	04910	12261	37566	80016	21245	69377
192	81453	20283	79929	59839	23875	13245
193	19480	75790	48539	23703	15537	48885
194	21456	13162	74608	81011	55512	07481
195	89406	20912	46189	76376	25538	87212
196	09866	07414	55977	16419	01101	69343
197	86541	24681	23421	13521	28000	94917
198	10414	96941	06205	72222	57167	83902
199	49942	06683	41479	58982	56288	42853
200	23995	68882	42291	23374	24299	27024

TABLE X. TABLE OF RANDOM NUMBERS

Table X 263

(7)	(8)	(9)	(10)	(11)	(12)	(13)	(14)
42564	59870	29399	67834	91055	89917	51096	89011
03216	78274	65863	37011	91283	33914	91303	49326
22940	24904	80523	38928	91374	55597	97567	38914
84426	59650	20247	19293	02019	14790	02852	05819
47939	91225	98768	43688	00438	05548	09443	82897
87261	30056	58124	70133	18936	02138	59372	09075
76439	61768	52817	81151	52188	31940	54273	49032
74541	07719	25472	41602	77318	15145	57515	07633
51941	84316	42067	49692	28616	29101	03013	73449
58624	17106	47538	13452	22620	24260	40155	74716
42637	45606	00011	20662	14642	49984	94509	56380
70752	05663	49081	26960	57454	99264	24142	74648
98059	67202	72789	01869	13496	14663	87645	89713
15412	81304	58757	35498	94830	75521	00603	97701
55417	41375	76886	19008	66877	35934	59801	00497
19069	67590	11087	68570	22591	65232	85915	91499
91407	49160	07379	34444	94567	66035	38918	65708
52721	73120	48025	76074	95605	67422	41646	14557
30550	76684	77366	32276	04690	61667	64798	66276
40162	89561	69199	42257	11647	47603	48779	97907
02945	00922	48189	04724	21263	20892	92955	90251
10954	10097	54243	06460	50856	65435	79377	53890
77701	99119	93165	67788	17638	23097	21468	36992
87310	69643	45124	00349	25748	00844	96831	30651
51736	33133	72696	32605	41569	76148	91544	21121
27830	45817	67867	18062	87453	17226	72904	71474
83432	49608	66520	06442	59664	20420	39201	69549
41436	25015	49932	20474	53821	51015	79841	32405
63520	31357	56968	06729	34465	70685	04184	25250
63988	98844	37533	08269	27021	45886	22835	78451
64630	34886	98777	75442	95592	06141	45096	73117
34548	93438	88730	61805	78955	18952	46436	58740
81969	92216	03568	39630	81869	52824	50937	27954
12060	44309	46629	55105	66793	93173	00480	13311
59929	95418	04917	57596	24878	61733	92834	64454
64068	98373	48971	09049	59943	36538	05976	82118
24580	90529	52303	50436	29401	57824	86039	81062
15634	79717	94696	59240	25543	97989	63306	90946
62263	68331	00389	72571	15210	20769	44686	96176
87368	29560	00519	84545	08004	24526	41252	14521
50420	85658	55263	68667	78770	04533	14513	18099
46808	74124	74703	35769	95588	21014	37078	39170
02861	86587	74539	65227	90799	58789	96257	02708
93551	72189	76261	91206	89941	15132	37738	59284
20748	12831	57166	35026	16817	79121	18929	40628
13305	94302	80703	57910	36933	57771	42546	03003
07423	57523	97234	63951	42876	46829	09781	58160
07460	69507	10600	08858	07685	44472	64220	27040
92196	20632	62045	78812	35895	51851	83534	10689
67460	94783	40937	16961	26053	78749	46704	21983

TABLE X. TABLE OF RANDOM NUMBERS

Table X

Col. Line	(1)	(2)	(3)	(4)	(5)	(6)
201	78994	36244	02673	25475	84953	61793
202	04909	58485	70686	93930	34880	73059
203	46582	73570	33004	51795	86477	46736
204	29242	89792	88634	60285	07190	07795
205	68104	81339	97090	20601	78940	20228
206	17156	02182	82504	19880	93747	80910
207	50711	94789	07171	02103	99057	98775
208	39449	52409	75095	77720	39729	03205
209	75629	82729	76916	72657	58992	32756
210	01020	55151	36132	51971	32155	60735
211	08337	89989	24260	08618	66798	25889
212	76829	47229	19706	30094	69430	92399
213	39708	30641	21267	56501	95182	72442
214	89836	55817	56747	75195	06818	83043
215	25903	61370	66081	54076	67442	52964
216	71345	03422	01015	68025	19703	77313
217	61454	92263	14647	08473	34124	10740
218	80376	08909	30470	40200	46558	61742
219	45144	54373	05505	90074	24783	86299
220	12191	88527	58852	51175	11534	87218
221	62936	59120	73957	35969	21598	47287
222	31588	96798	43668	12611	01714	77266
223	20787	96048	84726	17512	39450	43618
224	45603	00745	84635	43079	52724	14262
225	31606	64782	34027	56734	09365	20008
226	10452	33074	76718	99556	16026	00013
227	37016	64633	67301	50949	91298	74968
228	66725	97865	25409	37498	00816	99262
229	07380	74438	82120	17890	40963	55757
230	71621	57688	58256	47702	74724	89419
231	03466	13263	23917	20417	11315	52805
232	12692	32931	97387	34822	53775	91674
233	52192	30941	44998	17833	94563	23062
234	56691	72529	66063	73570	86860	68125
235	74952	43041	58869	15677	78598	43520
236	18752	43693	32867	53017	22661	39610
237	61691	04944	43111	28325	82319	65589
238	49197	63948	38947	60207	70667	39843
239	19436	87291	71684	74859	76501	93456
240	39143	64893	14606	13543	09621	68301
241	82244	67549	76491	09761	74494	91307
242	55847	56155	42878	23708	97999	40131
243	94095	95970	07826	25991	37584	56966
244	11751	69469	25521	44097	07511	88976
245	69902	08995	27821	11758	64989	61902
246	21850	25352	25556	92161	23592	43294
247	75850	46992	25165	55906	62339	88958
248	29648	22086	42581	85677	20251	39641
249	82740	28443	42734	25518	82827	35825
250	36842	42092	52075	83926	42875	71500

TABLE X.   TABLE OF RANDOM NUMBERS

Table X

265

(7)	(8)	(9)	(10)	(11)	(12)	(13)	(14)
50243	63423	69309	80308	49977	18075	43227	08266
06823	80257	44193	08337	47655	75932	29209	41954
60460	70345	37322	19987	67143	41129	89514	46892
27011	85941	01852	43096	31173	43730	48505	17958
22803	96070	10251	62711	66200	74330	13820	18966
78260	25136	62018	62919	73801	57195	83457	70597
37997	18325	88281	61091	97889	79977	04544	72963
09313	43545	43786	70443	41350	73369	42405	80516
01154	84890	04107	17469	59346	68651	97433	89491
64867	35424	25257	93844	39928	52519	34368	02114
52860	57375	52815	43539	18072	44270	27309	56535
98749	22081	52564	90431	35208	40323	87505	10227
21445	17276	90344	33199	02522	97883	09515	65930
47403	58266	52630	75573	91088	41118	27195	40650
23823	02718	28786	06121	29680	55295	67086	57574
04555	83425	46763	95315	23150	15116	18017	42730
40839	05620	62418	73374	92577	06755	21856	56272
11643	92121	22294	26648	69676	46198	00331	85186
20900	15144	26506	53770	76431	23861	71208	80694
04876	85584	78465	82182	03412	13217	14313	70593
39394	08778	38036	30140	89117	32054	44603	61849
55079	24690	84716	77732	35363	85525	17015	56344
30629	24356	05294	34236	65299	36922	46995	65765
05750	89373	79088	38088	65082	92504	80545	03090
93559	78384	99219	61747	96111	86965	33233	29812
78411	95107	10786	44886	44612	06830	27848	87597
73631	57397	08632	04762	69328	34926	07403	60916
14471	10232	19035	21695	07540	96447	20743	92472
13492	68294	87170	49468	40164	13374	23021	17006
08025	68519	95188	54788	32999	34374	05780	17506
33072	07723	87876	75258	22709	99869	11609	46666
76549	37635	91118	31062	89441	31839	88614	78168
95725	38463	03665	49189	46359	37401	73407	61817
40436	31303	79330	59083	34862	00540	21734	75535
97521	83248	52173	17636	77106	01044	22990	74874
03796	02622	78267	24503	73518	76545	99088	08369
66048	98498	46941	81427	44447	70357	18864	15525
60607	15328	09528	17277	84278	04463	12188	35359
95714	92518	10683	75617	78841	25315	74041	71554
69817	52140	03976	48795	60266	99592	68334	18790
64222	66592	67270	38593	18094	95095	08649	25047
52360	90390	73108	40475	80487	07787	35238	50990
68623	83454	49461	97707	12479	25041	40565	18313
30122	67542	54825	03274	02765	67162	40312	76127
32121	28165	21326	97375	44801	66977	08232	06807
10479	37879	21825	11453	29584	70067	09471	16319
91717	15756	78817	35541	01177	06869	10543	57652
65786	80689	49066	14456	91681	69371	18292	39377
90288	32911	79666	52959	01475	83321	24991	80102
69216	01350	92846	84792	87455	06842	22422	77379

TABLE X. TABLE OF RANDOM NUMBERS

Table X

Col. Line	(1)	(2)	(3)	(4)	(5)	(6)
251	89429	26726	15563	94972	78739	04419
252	43427	25412	25587	21276	44426	17369
253	58575	81958	51846	02676	67781	95137
254	61888	71246	24246	23487	78639	92006
255	73891	47025	40937	71907	26827	98865
256	40938	73894	40854	15997	55293	95033
257	98053	43567	17292	86908	71364	06089
258	59774	29138	46993	39836	99596	59050
259	09765	07548	63043	59782	81449	13652
260	38991	64502	24770	29209	82909	66610
261	25622	27100	56128	62145	82388	45197
262	31864	74120	66231	82306	91784	33177
263	81171	75639	60863	49562	28846	81581
264	69874	52803	28544	51569	56090	44558
265	27848	51107	05761	02159	53911	01952
266	69407	69736	75375	31488	67528	84234
267	29418	03091	06364	13151	40663	43633
268	38222	31231	79415	44558	62490	26936
269	94720	83796	93251	03568	62484	29140
270	45275	16852	02284	41361	73733	61486
271	97260	09352	82626	42915	45847	87401
272	01990	65259	60684	78175	43825	45211
273	24633	42314	81192	50253	67516	59076
274	98071	52677	74920	74461	52266	26967
275	34101	79442	28403	48541	13010	16596
276	77186	93967	25918	66403	73837	73445
277	23114	05481	42335	51396	60823	22680
278	59988	49944	41038	99977	16348	41119
279	11852	42254	82304	05588	75165	20179
280	59992	87922	56299	01700	07003	97507
281	42116	86593	22828	41422	18176	03250
282	39663	61401	21471	42702	70588	53144
283	53542	72009	96296	68908	58657	87117
284	25996	76108	98476	36397	89457	19577
285	91106	26450	14451	50328	29084	32332
286	37133	88924	27845	13024	90687	23726
287	13982	25736	10087	16762	02564	27250
288	26663	36187	81688	25005	46677	75851
289	62572	08275	16313	24936	81680	53829
290	65925	95455	08383	24643	72962	08172
291	97978	74676	08942	48919	51592	71196
292	01914	42524	67820	47985	91773	10383
293	68565	44811	39238	70394	78555	33539
294	54370	31672	03893	32423	54092	69375
295	79954	89601	23881	46951	69084	33477
296	55479	01069	44229	56975	06785	80930
297	38114	70330	42157	86699	46212	74692
298	29766	83452	66202	02488	72704	97821
299	31771	70640	34779	41831	33456	53194
300	77522	87188	83577	99067	83235	48662

TABLE X.   TABLE OF RANDOM NUMBERS

Table X 267

(7)	(8)	(9)	(10)	(11)	(12)	(13)	(14)
60523	31022	23728	37647	16476	11170	68376	56874
29010	45337	90245	92053	41447	14897	18753	68291
88430	78260	66962	31812	12759	06427	40337	50115
63846	92263	33212	26516	93662	72399	88244	33922
38882	25757	26662	91441	89357	87803	61521	80600
31736	75068	91314	75293	04895	39355	54837	57203
92394	73691	57883	09983	35643	79309	53449	95334
25419	04130	54632	17223	94604	22973	97731	99476
94420	74460	46707	94303	85523	95244	70995	10742
84418	66214	26001	78685	69117	72446	79783	22305
97609	83942	01120	71717	32858	58679	97165	02810
17681	18963	07216	49288	43185	62797	00735	27085
10249	23190	53440	32357	16472	99013	24328	93670
42095	92311	57915	13368	13719	15833	38744	56065
59273	32250	39647	29908	49075	23061	07795	95047
76462	13628	21286	13736	67478	45218	27867	93049
87954	69800	24773	62596	52476	60631	50503	94116
49682	16307	98535	44822	99574	58487	85020	68881
14152	37044	90398	92042	35099	31640	99753	44409
33189	08907	41159	08147	15472	33250	17361	79961
13339	53850	34931	00602	75307	99708	77863	04924
86287	78190	02431	66251	74970	50246	23975	80697
92006	65676	87343	89231	15760	73706	69426	01979
68284	31612	40335	28865	98949	64492	96905	29184
72001	38546	76305	22119	82668	84017	44111	40302
86663	15929	08237	05647	15785	70444	58670	95967
50459	05429	35227	92559	24136	13126	22099	52388
51548	19511	90142	65604	16147	63445	60525	10480
94198	25700	33473	59554	30974	69973	57629	38550
69260	53349	86947	27517	80159	01899	46890	53850
06079	85467	32052	56922	96804	51060	33157	83948
27087	05591	57759	51394	98873	45625	61069	78783
21483	28879	20480	57309	95552	09826	79928	17141
65877	04802	61938	25032	09190	74932	36925	82686
08635	25192	31337	20249	95073	93800	70022	99968
11212	30414	42185	49224	46560	80447	24334	74866
79316	83848	38684	20552	44402	85153	94526	41256
73938	73044	05132	61204	90354	90296	03182	36672
40412	01479	24241	58488	65341	93414	07135	43446
37824	87587	40698	34964	50166	74756	77033	41501
48534	16955	25759	95645	03148	10646	15660	86520
89514	07557	02084	16736	39198	69697	62485	61938
56310	40809	63204	14479	19635	97299	66947	58010
63308	08016	28407	98287	22874	57545	72695	01604
87968	15639	82409	34125	36864	52112	27102	87334
26443	44892	77561	51123	34495	31376	06238	15973
92603	91306	58558	57280	50639	80563	71370	81487
70614	53616	39050	30355	15340	97298	41795	35185
19602	74194	61154	51774	76822	73794	54182	45264
31503	54829	54723	13177	15307	26073	68915	88415

TABLE X.   TABLE OF RANDOM NUMBERS

Table X

Col. Line	(1)	(2)	(3)	(4)	(5)	(6)
301	64670	10396	82981	58320	71478	08143
302	25771	02205	73984	28436	88192	11470
303	27551	13537	54984	89406	88326	33993
304	91224	22417	44820	26189	57541	87558
305	75179	64320	71523	67868	38883	09674
306	64654	91085	65818	03313	39273	46384
307	98059	81123	67832	04102	66188	78200
308	38765	63585	18810	95805	11414	58096
309	01921	03564	71754	10213	80383	13473
310	16211	93671	27704	66778	96307	06732
311	70232	86076	61527	56123	48514	53935
312	22332	94265	67627	85815	00394	75271
313	81333	45965	64171	84367	15052	37965
314	39333	47453	66174	04546	10594	64271
315	29195	20825	50878	80273	26285	90070
316	74420	64037	06960	25109	08821	60143
317	22763	16508	24866	13177	07464	51730
318	72919	54618	40616	33287	51274	78491
319	92385	42402	15922	90033	21555	31647
320	85431	19857	97246	46118	71222	82744
321	40778	12451	14921	51464	45331	75822
322	88903	46592	60637	65231	08778	86813
323	29830	34899	85457	19548	83355	52479
324	22832	47422	08073	10107	46772	92299
325	75159	14809	11930	83531	51239	86298
326	99390	08217	56276	09263	82685	30451
327	68622	80897	08902	10867	91379	30068
328	92393	95901	41179	72129	72502	91097
329	53122	66033	38229	51879	29925	45574
330	43251	11941	86631	93264	53433	70281
331	16613	24901	34866	75002	55163	68308
332	12010	60852	92603	70393	17989	95755
333	85528	97879	27814	08219	02908	71582
334	32590	55079	33556	83169	92087	77939
335	92934	30650	16449	15805	61551	38689
336	80614	10150	09389	61892	79477	14522
337	62398	12034	90764	52872	22285	50592
338	02222	46811	05145	67916	15184	02636
339	08690	31785	61664	61322	24149	21471
340	61187	73897	66168	12885	73191	89432
341	12324	61149	85643	64999	63738	46671
342	47635	42279	98620	70677	52386	50904
343	70965	00390	08878	15373	70276	71889
344	58764	15262	96814	54548	00042	19721
345	07429	05609	31207	50254	68389	07714
346	15665	28659	54952	53217	76898	88931
347	64208	53232	99459	43605	04553	48451
348	17952	73276	52567	48489	64264	24220
349	60531	43217	39999	38615	97195	76928
350	76692	39999	43254	68110	88053	88727

TABLE X.   TABLE OF RANDOM NUMBERS

Table X 269

(7)	(8)	(9)	(10)	(11)	(12)	(13)	(14)
48294	42631	45464	58092	14187	12271	98179	87812
11775	67385	66360	59884	93873	29948	66302	82227
92324	13249	35271	60400	70762	08343	76456	90068
45835	28461	54835	92411	44369	47512	49508	02841
27645	76240	47587	01677	38342	85598	12482	30749
66677	14148	87552	38383	67435	21072	63866	74644
67466	46043	65406	22834	08620	17509	51424	25187
00295	82626	42683	44518	12209	83245	53771	95469
94128	62199	59411	46782	62871	51149	87146	40129
63750	04191	40003	51653	54228	14916	05361	08884
86784	42351	67586	07432	61499	01773	97463	58815
98385	53697	56378	50592	77441	88505	89791	16331
03122	81914	69381	70034	92563	61804	58326	97895
61026	39471	55981	18628	67943	35599	37209	34061
79586	12449	77293	36577	59192	03658	90056	83145
34485	19257	29417	72713	72326	41572	41553	46946
65802	95718	28560	11332	74272	59189	53167	13133
53604	66742	97777	64468	98224	45485	17257	31561
22288	75692	20592	84620	58679	24587	83517	55327
67892	77155	10785	00344	19641	98279	18716	13895
46859	66829	35803	27645	76095	41535	25508	53066
47819	19218	46837	89671	77661	08518	85216	62664
77801	01596	48890	56104	68733	40830	58611	59181
42975	86376	27869	52954	07900	75918	51398	87598
72661	63015	98804	98491	99565	42801	71816	84000
25742	41105	74711	42007	02082	93025	86641	28952
84289	45020	92459	03831	08531	63496	98230	42884
09488	84896	37720	68104	73817	67626	16221	63527
53938	72801	64067	76328	28941	43645	37181	95329
55000	24550	74751	32855	25399	95743	85393	20261
20070	36953	39378	71191	84510	47599	93608	24379
14672	58786	41996	02893	94163	36156	54203	94138
31439	00360	72264	87245	65903	42298	28061	81889
53792	78795	58159	86394	41749	91623	26973	81474
59179	85485	18537	70496	98694	19796	76804	03673
40270	45744	29582	29717	39590	10223	43049	78775
42505	80560	38213	18917	10015	03887	62589	15851
59078	57773	21259	86090	56705	65556	04487	95954
23328	03093	31266	14840	30703	01640	07874	16630
65414	41886	75911	35708	43208	59193	04727	31037
25408	69313	54455	04917	35047	09951	72776	84697
97403	03931	42090	28179	98028	47728	45696	74176
86953	37931	23286	20508	40100	22486	37323	35429
78869	85937	36639	29135	12633	67225	69588	74178
92268	64698	32823	60122	46213	05646	54742	98304
25786	55912	85269	29212	84976	08888	94332	58528
68154	49436	49891	65524	65133	55163	76765	26006
55498	97548	98437	26033	39026	17377	43519	27425
87688	99010	90189	12522	00675	01995	82781	95130
14187	98623	84225	78440	67082	37425	40559	16838

TABLE X.   TABLE OF RANDOM NUMBERS

Col. Line	(1)	(2)	(3)	(4)	(5)	(6)
351	06433	80674	24520	18222	10610	05794
352	39298	47829	72648	37414	75755	04717
353	89884	59651	67533	68123	17730	95862
354	61512	32155	51906	61662	64130	16688
355	99653	47635	12506	88535	36553	23757
356	95913	11085	13772	76638	48423	25018
357	55864	44004	13122	44115	01601	50541
358	35334	82410	91601	40617	72876	33967
359	57729	88646	76487	11622	96297	24160
360	86648	89317	63677	70119	94739	25875
361	30574	06039	07967	32422	76791	39725
362	81307	13114	83580	79974	45929	85113
363	02410	96385	79007	54939	21410	86980
364	18969	87444	52233	62319	08598	09066
365	87863	80514	66860	62297	80198	19347
366	68397	10538	15438	62311	72844	60203
367	28529	45247	58729	10854	99058	18260
368	44285	09452	15867	70418	57012	72122
369	86299	22510	33571	23309	57040	29285
370	84842	05748	90894	61658	15001	94055
371	56970	10799	52098	04184	54967	72938
372	83125	85077	60490	44369	66130	72936
373	55503	21383	02464	26141	68779	66388
374	47019	06683	33203	29608	54553	25971
375	84828	61152	79526	29554	84580	37859
376	68921	31331	79227	05748	51276	57143
377	36458	28285	30424	98420	72925	40729
378	95752	96065	36847	87729	81679	59126
379	26768	02513	58454	56958	20575	76746
380	42613	72456	43636	58085	06766	60227
381	95457	12176	65482	25596	02678	54592
382	95276	67524	63564	95958	39750	64379
383	66954	53574	64776	92345	95110	59448
384	17457	44151	14113	62462	02798	54977
385	03704	23322	83214	59337	01695	60666
386	21538	16997	33210	60337	27976	70661
387	57178	16739	98310	70348	11317	71623
388	31048	40058	94953	55866	96283	46620
389	69799	83300	16498	80733	96422	58078
390	90595	65017	59231	17772	67831	33317
391	33570	34761	98939	78784	09977	29398
392	15340	82760	57477	13898	48431	72936
393	64079	07733	36512	56186	99098	48850
394	63491	84886	67118	62063	74958	20946
395	92003	76568	41034	28260	79708	00770
396	52360	46658	66511	04172	73085	11795
397	74622	12142	68355	65635	21828	39539
398	04157	50079	61343	64315	70836	82857
399	86003	60070	66241	32836	27573	11479
400	41268	80187	20351	09636	84668	42486

TABLE X.   TABLE OF RANDOM NUMBERS

Table X 271

(7)	(8)	(9)	(10)	(11)	(12)	(13)	(14)
37515	48619	62866	33963	14045	79451	04934	45576
29899	78817	03509	78673	73181	29973	18664	04555
08034	19473	63971	37271	31445	49019	49405	46925
37275	51262	11569	08697	91120	64156	40365	74297
34209	55803	96275	26130	47949	14877	69594	83041
99041	77529	81360	18180	97421	55541	90275	18213
00147	77685	58788	33016	61173	93049	04694	43534
73830	15405	96554	88265	34537	38526	67924	40474
09903	14047	22917	60718	66487	46346	30949	03173
38829	68377	43918	77653	04127	69930	43283	35766
53711	93385	13421	67957	20384	58731	53396	59723
72268	09858	52104	32014	53115	03727	98624	84616
91772	93307	34116	49516	42148	57740	31198	70336
95288	04794	01534	92058	03157	91758	80611	45357
73234	86265	49096	97021	92582	61422	75890	86442
46412	65943	79232	45702	67055	39024	57383	44424
38765	90038	94209	04055	27393	61517	23002	96560
36634	97283	95943	78363	36498	40662	94188	18202
67870	21913	72958	75637	99936	58715	07943	23748
36308	41161	37341	81838	19389	80336	46346	91895
56834	23777	98392	31417	98547	92058	02277	50315
69848	59973	08144	61070	73094	27059	69181	55623
75242	82690	74099	77885	23813	10054	11900	44653
69573	83854	24715	48866	65745	31131	47636	45137
28504	61980	34997	41825	11623	07320	15003	56774
31926	99915	45821	97702	87125	44488	77613	56823
22337	48293	86847	43186	42951	37804	85129	28993
59437	33225	31280	41232	34750	91097	60752	69783
49878	06846	32828	24425	30249	78801	26977	92074
96414	32671	45587	79620	84831	38156	74211	82752
63607	82096	21913	75544	55228	89796	05694	91552
46059	51666	10433	10945	55306	78562	89630	41230
77249	54044	67942	24145	42294	27427	84875	37022
48349	66738	60184	75679	38120	17640	36242	99357
97410	55064	17427	89180	74018	44865	53197	74810
08250	69599	60264	84549	78007	88450	06488	72274
55510	64756	87759	92354	78694	63638	80939	98644
52087	80817	74533	68407	55862	32476	19326	95558
99643	39847	96884	84657	33697	39578	90197	80532
00520	90401	41700	95510	61166	33757	23279	85523
93896	78227	90110	81378	96659	37008	04050	04228
78160	87240	52716	87697	79433	16336	52862	69149
72527	08486	10951	26832	39763	02485	71688	90936
28147	39338	32169	03713	93510	61244	73774	01245
88643	21188	01850	69629	49426	49128	14660	14143
52594	13287	22531	04388	64693	11934	35051	68576
18988	53609	04001	19648	14053	49623	10840	31915
35335	87900	36194	31567	53506	34304	39910	79630
94114	81641	00496	36058	75899	46620	70024	88753
71303	19512	50277	71508	20116	79520	06269	74173

TABLE X.   TABLE OF RANDOM NUMBERS

Table X

Col. Line	(1)	(2)	(3)	(4)	(5)	(6)
401	05073	90103	85167	53900	19720	41488
402	93320	80269	56684	39192	53220	74539
403	18806	70257	96424	13606	14356	76599
404	22253	45923	29815	18578	23316	30896
405	93640	45982	40011	74142	29106	45729
406	47630	45980	76619	57138	57492	00030
407	01781	55061	07455	47083	71870	90597
408	69694	45054	33587	03664	95007	31567
409	51236	05052	26503	94651	29874	73492
410	89445	51039	73837	26720	38650	47322
411	40867	96834	02162	41517	88937	26099
412	92946	56944	93407	05010	54896	33173
413	75898	02275	90768	31902	52114	36634
414	22729	21695	90824	80500	09332	54667
415	28733	62663	23644	16416	47135	39137
416	51323	37770	42114	79742	59905	38480
417	69325	65551	49927	68073	56979	49454
418	11333	60801	36992	76128	27959	41306
419	86347	03703	36778	72501	95229	65735
420	73452	36179	82893	92262	43850	31888
421	75483	74009	73699	05870	36804	89338
422	73302	84917	75128	34085	86208	98399
423	42785	24350	05933	65282	12832	75382
424	40429	33209	58622	09308	38098	55947
425	92876	58271	99325	12301	72957	22690
426	32951	39844	99126	94838	48715	36586
427	09772	28139	48130	73301	35915	90923
428	78459	91322	50072	77941	65046	78363
429	14419	96517	99075	43664	81119	63487
430	97769	50967	24427	21011	92226	44380
431	09175	37545	39088	06879	21277	05153
432	52062	95519	54087	14072	50953	63477
433	70558	85169	01086	97202	10390	01819
434	22553	61317	08968	67521	16627	48855
435	95216	75263	60351	02643	00063	20824
436	49087	61399	47781	32173	96672	04528
437	24808	79068	70787	43106	97133	37236
438	89879	79942	43781	05069	80143	59176
439	61178	79295	58926	21977	28435	32631
440	37444	56047	23208	34710	12147	28558
441	99633	00363	16853	20789	87674	03938
442	87363	59239	42023	78056	51254	95644
443	23923	87269	85277	34727	78036	74471
444	45610	26370	13094	34500	36750	54517
445	44166	80095	08286	38126	48834	73423
446	81875	27486	53925	22330	37168	97954
447	79400	83852	52174	42577	18553	14023
448	42799	46647	36718	49704	17150	07935
449	09302	36408	64569	93033	95645	56791
450	88078	81456	17242	84590	93660	34619

TABLE X.   TABLE OF RANDOM NUMBERS

Table X 273

(7)	(8)	(9)	(10)	(11)	(12)	(13)	(14)
57476	39458	16621	69774	47953	35039	39283	21573
26393	00787	94490	23386	38454	33466	32159	77439
25390	63236	04513	16358	30504	10551	32498	18685
64771	11220	86218	75956	22399	36234	61644	80682
43406	21457	04301	39651	76025	73819	11462	97385
77897	76236	64990	35985	57748	11606	72081	18359
10151	59606	96919	31174	99872	15843	99173	79512
25334	26433	75002	67607	33135	07076	82984	82675
88941	08488	09418	08173	63380	82067	58143	64983
68474	95047	20404	41577	46865	39849	78735	99192
56047	49164	35127	64916	75451	79160	14014	00445
30548	23667	43171	47849	40449	91072	91092	17613
46803	97970	92216	55398	75320	70475	82931	20172
46696	38166	02005	24615	85613	25948	75389	25765
62190	31032	58702	03805	67252	23712	92697	19071
25293	32993	36946	62701	51198	72941	52215	85257
79451	60753	70872	07422	06399	75240	80847	78231
93543	15926	99159	27102	98684	80175	98732	45405
14269	50220	77270	68604	05677	23347	43686	31584
71151	40682	49775	63628	45415	96270	31735	01509
73891	40740	98753	74566	74733	34777	05786	38294
79433	61960	01720	87458	24023	89971	09532	68155
29826	33197	81781	53542	63985	57022	22712	61343
12001	73526	23170	13721	37856	86502	74299	01346
62705	73892	01974	77759	92733	11331	08323	86196
42076	15283	19280	29166	24522	73131	83401	38920
19255	75242	84655	30163	75510	83315	98529	93805
21951	42319	46472	67617	34134	05905	61251	51040
95589	51785	07398	23245	10086	49097	46173	00507
23422	10654	43617	80504	90663	60751	79728	41132
81855	84043	35307	59465	75395	74758	09427	84460
64635	34552	75243	70222	75023	81454	70606	31861
88167	21851	87837	85287	69883	08289	74968	46947
97263	94242	93354	72446	28840	88195	82751	94352
67468	89441	84055	47035	29741	47972	61914	66864
15881	46764	20115	03226	79308	31970	49804	85150
77888	48451	20788	44648	70350	54965	57715	94826
47392	70372	26899	16228	71205	14564	97087	95690
23062	31822	70462	05965	22312	33013	74612	23733
58817	98807	56775	08129	08794	23646	92846	61706
36077	41012	08813	51168	78822	37353	61281	31172
90527	41398	74996	94977	22149	96616	54435	52469
12157	11655	25194	47557	26181	67825	80224	41490
85011	26567	01021	32485	58903	43529	24191	91832
13617	08853	16286	16023	77901	39118	14288	39385
11967	03309	97096	64221	11318	98720	01100	13651
69629	61913	41050	69689	57284	38160	57756	16762
62372	39933	20838	27652	54801	41067	08240	35163
14830	81699	45057	85796	63756	93944	60649	84847
51965	85618	36558	54410	68456	98504	83011	19393

TABLE X.   TABLE OF RANDOM NUMBERS

Table X

Col. Line	(1)	(2)	(3)	(4)	(5)	(6)
451	85018	23508	91507	76455	54941	72711
452	11904	73678	08272	62941	02349	71389
453	75344	98489	86268	73652	98210	44546
454	65566	65614	01443	07607	11826	91326
455	51872	72294	95432	53555	96810	17100
456	03805	37913	98633	81009	81060	33449
457	21055	78685	71250	10329	56135	80647
458	48977	36794	56054	59243	57361	65304
459	93077	72941	92779	23581	24548	56415
460	84533	26564	91583	83411	66504	02036
461	11338	12903	14514	27585	45068	05520
462	23853	68500	92274	87026	99717	01542
463	94096	74920	25822	98026	05394	61840
464	83160	82362	09350	98536	38155	42661
465	97425	47335	69709	01386	74319	04318
466	83951	11954	24317	20345	18134	90062
467	93085	35203	05740	03206	92012	42710
468	33762	83193	58045	89880	78101	44392
469	49665	85397	85137	30496	23469	42846
470	37541	82627	80051	72521	35342	56119
471	22145	85304	35348	82854	55846	18076
472	27153	08662	61078	52433	22184	33998
473	00301	49425	66682	25442	83668	66236
474	43815	43272	73778	63469	50083	70696
475	14689	86482	74157	46012	97765	27552
476	16680	55936	82453	19532	49988	13176
477	86938	60429	01137	86168	78257	86249
478	33944	29219	73161	46061	30946	22210
479	16045	67736	18608	18198	19468	76358
480	37044	52523	25627	63107	30806	80857
481	61471	45322	35340	35132	42163	69332
482	47422	21296	16785	66393	39249	51463
483	24133	39719	14484	58613	88717	29289
484	67253	67064	10748	16006	16767	57345
485	62382	76941	01635	35829	77516	98468
486	98011	16503	09201	03523	87192	66483
487	37366	24386	20654	85117	74078	64120
488	73587	83993	54176	05221	94119	20108
489	33583	68291	50547	96085	62180	27453
490	02878	33223	39199	49536	56199	05993
491	91498	41673	17195	33175	04994	09879
492	91127	19815	30219	55591	21725	43827
493	12997	55013	18662	81724	24305	37661
494	96098	13651	15393	69995	14762	69734
495	97627	17837	10472	18983	28387	99781
496	40064	47981	31484	76603	54088	91095
497	16239	68743	71374	55863	22672	91609
498	58354	24913	20435	30965	17453	65623
499	52567	65085	60220	84641	18273	49604
500	06236	29052	91392	07551	83532	68130

TABLE X.   TABLE OF RANDOM NUMBERS

Table X 275

(7)	(8)	(9)	(10)	(11)	(12)	(13)	(14)
39406	94620	27963	96478	21559	19246	88097	44926
45605	60947	60775	73181	43264	56895	04232	59604
27174	27499	53523	63110	57106	20865	91683	80688
29664	01603	23156	89223	43429	95353	44662	59433
35066	00815	01552	06392	31437	70385	45863	75971
68055	83844	90942	74857	52419	68723	47830	63010
51404	06626	10042	93629	37609	57215	08409	81906
93258	56760	63348	24949	11859	29793	37457	59377
61927	64416	29934	00755	09418	14230	62887	92683
02922	63569	17906	38076	32135	19096	96970	75917
56321	22693	35089	07694	04252	23791	60249	83010
72990	43413	59744	44595	71326	91382	45114	20245
83089	09224	78530	33996	49965	04851	18280	14039
02363	67625	34683	95372	74733	63558	09665	22610
99387	86874	12549	38369	54952	91579	26023	81076
10761	54548	49505	52685	63903	13193	33905	66936
34650	73236	66167	21788	03581	40699	10396	81827
53767	15220	66319	72953	14071	59148	95154	72852
94810	16151	08029	50554	03891	38313	34016	18671
97190	43635	84249	61254	80993	55431	90793	62603
12415	30193	42776	85611	57635	51362	79907	77364
87436	37430	45246	11400	20986	43996	73122	88474
79655	88312	93047	12088	86937	70794	01041	74867
13558	98995	58159	04700	90443	13168	31553	67891
49617	51734	20849	70198	67906	00880	82899	66065
94219	88698	41755	56216	66852	17748	04963	54859
46134	51865	09836	73966	65711	41699	11732	17173
79302	40300	08852	27528	84648	79589	95295	72895
69203	02760	28625	70476	76410	32988	10194	94917
84383	78450	26245	91763	73117	33047	03577	62599
98851	50252	56911	62693	73817	98693	18728	94741
95963	07929	66728	47761	81472	44806	15592	71357
77360	09030	39605	87507	85446	51257	89555	75520
42285	56670	88445	85799	76200	21795	38894	58070
51686	48140	13583	94911	13318	64741	64336	95103
55649	36764	86132	12463	28385	94242	32063	45233
04643	14351	71381	28133	68269	65145	28152	39087
78101	81276	00835	63835	87174	42446	08882	27067
18567	55524	86088	00069	59254	24654	77371	26409
71201	78852	65889	32719	13758	23937	90740	16866
70337	11861	69032	51915	23510	32050	52052	24004
78862	67699	01009	07050	73324	06732	27510	33761
18956	50064	39500	17450	18030	63124	48061	59412
89150	93126	17700	94400	76075	08317	27324	72723
52977	01657	92602	41043	05686	15650	29970	95877
00010	13800	76690	75133	60456	28491	03845	11507
51514	98135	42870	48578	29036	69876	86563	61729
93058	08313	99293	00990	13595	77457	79969	11339
47418	90974	83965	62732	85161	54330	22406	86253
56970	33273	61993	88407	69399	17301	70975	99129

TABLE X.   TABLE OF RANDOM NUMBERS

Table X

Col. Line	(1)	(2)	(3)	(4)	(5)	(6)
501	88188	99345	94118	40373	50387	24802
502	05200	50533	59428	02797	16833	10038
503	82828	41316	92617	31346	89263	06589
504	71006	99318	19269	35233	79183	78538
505	05937	00875	32264	82808	00229	03868
506	06021	04370	93070	90737	05354	68427
507	54789	10960	44023	57857	56556	83993
508	90400	05707	29128	14859	84117	72206
509	51424	01651	99970	73521	82356	03297
510	79743	88757	43370	86536	07166	06401
511	77418	00322	98854	51507	00565	33066
512	17580	49302	16408	05678	75532	46218
513	15489	45559	28548	64330	42126	43145
514	56342	66773	18536	32600	73958	75993
515	20202	19216	23762	47856	04623	70728
516	84877	51708	69357	67914	55372	97225
517	01647	00311	44989	21900	96079	15793
518	45652	89311	45302	14539	32045	86727
519	79975	06153	08932	59185	71386	19070
520	49744	54713	37053	77467	15348	03383
521	40922	94903	29638	46870	14108	84391
522	53319	48020	77444	51447	07916	99506
523	76682	10559	85446	56236	85919	76388
524	48869	97229	69581	84581	71728	45150
525	95961	19279	38078	17473	43945	21562
526	16521	25945	94076	91281	92272	41233
527	78282	26332	44072	55104	16895	98311
528	43473	39179	53174	43498	72674	13087
529	06513	31352	09177	21367	64725	23784
530	48734	39737	03448	99009	98136	34562
531	54832	70111	48339	75270	11652	41697
532	55844	69515	22658	75438	83086	41325
533	42829	54398	93338	90705	00626	97752
534	81128	63461	10925	44382	73365	98875
535	62885	26354	10368	78026	00186	46783
536	19525	10375	27010	42791	49471	90607
537	26570	99202	73924	59888	01827	93314
538	04772	17749	01537	96036	02102	02622
539	49129	12491	62552	64323	44856	29045
540	19937	75104	57780	95871	94547	53541
541	52571	67962	72775	28480	87411	12075
542	54943	80723	81195	84069	28144	48106
543	16375	88048	29625	08111	92924	53335
544	38745	91458	30363	95005	55854	38628
545	09937	17776	86425	88916	80594	28347
546	30097	47192	27960	15937	42080	61048
547	02410	60124	62825	42947	74590	89730
548	44804	80165	19442	72194	76910	40274
549	37352	79142	51032	58844	03167	57351
550	60640	14199	48263	71533	94235	42431

TABLE X. TABLE OF RANDOM NUMBERS

Table   X 277

(7)	(8)	(9)	(10)	(11)	(12)	(13)	(14)
81352	61640	56614	71506	75541	37818	88047	94144
18901	40743	99449	49825	44637	72724	42649	67052
07121	07151	23905	98435	50453	12983	04738	76421
06326	62715	28701	52809	56581	05925	85210	17745
71072	11519	44876	34508	07859	62424	54319	32842
25554	11165	00123	80338	03876	85648	24978	01687
70787	28193	65872	33723	00125	99818	85571	69509
53740	00464	51853	78852	83593	82926	48985	64355
36288	93531	69269	84798	78962	06336	95618	89718
14413	23643	21527	91902	91384	31444	54783	38760
65791	47857	32483	38493	52606	91078	13631	67863
74359	77556	82242	00134	70154	09027	79459	18730
81287	73884	69312	03395	06879	49662	40000	61598
84250	19254	06677	54192	53422	58200	74464	73949
86657	70801	53719	25214	65635	07565	49977	45525
52837	46723	00256	96221	26641	00309	36009	48392
13148	01433	78721	02647	25454	53913	97554	41578
40595	55953	93448	07805	53622	27330	18749	57867
87098	19392	13899	56096	83645	45871	35950	52272
96086	93295	12413	55774	97318	66402	11209	52495
87313	65969	43349	85142	25650	01896	48680	51236
83504	22290	63835	45589	04884	92760	70462	00538
59850	03262	60347	31077	07165	26588	31296	56112
16901	88717	62688	24828	89469	35483	76532	30256
90937	52140	73771	56084	08775	94820	78139	25987
58614	18912	58454	34011	85969	83621	92099	19131
56005	23331	21939	03463	53828	78930	30987	40988
54261	01844	45738	93150	13240	16694	59155	67589
18125	74873	83971	92678	96950	69821	41119	43312
30339	93143	07350	94289	76144	47238	08110	00037
43277	58089	70520	96997	71007	87803	52458	06637
04694	40359	28351	53492	73134	02370	72313	53039
93482	27726	51835	23966	50279	26329	25754	43530
77605	27351	49177	36914	50258	62361	38229	89608
02059	98892	98061	15330	31705	71923	29266	72716
98103	31752	04842	13693	84292	48485	76178	41716
63949	35394	12989	05867	11568	45056	16609	20470
06007	52239	61201	57415	35609	38761	19589	24238
76871	80449	81351	73642	48643	23848	48390	56829
77723	54114	90290	62627	65151	15687	81062	06729
45177	08796	99297	48807	88310	75454	45456	85394
04169	16575	62665	97861	71650	56981	61794	94285
09525	88290	17679	08945	25816	11848	95106	22031
13599	73065	40870	82576	37089	86738	16284	44725
08092	64255	55604	78635	13197	72213	95102	36723
14358	44508	72683	51088	55368	85587	27046	11198
16073	28184	30078	92578	83789	08044	76238	47599
93861	06568	92482	70037	66779	63312	00619	94053
51850	92810	35331	78995	44221	41532	51606	26430
44114	90993	41149	06159	39242	11163	14764	19246

TABLE X.   TABLE OF RANDOM NUMBERS

Col. Line	(1)	(2)	(3)	(4)	(5)	(6)
551	31630	67734	78201	94545	80152	62327
552	25101	98983	36993	40028	58036	14075
553	86207	09805	46240	70644	76012	37000
554	31611	47643	28795	48115	17223	63161
555	10649	89132	59781	12373	35999	30832
556	68210	16228	34801	40972	22887	89759
557	32367	69587	66162	44358	69844	73042
558	14684	42446	01751	37459	31945	03627
559	64260	04661	39957	01200	84800	27930
560	66035	77943	70861	32037	96699	56314
561	20966	71492	32323	11867	47523	24094
562	20498	68176	02027	22358	15907	13247
563	56320	79875	60634	17556	52153	63549
564	50559	90270	33571	88091	34749	56784
565	49366	90095	73459	12225	28483	35358
566	29022	19268	03003	96622	30239	20482
567	77212	76531	68842	37777	20085	38703
568	41121	90499	83459	71424	27596	74645
569	06613	95412	94751	60763	56611	73508
570	66430	95324	60108	42377	56350	67861
571	39380	09648	47285	62864	03421	34292
572	27595	63289	75149	03348	91237	28372
573	43525	45549	58819	48478	14007	11384
574	60024	79858	72015	01236	27444	47010
575	35914	73076	05158	40190	41294	72776
576	58253	32995	54370	34437	98365	17630
577	85887	54618	23532	73821	80904	05950
578	88988	60426	34636	40601	57718	93925
579	65381	17333	32358	49608	36893	43453
580	20214	88406	06098	07770	51679	64857
581	04970	56425	74303	94793	55055	44762
582	87556	31521	19669	16311	01767	67301
583	34615	85796	30299	56090	41453	82886
584	95769	07371	49569	42262	74097	44317
585	57338	58325	53918	08075	69395	08189
586	08882	30762	14602	54767	67683	57818
587	62644	32814	31337	06011	80623	36021
588	55358	48010	24440	51092	89263	82352
589	02254	43319	79888	86311	57615	69666
590	61023	48494	57279	47133	89534	37085
591	41900	90881	96080	35258	70734	59465
592	94500	19764	94752	73077	74726	86176
593	66375	34653	37125	84780	92759	09781
594	27141	20959	02318	57546	45467	06653
595	25670	88933	19316	57014	52797	83779
596	06590	67256	21117	69351	98168	81043
597	36429	99750	38004	64992	25021	45680
598	52011	93461	06334	23801	34422	13728
599	09443	77694	26882	15663	45983	29425
600	25862	81855	15254	28462	95680	42433

TABLE X.   TABLE OF RANDOM NUMBERS

Table X 279

(7)	(8)	(9)	(10)	(11)	(12)	(13)	(14)
83165	31035	82295	11824	06765	29501	62849	50419
05980	57094	45527	18766	77741	12985	14112	65058
98321	97197	30645	56169	09363	44394	29087	96569
29677	69820	77159	20762	94296	94528	82984	71418
02508	93055	57173	79848	25439	18861	26742	54970
09095	00587	03998	13659	64179	98567	69313	84637
88091	07288	74971	47066	36927	53520	58309	58605
47690	97813	45272	42789	99315	26662	15833	37246
98937	76108	11043	29101	01767	78894	92922	66537
75755	68667	04730	15256	13957	52743	42306	87515
23334	78839	81588	67374	43855	24512	81956	75721
44784	82957	53009	73379	44093	58405	38515	85531
03661	78290	96447	04192	30157	63198	75932	02367
98486	06018	27447	00884	29564	51522	35571	69208
99941	63054	54358	80748	54049	85937	64718	21466
50028	16632	57708	78559	70241	47977	78645	48550
31753	18608	52524	08585	91711	63572	57007	11379
68790	66478	78885	53799	02026	24596	88692	55936
88479	86151	78563	31633	92321	23304	77153	86639
47478	11961	78516	95316	64393	52020	44994	88205
36084	39604	89838	03635	30064	72710	26327	65521
65330	00966	73904	17477	34953	08975	83142	48425
19576	68138	30774	51898	24711	72537	18360	95682
80938	15828	86484	92753	04322	27171	41828	79025
38528	22272	34709	34561	65554	53461	61776	03585
42196	49736	39619	16731	71792	38047	85559	56700
63441	86109	79900	14063	03152	39235	74289	42342
67763	67671	21739	87534	83385	91492	45796	04621
38580	27639	52832	01522	11108	59992	23168	04414
36998	95796	46745	36780	56791	34690	61634	85411
22800	02663	91182	13102	07408	50545	21312	11365
63016	01227	29273	79256	50368	88653	23329	86500
33077	99791	86553	64640	61529	88660	66941	15936
40304	05346	90342	73324	81555	82769	23559	55238
66668	11663	84852	93146	27182	34936	97267	63996
98756	49119	91258	71916	65948	24841	38607	39412
81846	71868	98242	72307	25917	56240	73499	45106
95072	68828	28001	48991	19201	90963	34192	63336
60651	49084	39681	66415	10201	53931	44245	42850
28495	39162	89121	52021	23143	14829	55792	84641
49227	01431	90694	96186	57811	54512	30108	01261
31623	14569	89225	09606	73432	95276	21237	36807
06912	96802	92502	97497	67702	49763	25950	49924
99212	60612	26046	53553	59757	70491	69632	46009
56495	33104	26858	07662	41253	97688	76883	29444
26682	00063	96223	30436	21987	15450	56574	45011
46501	92943	99165	20707	43410	53746	54716	17090
35016	80605	86628	21689	34082	26035	11928	41817
12085	96233	29036	91135	28258	27709	90674	09705
70311	09702	71615	72688	85922	58369	89734	08750

TABLE X.   TABLE OF RANDOM NUMBERS

Table X

Col. Line	(1)	(2)	(3)	(4)	(5)	(6)
601	60359	07603	81594	66235	48154	61257
602	34992	97880	79115	47587	76167	47086
603	04887	64208	71842	97885	32616	23280
604	09332	86232	88199	66094	72594	30100
605	42326	62962	06485	04978	96639	96214
606	49187	42836	17042	35179	31880	48444
607	09228	57404	42180	07949	98750	31506
608	69720	73477	91252	48009	81393	76401
609	82222	13787	98611	95257	34753	36674
610	30703	00513	54586	05623	43999	55387
611	86369	62151	70713	41166	79321	52215
612	83331	99035	68506	96734	91074	24356
613	43053	60600	98921	43720	77342	26186
614	57104	49148	18487	01775	71782	04679
615	33177	11409	13925	18130	54242	13460
616	05424	76714	05732	29415	01183	45054
617	92950	58665	41191	69259	50244	55322
618	54925	20502	71767	82737	64847	04496
619	41980	43710	55304	57526	29616	92314
620	83825	70977	67987	61545	92066	71215
621	84047	83627	37763	07081	33048	57895
622	12776	69127	67921	57611	85876	30744
623	81419	55440	69506	09115	45032	48343
624	59844	03603	96297	58028	93069	35674
625	18350	74940	07044	11210	53622	00779
626	79960	18784	13376	03415	84450	78874
627	45420	24157	16374	22384	56892	84941
628	13945	09559	68152	56960	39453	51654
629	91206	33871	60730	96821	95808	29763
630	24847	08724	81499	72905	95102	63004
631	94303	08209	27804	49372	66392	50578
632	22732	95331	60954	93333	71142	38827
633	82809	24004	65983	01091	70431	91145
634	62700	79965	09610	97213	48579	43574
635	89870	73755	48525	32765	50818	71468
636	81493	24124	67928	12735	41249	24180
637	43630	32189	08532	43055	08080	84208
638	60234	18992	13283	96334	39746	07272
639	00107	21861	60367	48999	71634	34053
640	09657	36088	05976	88267	62683	57675
641	93948	38350	63464	08008	96607	73505
642	42746	29761	72298	48186	88584	90141
643	12939	04181	27698	48297	20574	30169
644	71032	55283	94804	00202	12254	22920
645	09188	78876	95736	70659	32725	23024
646	79236	54729	47052	49717	22312	06735
647	41337	52635	48056	43317	11599	26382
648	73732	99966	30485	45994	30195	40239
649	92113	55625	03726	76886	64237	33300
650	63797	22667	74860	99731	06975	63055

TABLE X. TABLE OF RANDOM NUMBERS

Table   X   *281*

(7)	(8)	(9)	(10)	(11)	(12)	(13)	(14)
27978	64695	63165	44593	08210	16863	09655	00855
57064	16730	74172	60317	83215	38133	06303	05466
11783	19852	64266	24446	14189	77419	30991	92130
23673	68705	66989	42666	81857	34651	36167	24221
91478	12408	21457	19862	99102	91426	10181	51762
89877	50915	37426	21556	25999	84256	82314	18813
78442	45809	12725	49774	11276	46371	81681	00623
48168	25967	33372	84414	21506	46131	46046	12354
44326	66070	61131	70620	42865	89251	54844	04013
67189	95058	91174	13121	27557	16512	77963	40635
94358	28962	35868	22796	87221	40014	68875	71420
03035	66926	32197	54944	76781	86722	11769	27368
64554	46226	64244	10703	49564	69737	32948	43060
64369	06208	71669	63046	10470	54194	96709	86502
39174	63528	22670	31810	04313	50669	20653	31779
13493	44006	61641	80304	96504	52181	05359	72203
75137	90193	31989	17381	43795	26981	15326	02303
35921	42670	08584	54090	52907	75331	09155	54187
85883	21584	55045	81997	62277	58884	01590	13532
93967	63071	69928	98917	05699	35957	04679	58769
42182	73279	08032	19165	01701	35656	03328	81785
40886	68396	79787	76434	71221	86769	15104	19062
78352	39075	31689	76469	64918	15149	88457	97144
38479	54639	54455	10300	73946	94827	53164	07458
36027	51496	01694	57895	84570	18271	54461	42210
22050	19730	92598	54291	60658	73188	03446	49864
97157	99656	33978	81436	10955	98991	10456	35727
10617	55628	47933	85161	52998	75414	59552	03546
39678	73104	43398	38181	44314	58343	28884	94613
22223	19808	90777	54986	97234	18458	22889	83960
02966	90907	33164	83044	97985	78526	00983	29271
48222	21779	35598	95957	58844	82319	19780	08330
88207	52216	94846	75303	85105	89486	08182	56504
37652	12447	80233	42473	94585	84840	99926	74778
37876	28334	07762	16180	45346	78324	20422	85784
54740	44290	58903	38681	04066	69393	84595	42173
06295	07813	24068	67549	43051	78581	02095	03471
25295	07871	34201	49620	52178	07290	89767	63890
28265	02064	06290	10620	17941	81086	51759	57028
85265	10856	06525	37911	52332	55752	25054	30436
75513	91238	11042	40972	62837	30260	84002	99947
72879	54531	99127	60063	22374	76895	63812	94877
45545	04462	91067	43847	62739	31141	30385	30098
73225	51484	73943	08431	45681	32663	67097	15644
04656	48102	15904	19019	09882	87431	16879	61253
58347	04402	03838	97049	35378	38579	24489	86899
41305	04589	92877	52732	53130	45275	30183	15962
52751	64124	67778	60982	12167	63134	10730	11350
48004	37440	76329	80441	74766	70630	97855	88039
72287	81976	43983	97018	25559	96618	93350	67143

TABLE X.   TABLE OF RANDOM NUMBERS

Table X

Line \ Col.	(1)	(2)	(3)	(4)	(5)	(6)
651	98707	41348	74106	43550	16638	72858
652	65496	87176	71726	09722	85667	37498
653	17617	18337	83583	73510	52998	02570
654	85006	32658	83348	38599	77549	70275
655	77279	66357	38044	75041	06698	16798
656	84133	32224	55350	73251	64120	86718
657	86535	01806	18470	94806	18228	71262
658	58459	99005	64939	57060	34609	06739
659	38783	61870	02744	23773	74163	30029
660	49454	76832	17745	75922	40087	51566
661	71219	82115	51031	28586	91328	59635
662	57770	63699	49430	82846	13285	69178
663	94539	92056	65825	27167	69783	48976
664	94992	45377	75471	45547	34348	61707
665	60717	63093	60684	08910	30296	12345
666	28040	08603	43675	64665	40567	76211
667	69841	63836	53186	49970	05485	65171
668	80931	69365	75077	19929	67852	63937
669	94000	24074	95072	32602	89373	78009
670	70609	17467	26861	51035	44534	49774
671	24016	76531	77282	21820	61170	37198
672	36008	76472	60324	45386	67266	98846
673	65769	02131	95850	61875	52572	82826
674	71033	43026	99601	18102	33654	29885
675	25152	11755	33527	05149	71696	63492
676	81251	78313	53908	66912	84868	51859
677	37401	18809	83908	01638	72548	80521
678	01383	28893	13771	99613	39531	20895
679	03921	79304	91058	03175	33155	58344
680	19882	28056	10093	45334	89019	23914
681	55339	29655	37282	26504	70059	91780
682	01725	18314	40197	00964	44112	60484
683	45073	49109	77778	46092	30928	16111
684	60861	39467	17392	05446	26083	65352
685	88775	27006	69664	22246	77064	40312
686	20497	65297	17965	97094	26451	47473
687	76692	21880	39663	77289	36681	47935
688	00528	63679	57762	88094	26402	25925
689	57718	09889	94003	43899	32989	87707
690	85652	55278	23342	30372	47987	19936
691	16543	93477	41421	34710	59498	55069
692	50315	17150	14006	10453	96007	41512
693	73876	27245	79551	56241	53843	66714
694	26749	07215	99485	51013	50210	69934
695	89438	39457	58255	75693	72570	17885
696	24388	81801	22581	26331	47945	11717
697	19688	39199	20531	96205	50355	14725
698	07943	07018	03516	05747	94259	73583
699	90391	97135	97952	83722	35578	82905
700	61564	65108	76451	71430	08671	84975

TABLE X.   TABLE OF RANDOM NUMBERS

Table X 283

(7)	(8)	(9)	(10)	(11)	(12)	(13)	(14)
32375	50634	80903	43867	54215	65017	70776	09596
03926	72585	15193	45615	67073	65110	42404	28419
37334	82042	07261	97361	93327	81104	17979	15192
85147	08030	20980	86223	58024	56948	10793	81858
83023	84686	53539	86491	70402	31779	09591	22398
61061	66146	36275	48583	74783	85824	87328	39207
39195	99049	75632	67645	26117	89606	83976	53307
40489	08146	74014	24803	50699	05588	73658	24352
64988	67851	68279	53341	88122	03631	97916	27003
38620	92988	84531	99883	40785	53456	03864	20593
15746	09832	53937	07006	00272	00914	31619	73222
66391	44979	69923	02920	98234	68478	52861	04824
07466	76421	71439	98358	21989	77098	86413	62724
77657	29541	67815	56777	91185	00390	17361	18119
22178	71824	11973	61293	25937	61542	76603	48032
18715	62133	75336	98118	88984	44341	11202	34471
40135	67168	42593	37922	55146	40354	32803	63430
67390	94560	76131	58351	72302	68973	72689	28187
38009	60982	27122	29283	08881	91436	80644	09688
88833	15577	80076	68625	42230	94087	97450	13569
16941	14009	97386	61256	11687	70360	18405	85586
17763	98235	53433	01295	74947	71106	33976	55405
09571	61125	81326	72388	40377	93419	38458	13301
48692	09691	95837	77697	69896	47214	85897	59723
32871	48386	37657	82062	02643	34598	36128	42411
15959	31119	92964	94054	57594	61310	42452	76262
02945	29410	37736	96301	32848	16332	63069	35975
92322	01032	66104	86756	84812	10692	73677	58045
13242	02686	57998	84750	88695	99642	86560	79808
12549	69492	09859	46366	22224	63308	95897	20408
59955	01729	92886	28616	52061	47284	02989	20385
07486	10883	23106	18254	23711	26458	09878	12981
15372	76016	15402	98293	24495	79618	69649	19768
24501	20502	13346	11151	37472	05548	35893	47964
14388	70861	47198	24261	69913	07368	78156	07858
07292	03191	47612	36599	67827	74007	48783	12329
76260	96157	44019	73403	83985	84210	58091	88097
10598	09606	41596	90916	25311	91501	63286	87423
46639	53512	25509	92751	60744	24239	13608	90176
89623	38221	45073	67291	57082	13194	39749	75753
70504	08544	80118	77512	88070	03943	84969	02116
36027	60557	57141	50554	80556	47441	56506	62857
24050	26354	14915	62391	75724	57468	34621	22325
73613	95543	94599	13955	29586	62507	90618	52247
07259	10273	25229	52788	55762	14772	50200	55909
42187	06031	02488	09199	74752	65757	27989	81532
38656	79167	44771	43966	73425	79632	77181	84817
43619	99509	02102	08279	57372	53487	10885	95017
41761	66670	43482	34931	94438	93341	60927	31368
36182	70787	79442	29461	01209	20022	93055	28312

TABLE X. TABLE OF RANDOM NUMBERS

Table   X

Col. Line	(1)	(2)	(3)	(4)	(5)	(6)
701	79681	63467	02907	86515	71330	04490
702	30305	20743	10302	71391	18138	23412
703	32763	33847	58250	64362	87550	94978
704	59166	21978	40556	13084	31782	00518
705	55843	94845	30006	51045	17428	50657
706	33537	82468	52422	32155	54419	61661
707	13533	29605	31430	07663	95274	11484
708	24626	54219	12284	06890	05239	42846
709	48002	32024	17230	37523	47488	31080
710	03742	00004	98249	12256	94253	95378
711	17749	89193	37944	53702	49918	65397
712	34837	36219	22048	97047	68804	09633
713	99451	37922	90191	39229	07564	41077
714	74045	00036	53137	15250	19646	20451
715	98998	98774	98159	00032	97323	81490
716	61513	02266	36871	85993	23028	67082
717	67056	19960	53863	63917	68283	31123
718	83036	04625	93284	14368	10979	95800
719	71901	25497	76987	74388	41605	39295
720	46484	77860	02062	92917	70275	40593
721	18312	05137	64361	86541	17794	32313
722	63093	94089	17729	19607	19340	19022
723	38109	69439	62094	49578	37728	17809
724	41421	22003	36770	32741	10325	30892
725	92320	12828	57972	83551	63054	95028
726	42226	72413	67949	96906	17848	21446
727	01094	08525	21349	41981	55232	76652
728	75760	51119	37218	16828	89127	42801
729	62568	56665	42394	67135	03069	93275
730	77151	67677	85258	46925	92504	87860
731	71920	39074	15464	36753	86550	24330
732	99411	04216	66076	90718	67214	03688
733	05654	88507	03119	93043	06951	35126
734	65937	81013	09884	97787	85851	00011
735	87649	70531	88258	21822	97418	67341
736	57827	19642	95661	23788	19164	78112
737	78911	81376	22392	42570	33512	17996
738	91302	54963	94112	60597	31843	40120
739	82950	87509	65702	14385	86299	23769
740	21888	66504	85577	67163	46317	92073
741	73799	60026	87226	26744	12037	98558
742	48237	10339	99550	86134	30229	39131
743	24293	69496	20243	17738	55798	96178
744	18748	01580	73315	84924	81621	67021
745	94470	36824	89203	23689	37016	18462
746	87639	11791	63380	25952	20838	13638
747	65676	78482	33343	65797	56005	15782
748	94357	62236	54083	37960	43467	79372
749	06595	83512	74524	10051	97759	64738
750	34033	69035	18588	88893	83679	27789

TABLE X.   TABLE OF RANDOM NUMBERS

Table X 285

(7)	(8)	(9)	(10)	(11)	(12)	(13)	(14)
47372	68791	27576	02044	03784	94581	60105	75131
11858	47818	22324	52031	10600	49892	34101	71430
22888	78355	34651	41604	51892	89533	81610	72641
18621	60508	93095	74017	17416	76900	25261	63227
68237	02969	30500	43569	28051	22505	07159	78162
64835	06496	16377	92607	86248	89492	49306	95414
42579	15718	54485	08857	93691	02973	26687	90437
24773	15025	15161	51340	54739	99433	54328	47800
29352	61444	04011	56275	19259	72475	53451	43397
88918	98167	46646	19727	24181	83358	28999	11769
72597	63520	77429	68355	21003	00657	02157	68031
28689	80484	59331	77577	30376	10021	78267	78049
91554	46657	74652	84677	49671	94805	82406	99797
46677	53620	74712	17246	96626	28587	33618	46845
21552	35001	10913	48910	91005	62408	83253	19770
93486	45110	86288	34493	66710	04268	04955	49074
17443	32019	19695	85622	46808	03535	91566	36785
72182	77004	07320	79516	00915	53209	00884	65464
75622	41203	87987	09672	81312	08728	49867	95245
93265	92722	39193	47099	39046	85989	24607	72287
52847	08862	36752	32624	11035	92500	35016	18519
50080	21998	49864	07107	64287	41647	75264	09230
11563	10073	17299	69238	88068	04754	51698	11641
14112	34880	92387	45169	96668	89183	10099	44382
50857	40315	04962	36431	54964	33961	89397	70359
35722	10376	84226	16403	14642	23253	07162	57664
00857	77173	63362	64936	96601	95816	14729	35398
01084	78402	28359	41533	83339	69672	64909	55192
11662	23607	00878	53800	14840	28975	51693	97620
85299	45952	76388	72989	12170	03449	28315	95994
17873	20798	15221	80763	69974	95552	66857	85097
71088	92479	27623	97466	27560	63689	35799	28078
26154	26820	68861	37807	17485	58902	92005	57597
66801	02686	73801	19522	67200	80477	73121	88815
54342	80836	46142	04718	08348	75316	53030	90533
07304	32337	12845	12588	70054	12267	96360	43459
29406	47329	66928	89312	16994	87633	71914	52038
00386	35486	27379	02873	69868	03803	29671	74165
87183	66267	13819	58266	16843	06546	00521	50286
68325	46664	03841	58572	46048	04635	78350	53233
66640	32882	11415	64686	78236	05844	76840	80802
76976	56296	89453	79556	48059	70552	45852	06593
86022	42073	16407	53031	48671	82154	04054	90741
99065	43590	82522	71919	25097	49536	03140	63610
59404	27230	78689	88837	41119	95462	94394	13374
32782	23841	16936	91384	20472	81876	85484	35003
27311	64066	28230	36207	06446	09976	09463	56698
26370	08273	18180	84100	30757	33315	28583	53434
10370	07874	28301	08201	07624	63508	82486	82993
22340	12208	00381	06023	77844	28666	24220	19220

TABLE X.   TABLE OF RANDOM NUMBERS

Table X

Col. Line	(1)	(2)	(3)	(4)	(5)	(6)
751	03548	52011	53722	62927	01693	90948
752	05066	74263	11659	84658	52063	42299
753	19814	89956	36256	19896	56654	69424
754	44928	11206	21377	35086	62233	93761
755	52158	08923	94812	04443	32028	96465
756	53211	58616	05135	52204	51079	11341
757	27334	43976	14685	38119	29486	57290
758	66166	06709	69495	42150	38018	43875
759	01837	16750	96491	33095	30383	48804
760	38295	51093	63495	13203	93562	94132
761	72510	80830	39948	94133	00780	14167
762	16354	00336	06494	30078	46134	62486
763	45168	70700	04592	35281	47737	28881
764	68137	13619	84666	78104	83546	72551
765	31848	95753	76858	89517	91138	62356
766	17216	37292	05495	50885	98994	32966
767	73211	42922	57386	95490	56100	08977
768	29217	56753	90171	87554	57421	35839
769	66780	34571	71684	28798	47123	25232
770	55780	48081	93674	70837	92534	65892
771	05788	64237	58140	63279	60170	17229
772	21911	51065	51525	65122	52608	52836
773	32645	38561	25181	18042	31903	46525
774	72304	15382	01151	63162	23656	69649
775	27637	04122	86132	22538	98976	90718
776	66305	85906	87925	50081	37585	63674
777	53470	40332	81044	00558	50403	48029
778	01355	77096	64828	02445	94511	09503
779	66954	04728	86153	10933	86557	10877
780	53734	15628	08080	24011	04187	65722
781	23114	08743	07186	23825	04298	72839
782	16803	91335	64192	13631	20332	74852
783	57233	99034	38028	32038	81270	77809
784	83203	59665	72314	67942	01320	24467
785	79299	97340	04568	65386	04876	31514
786	99442	63865	14360	83898	98873	17471
787	77773	19621	81557	84629	18808	89056
788	18581	12572	15185	57989	87644	88902
789	80978	13327	19682	53353	96223	04775
790	96884	95522	04791	93463	20316	84054
791	61113	82511	44196	73740	16111	73200
792	01272	75657	28365	17431	93603	32457
793	82357	77572	75628	93073	38281	72103
794	20434	70899	44243	19741	59954	73617
795	38580	10399	44894	25476	04984	64543
796	20211	12980	45261	82527	87534	39405
797	98954	03124	09433	19894	01380	18962
798	99749	02140	46641	56354	78746	89410
799	47167	24716	84417	40097	46608	11667
800	65812	77947	27864	98144	01818	28214

TABLE X.   TABLE OF RANDOM NUMBERS

Table X

(7)	(8)	(9)	(10)	(11)	(12)	(13)	(14)
75340	16660	00939	77148	52778	20615	97851	58353
94340	20391	00080	43359	44231	06891	86588	88565
84446	95294	00919	60267	34349	64353	92469	01606
56000	42066	35898	48944	32352	33177	16239	76624
94430	42834	22836	88818	44467	18329	30144	31536
20366	86248	35160	31485	92436	69726	12722	52722
34141	18058	91299	58001	17944	00296	13949	94904
44534	48712	98089	76889	55720	09038	95667	57007
21228	10863	08350	25610	80866	00115	27666	94607
54810	11410	73776	55752	36138	52297	31528	77790
83322	32747	97291	66126	64996	21354	64615	64234
70923	17400	28797	83599	78655	67488	09715	69232
57221	79935	99756	61564	36936	55668	87148	95939
40848	52138	36343	93975	09556	58888	08125	80961
66405	98171	66239	49761	88800	29069	30175	31267
83496	54614	75214	34959	15500	32107	81638	46696
78499	53540	41745	09805	67110	53329	92776	32312
60961	87828	24777	19688	42464	35569	84904	13130
67995	39562	35855	61645	64448	65170	46792	77081
32422	02887	03170	05372	48068	65758	62376	94055
69038	02500	21071	26294	95005	47815	87991	93358
86157	19943	63173	34921	11943	36931	74722	06193
96290	24323	68269	45841	73918	13720	52336	48416
54580	33479	62899	33716	53668	86735	36746	21939
81840	33461	28526	96231	28082	89710	16038	26648
65144	75814	87596	04642	07590	82276	59336	84262
07719	83340	20258	49737	45499	72241	90103	96141
12283	15264	32845	55594	03668	05664	67999	06001
97049	40124	87071	03864	29526	42888	09057	82966
82863	59616	60664	76761	84417	72560	77869	42603
43963	52136	60923	88568	55458	87335	81939	67361
67813	68131	56695	93253	71650	81735	13302	30319
99813	67091	56997	62005	95516	34011	52035	79500
27813	74783	56696	84101	93852	12016	20869	57009
78813	77887	66697	51048	28395	62626	42911	99130
01813	81541	66913	52874	10563	59450	16885	16009
48194	65899	26244	74431	24232	33299	58332	18314
49946	76235	92633	10232	84241	66467	86018	92272
05590	55237	21695	31870	48893	94058	88943	77040
68980	56907	77188	54803	67622	57968	78532	66688
65980	66320	77327	14715	24759	08145	28077	32303
57980	66157	27797	55887	57788	29337	73409	35682
39980	71150	87282	01585	25850	46138	21777	97555
11661	81574	05903	25909	38638	32196	41345	22152
14141	97911	89418	97980	49571	60483	94714	51753
03701	94321	23777	61546	21541	69603	61768	40328
40701	05185	23280	04945	49801	98051	77848	62320
35701	00250	23508	51246	09460	02313	71279	74223
40701	06155	23784	68961	34289	04531	12752	08516
83701	10257	33945	08795	30976	14017	59161	31388

TABLE X.   TABLE OF RANDOM NUMBERS

Table X

Line\Col.	(1)	(2)	(3)	(4)	(5)	(6)
801	33993	51249	78123	16507	57399	77922
802	39041	05779	74278	75301	01779	60768
803	56011	26839	38501	03321	43259	73148
804	07397	95853	45764	43803	76659	57736
805	74998	53337	13860	89430	95825	65893
806	59572	95893	69765	43597	90570	60909
807	74645	13920	28640	00127	04261	17650
808	42765	23855	38451	11462	32671	52126
809	66561	56130	30356	54034	53996	98874
810	50670	13172	31460	20224	34193	59458
811	53971	08701	38356	36149	10891	05178
812	47177	03085	37432	94053	87057	61859
813	41494	89270	48063	12253	00383	96010
814	07409	32874	03514	84943	74421	86708
815	03097	12212	43093	46224	14431	15065
816	34722	88896	59205	18004	96431	41366
817	48117	83879	52509	29339	87735	97499
818	14628	89161	66972	19180	40852	91738
819	61512	79376	88184	29415	50716	93393
820	99954	55656	01946	57035	64418	29700
821	61455	28229	82511	11622	60786	18442
822	10398	50239	70191	37585	98373	04651
823	59075	81492	40669	16391	12148	38538
824	91497	76797	82557	55301	61570	69577
825	74619	62316	80041	53053	81252	32739
826	12536	80792	44581	12616	49740	86946
827	10246	49556	07610	59950	34387	70013
828	92506	24397	19145	24185	24479	70118
829	65745	27223	22831	39446	65808	95534
830	01707	04494	48168	58480	74983	63091
831	66959	80109	88908	38759	80716	36340
832	79278	02746	50718	90196	28394	82035
833	11343	22312	41379	22297	71703	78729
834	40415	10553	65932	34938	43977	39262
835	72774	25480	30264	08291	93796	22281
836	75886	86543	47020	14493	38363	64238
837	64628	20234	07967	46676	42907	60909
838	45905	77701	98976	70056	80502	68650
839	77691	00408	64191	11006	39212	26862
840	39172	12824	43379	57590	45307	72206
841	67120	01558	99762	79752	17139	52265
842	88264	85390	92841	63811	64423	50910
843	78097	59495	45090	74592	47474	56157
844	41888	69798	82296	09312	04150	07616
845	46618	07254	28714	18244	53214	39560
846	29213	42101	25089	11881	77558	72738
847	38601	25735	04726	36544	67842	93937
848	92207	10011	64210	77096	00011	79218
849	30610	1·3236	33241	68731	30955	40587
850	74544	72806	62226	65685	37996	00377

TABLE X.   TABLE OF RANDOM NUMBERS

Table X 289

(7)	(8)	(9)	(10)	(11)	(12)	(13)	(14)
38198	63494	00278	30782	33119	64943	17239	69020
22023	07510	67883	55288	67391	54188	31913	29733
43615	49093	91641	77179	50837	48734	85187	41210
44801	45623	23714	69657	87971	24757	94493	78723
96572	73975	19577	87947	23962	78235	64839	73456
06478	77692	30911	08272	81887	57749	02952	51524
34050	78788	57948	36189	88382	72324	59253	30258
23800	02691	57034	34532	19711	71567	90495	55980
78001	29707	91938	72016	16429	69726	41990	33673
24410	01366	68825	22798	52873	18370	15577	63271
55653	31553	20037	39346	28591	13505	04446	92130
97943	81113	62161	11369	54419	58886	89956	12857
41457	54657	46881	75255	29242	07537	53186	95083
34267	66071	62262	99391	61245	95839	75203	93984
18267	60039	62089	38572	70988	17279	05469	28591
50982	92400	59369	43605	26404	04176	05106	08366
42848	81449	80024	81312	59469	91169	70851	90165
23920	75518	32041	13411	61334	52386	33582	72143
96220	82277	64510	43374	09107	28813	41848	08813
99242	42586	11583	82768	44966	39192	82144	05810
36508	98936	19050	57242	33045	54278	21720	87812
67804	84062	27380	75486	63171	24529	60070	66939
73873	68596	25538	83646	61066	45210	24182	18687
23301	31921	09862	73089	69329	41916	41165	34503
65201	92165	93792	30912	59105	76944	70998	00317
41819	85104	25705	92481	95287	61769	29390	05764
64460	96719	43056	24268	23303	19863	43644	76986
42708	54311	95989	08402	77608	98356	47034	01635
03348	11435	24166	62726	99878	59302	81164	08010
81027	72579	67249	48089	34219	71727	86665	94975
30082	43295	37551	18531	43903	94975	31049	19033
03255	39574	41483	12450	32494	65192	54772	97431
65082	57759	79579	41516	46248	37348	34631	88164
95828	98617	27401	50226	17322	44024	23133	57899
51434	66771	20118	00502	07738	31841	90200	46348
16322	45503	90723	35607	43715	85751	15888	80645
73293	38588	31035	12226	37746	45008	43271	32015
24469	15574	40018	90057	96540	47174	03943	37553
99863	58155	66052	96864	61790	11064	49308	94510
53283	75882	93451	44830	06300	45456	49567	51673
97997	66806	55559	62043	51324	32423	88325	99634
38189	88183	56625	22910	52250	70491	71111	37202
88287	47032	66341	38328	70538	91105	12056	36125
34572	83202	58691	27354	37015	11278	49697	65667
68753	16825	48639	38228	95166	53649	05071	26894
57234	28458	74313	29665	97366	94714	48704	07033
68745	62979	97750	28293	75851	08362	71546	17993
52123	29841	76145	82364	55774	15462	44555	26844
45206	11949	28295	12666	98479	82498	49195	46254
59917	91100	07993	15046	51303	19515	25055	56386

TABLE X. TABLE OF RANDOM NUMBERS

Table X

Line \ Col.	(1)	(2)	(3)	(4)	(5)	(6)
851	76385	05431	82252	79850	31192	86315
852	08059	15958	10514	86124	29817	19044
853	30636	03463	50326	69684	38422	59826
854	23794	51463	67574	48953	73512	46239
855	01117	60216	29314	65537	84029	00741
856	29527	19577	01414	35290	70174	37019
857	64236	24229	17970	92022	64164	17873
858	92331	30325	61918	71623	38040	51375
859	93454	37190	23790	40058	03758	01774
860	17101	42181	45798	68745	24190	16539
861	30742	93358	95730	52535	34404	76057
862	02472	01280	67106	47893	93551	76697
863	80718	72187	67178	77179	06212	37409
864	85406	73687	02116	57637	94701	46754
865	00563	67156	88141	13491	92592	35746
866	89190	58965	55213	24337	58807	36123
867	01438	81590	83758	45361	76209	65081
868	79127	53282	50510	80129	23960	78423
869	33952	92823	32840	94420	51193	69652
870	57146	14126	71734	56942	83371	31526
871	33158	61761	73207	01764	81696	55137
872	63615	69083	00118	47991	99521	88655
873	89010	46915	70186	55657	76955	25430
874	32547	43398	30909	62599	53105	27460
875	61992	28258	27359	61002	16882	44018
876	78326	74541	22198	48380	45919	76160
877	35493	53008	78622	38329	27611	12327
878	19130	59917	28850	76593	02389	80759
879	00317	05769	03497	42174	32653	23663
880	84122	36454	70776	17000	83017	07027
881	76320	32120	91585	39640	23470	86000
882	09234	36233	94404	42812	39210	25967
883	16206	70598	95378	70573	42636	53862
884	04071	51662	67884	73911	08708	66287
885	97545	87732	23795	38027	90239	80044
886	53253	56120	42720	25660	36921	30891
887	66817	18439	53188	35155	24309	88284
888	28077	26409	11443	22200	23129	32407
889	18889	00291	13701	12401	26466	67700
890	10598	64974	66296	33329	30560	73380
891	18656	81152	45498	14400	92435	67664
892	79044	10440	25777	05486	65659	22183
893	74042	20365	42672	34850	60670	56980
894	87249	06640	09090	03242	68467	85678
895	82839	52537	00518	60559	43669	44297
896	07749	62249	01611	43795	17129	44447
897	37171	34598	87234	28324	85927	23465
898	55432	45030	85336	49128	40487	63959
899	43658	35437	83506	11209	24770	87123
900	59536	79475	04874	50831	16996	04750

TABLE X.   TABLE OF RANDOM NUMBERS

Table X

(7)	(8)	(9)	(10)	(11)	(12)	(13)	(14)
75612	59985	76421	39300	64976	27951	17855	02220
03555	80725	67857	31395	68780	16560	79952	41739
47858	90601	50834	88109	43882	15687	06212	19886
10953	04622	60650	35048	34705	90502	31011	81004
40851	96344	13861	43421	57107	60813	06877	52161
80223	62206	22928	63414	03940	02188	20345	13183
41189	28240	60697	26495	87634	75899	09741	84939
91127	93903	83715	93244	04366	57679	70829	90088
90696	81674	53791	15559	42798	46892	57960	06575
32330	21732	65547	94356	38651	35102	16327	17886
21325	87526	93020	94861	83865	61393	89645	00773
56598	67982	77376	33312	58893	69370	59118	95277
48788	68930	21672	88783	59304	82369	19410	93050
54019	96344	93343	47764	57490	21321	29075	40086
72117	37593	72780	78271	75915	85972	58615	71755
09235	95541	96979	03336	34380	66288	98659	46572
34785	68423	04408	73827	78494	02765	46174	83192
31988	78571	79458	95043	23997	97528	21631	63898
04332	81675	64644	68673	33718	02256	34414	87710
13444	11912	03152	66411	42853	08437	35667	26251
41834	81860	81310	14711	36599	78042	62086	41752
94451	67445	99377	75528	40794	30140	82298	85868
91951	56473	34225	68103	25353	89595	04715	20102
56734	41954	47696	82113	38508	88941	49983	36899
85376	66756	14395	66865	67036	78374	43612	44134
10974	03127	58980	18350	22089	54977	94019	84739
52541	00861	62380	65890	79729	99710	64836	43706
18481	02724	57578	35705	89265	25033	13767	95888
29569	36342	85908	29572	60063	41170	59957	14755
98058	41274	22476	27436	30798	62287	21235	00249
68204	23980	17625	53197	35128	76385	02848	61680
12232	38195	16649	96739	64610	96067	89561	15772
81334	65439	28858	97619	59608	61460	00581	43226
89261	73451	81146	77733	70162	42449	44755	56401
64677	47912	82144	85918	93508	05816	57549	74831
42042	80370	97880	62507	01218	19202	22323	81363
74644	25454	19606	61460	52684	36568	68108	45653
52401	78416	63693	35633	77724	86835	89829	81383
55805	63818	16067	95185	97241	66126	16774	39342
94905	04959	80213	14228	97242	94826	64276	37466
86229	74358	76537	87066	42293	91743	49462	42808
82080	04351	19530	49941	29181	34667	28910	70927
88333	75288	64996	26913	62379	55068	91239	85752
23411	53443	19526	03205	29261	36061	34325	03761
75071	17146	35492	60718	38106	06409	75657	66013
95197	25088	22150	54427	11578	77560	26460	55002
80833	62872	40826	10066	64858	33605	24848	30881
25879	60415	26744	38584	51543	17333	47300	13834
21494	85056	56630	75919	26005	90077	50380	09261
02246	08846	82410	50997	45824	55547	08168	16679

TABLE X.   TABLE OF RANDOM NUMBERS

Table X

Col. Line	(1)	(2)	(3)	(4)	(5)	(6)
901	77583	63193	14378	66314	92154	31173
902	41435	82033	40363	74800	79198	03991
903	17163	74517	11281	83105	16146	27577
904	47909	53647	99896	13016	53708	18549
905	88024	64045	70954	56434	73860	50310
906	51540	26148	65528	66246	74154	47254
907	29122	04072	85663	33914	63587	20151
908	34873	53537	85834	11394	44898	92403
909	26132	07856	14223	02824	66094	88093
910	26892	74387	97124	32851	83465	51300
911	73485	32215	29894	87362	47818	52494
912	45678	13831	93961	93558	34050	42936
913	71687	79185	46096	22818	19051	39123
914	11833	28572	61695	36928	50851	98599
915	90529	59895	55565	57955	75623	55721
916	31697	92763	05134	45963	15725	39771
917	47212	92010	40601	23364	72705	15989
918	56686	61285	36139	18867	12877	89109
919	50669	06740	74638	62054	53083	22790
920	16821	64195	64269	62235	41829	93374
921	30608	94750	38092	42230	85443	34743
922	83211	45796	33370	31291	76074	40065
923	51774	53905	66945	65882	70352	38447
924	89095	90943	76279	78323	37210	77141
925	54047	41638	45626	98549	20648	89365
926	81549	98773	68979	28308	53686	57192
927	72335	08233	56777	99798	57340	54203
928	02540	14032	90858	26020	28174	91563
929	09226	10658	79614	84200	53035	16839
930	18050	79976	70461	19890	08214	77015
931	44378	42627	16636	69923	89122	92792
932	92101	13475	15373	53642	77422	26662
933	27926	69613	88880	95537	08518	66627
934	06900	75222	25385	63569	03483	90973
935	08717	31244	56529	97336	17309	32629
936	79966	02658	18012	59422	86272	69447
937	85668	03060	61274	14020	16594	72642
938	29530	85435	54205	01781	88806	28369
939	69284	91923	11744	13051	69350	55403
940	41503	48684	36452	06825	80218	76395
941	47063	45686	78929	82808	67899	16936
942	90503	44995	53753	88362	99077	14510
943	08367	68828	82856	56380	26969	16987
944	05207	94071	56234	12584	21319	90050
945	15445	53683	92875	02502	48973	00564
946	99603	19507	68314	12524	57700	39240
947	82736	02198	84682	18512	42010	69026
948	04925	66489	08883	69176	95712	25057
949	48692	63679	55426	75873	21338	78242
950	23766	44510	30176	02784	65111	89090

TABLE X. TABLE OF RANDOM NUMBERS

Table X 293

(7)	(8)	(9)	(10)	(11)	(12)	(13)	(14)
75223	92947	81041	91385	89091	83989	34982	66565
01635	43666	02630	46039	53273	86262	91450	79883
33565	39621	27321	06387	51838	37591	54290	54527
75241	22312	53261	80185	69888	12646	64464	98505
29601	52204	93295	05735	15486	39857	83584	59587
65546	75309	90415	17346	20559	29568	30124	11974
36720	25827	13447	17082	74817	92523	81519	65724
01979	22785	94354	33271	44579	40756	70108	30514
35794	69015	67498	97087	79351	52239	07472	74585
58546	12579	43451	48661	79351	58231	11824	77723
27684	83573	91237	03585	13266	63905	94600	43823
00860	66594	36783	56939	47057	55257	16555	65642
08793	18798	79528	05621	12131	88365	51154	46175
76466	73848	76675	68072	50077	27709	60895	55306
43773	98277	38842	05069	31590	31482	38040	71541
54430	76890	52975	52037	06071	14710	84160	05126
02965	78394	53956	48542	35568	09897	20287	54478
48963	58636	18825	47664	74509	53250	61778	43202
82404	85132	17634	35664	26876	54191	61412	08062
74377	17728	84164	32762	18275	91425	96275	97711
00348	36016	74853	00499	03387	07739	96416	28578
50006	74244	44355	96978	79518	51941	26627	39503
50881	25003	88424	98825	47209	16332	48104	50130
50200	95786	00463	82949	55660	93650	41134	43309
10934	10799	37862	71684	98288	94592	70961	51028
65423	47225	27490	78045	57512	11431	50502	11301
12334	16800	91872	82213	79086	33896	31707	48739
65006	14401	58189	32658	13812	72382	19169	24151
83058	55540	32716	03834	05299	58331	90479	34808
86657	29965	21658	72049	74959	95217	71503	74950
95380	20397	06649	02752	46279	99681	05154	56400
58069	86649	38783	30050	20087	26681	05859	41218
61173	27188	55705	29607	24361	83286	00052	41181
86572	53485	23353	18345	24553	43494	84651	48483
34184	47240	07490	86150	78423	82309	01762	33463
45731	19344	49685	95236	60922	09665	50176	98674
59596	70634	58195	78404	80798	60696	00686	75274
69095	26496	77711	62699	92045	74777	97537	69768
75969	64260	42303	54467	61374	22426	70625	96732
85109	67756	71072	04047	38520	31872	00118	84181
94990	90513	44712	51443	78685	61446	11529	64531
95018	53155	71938	99420	15408	44461	27981	90368
20563	11277	42307	32736	99532	65361	64300	76678
20768	27895	83856	69778	39159	68796	17739	30795
70805	92775	34093	00317	34349	79168	87667	98240
96194	11150	67555	95582	47540	91542	35023	35235
15920	14305	46477	35022	30006	35516	48513	09493
80789	87303	24848	46375	92999	26747	77435	49786
14907	30788	79299	34428	88050	39956	35155	55514
99467	74257	98806	10498	63125	99067	24489	32313

TABLE X. TABLE OF RANDOM NUMBERS

Col. Line	(1)	(2)	(3)	(4)	(5)	(6)
951	17222	24847	14225	43238	39943	20269
952	51117	86027	85128	33442	59152	67511
953	98122	85707	47835	60583	81100	24151
954	07079	03751	09486	74453	95050	28861
955	91827	97079	74080	85168	67841	92785
956	63924	46330	97808	84837	61032	98609
957	27527	61690	11334	66616	74361	26430
958	62640	79415	84174	76251	22673	44026
959	24485	14761	49553	21948	82300	95215
960	31163	12555	57763	73074	54090	95819
961	75829	90951	85627	98027	69242	44145
962	50265	07807	43138	62890	82977	38769
963	71200	80801	81289	44767	02901	61530
964	88107	27970	99234	22295	36871	14761
965	61197	02791	14924	93662	36173	00145
966	83405	26615	87854	03164	98061	64144
967	02276	39832	64484	17204	96556	46324
968	13020	23251	09546	06687	39893	85402
969	50766	20504	47573	87991	62684	10921
970	74813	19126	38598	77307	59250	62431
971	34150	68179	87177	64247	40979	62952
972	95035	52598	27655	90383	76468	08022
973	76846	66076	01650	77548	03009	48350
974	24146	49880	75124	54340	74297	31307
975	27427	83596	07262	58388	43212	81674
976	62277	00386	94965	57568	04965	52742
977	26218	09494	96963	28005	15942	13443
978	95828	98777	58855	21092	97414	28481
979	06052	32502	47874	51676	13316	43804
980	79103	19996	36294	42929	95081	45359
981	69040	75170	31286	93852	19394	43743
982	02688	26065	29875	54689	12247	67229
983	59669	62317	56242	90656	43992	64306
984	03409	78084	23818	21489	87944	11732
985	05188	62591	48883	62080	49460	32772
986	91857	24665	14441	83430	90899	23068
987	78853	88643	78621	02377	02253	06181
988	57803	03002	43966	52273	88993	47814
989	20414	62347	23572	02107	86764	95612
990	76748	64133	70898	44429	28121	56265
991	98506	47155	93726	32281	64505	56978
992	54242	81976	67665	70890	07185	20131
993	81990	00775	05836	34512	64536	69060
994	44250	04275	16365	24813	43750	62646
995	27942	19415	30541	10707	42525	84144
996	64300	38752	56657	85280	28224	84314
997	14653	08589	09404	81557	16276	63464
998	95087	67947	29851	06124	04054	73563
999	08774	32723	44960	29042	94665	36955
1000	37039	97547	64673	31546	99314	66854

TABLE X.   TABLE OF RANDOM NUMBERS

Table X 295

(7)	(8)	(9)	(10)	(11)	(12)	(13)	(14)
52593	99929	69993	66806	16035	74175	04974	75631
36916	93946	25402	53934	83623	18774	21413	37540
58693	46274	83783	68899	42100	14798	30325	01988
19001	56932	51925	60421	43291	13854	87524	26618
85584	97685	18369	02716	04330	45215	67343	98098
67177	96231	44689	05224	72408	37347	30369	45386
06353	29841	87500	03627	42526	26924	64139	45052
59218	78743	90058	54715	88307	88689	51172	47812
58519	81521	21773	21559	07775	96603	12156	58338
02558	92149	15685	15315	84154	89401	53301	70641
74974	58554	71020	36095	41778	51657	99796	93408
11754	81428	56568	40473	43510	66127	10806	57208
67617	50078	88909	40539	31521	68964	47991	59963
74494	02770	94610	89209	52688	92319	56341	04119
81019	63834	09891	16660	20541	48016	48620	66704
81948	49844	07816	04927	68917	11008	96965	74719
12068	68780	30396	01443	36666	76408	18082	19438
96369	71504	04242	80485	66151	39141	52873	13346
29624	44384	82938	30828	72266	20455	32794	18715
25603	49204	72792	87154	34140	45891	44767	55879
88828	81184	83310	49826	78268	84810	34755	76708
90814	78356	95077	65538	77477	12978	72000	34428
61844	17850	10898	75323	23270	74928	29066	50899
87627	14922	55793	78307	61406	73263	29927	21784
09745	74206	19310	11767	88334	52762	50280	77990
85423	87258	25805	97116	03662	40715	26670	51038
96330	50222	50856	27308	78353	83501	28171	69667
39244	80530	45700	78547	36536	86420	71140	25607
85364	25972	77144	32777	81828	16275	99252	28637
82772	10679	34236	27171	90309	42230	97146	91977
79837	38978	99397	91989	25596	24940	91964	97249
69268	72790	59095	68115	52432	03486	47945	00001
01038	22952	49953	72014	05599	06801	92765	39212
69494	98213	41356	65383	25792	96224	57038	19936
09094	13547	94514	38094	93755	39313	56688	30338
39436	65648	90795	04028	44670	88737	54796	80249
15150	41361	65890	86631	08289	15996	31647	89466
41363	97015	32977	09224	77829	77519	42395	21926
22638	02471	49752	86400	21017	92299	36523	28074
53054	85704	25631	56706	54899	25569	71394	48333
48469	71323	70581	72004	67535	79545	88388	46777
73487	53010	31677	34437	88622	03580	41811	40951
23223	57511	61764	85828	21119	66098	83996	70640
97455	01174	23358	71401	28580	92308	62973	72123
17237	38447	76318	03660	59959	76447	37567	66990
58118	97734	35701	06593	03690	50302	97271	97034
79223	12993	51056	57233	75099	31082	01867	25094
67625	44199	40748	52598	76137	26030	34138	76445
46376	14300	13280	88986	81039	34211	38168	67630
97855	99965	26373	61701	24760	52014	96547	65403

TABLE X.   TABLE OF RANDOM NUMBERS

Table X

Col. Line	(1)	(2)	(3)	(4)	(5)	(6)
1001	25145	84834	23009	51584	66754	77785
1002	98433	54725	18864	65866	76918	78825
1003	97965	68548	81545	82933	93545	85959
1004	78049	67830	14624	17563	25697	07734
1005	50203	25658	91478	08509	23308	48130
1006	40059	67825	18934	64998	49807	71126
1007	84350	67241	54031	34535	04093	35062
1008	30954	51637	91500	48722	60988	60029
1009	86723	36464	98305	08009	00866	29255
1010	50188	22554	86160	92250	14021	63859
1011	50014	00463	13906	35936	71761	95755
1012	66023	21428	14742	94674	23308	58533
1013	04458	61862	63119	09541	01715	87901
1014	57510	36314	30458	09712	37714	95482
1015	43373	58939	95848	28288	60341	52174
1016	40704	48823	65963	39359	12717	56201
1017	07318	44623	00843	33299	59872	86774
1018	94550	23299	45557	07923	75126	00808
1019	34348	81191	21027	77087	10909	03676
1020	92277	57115	50789	68111	75305	58289
1021	61500	12763	64433	02268	57905	72347
1022	78938	71312	99705	71546	42274	23915
1023	64287	93218	35793	43671	64055	88729
1024	35314	29631	06937	54546	04470	75463
1025	96864	11554	70445	24841	04779	76774
1026	96093	58302	52236	64756	50273	61566
1027	16623	17849	96701	94971	94758	08845
1028	50848	93982	66451	32143	05441	10399
1029	48006	58200	98367	66577	68583	21108
1030	56640	27890	28825	96509	21363	53657
1031	01554	40592	26557	79189	81099	03951
1032	63799	48415	28087	17196	42784	58377
1033	77654	05943	43283	63181	72300	50360
1034	51058	95654	72648	60349	59563	59021
1035	14982	68789	41999	84797	17743	18775
1036	22305	50900	77277	32383	41723	35868
1037	51375	62050	18718	88109	36767	27440
1038	14590	70151	63862	02246	33710	46456
1039	60312	56009	52588	65722	46072	26273
1040	48513	62172	46711	56744	52250	95301
1041	19052	03511	71346	50933	10108	19747
1042	11246	92882	40354	14700	66716	93470
1043	34268	97508	69711	07165	69844	07591
1044	22931	31857	95595	48438	78029	07045
1045	11476	16553	10297	23643	06192	72478
1046	57005	88551	02117	87827	00151	39027
1047	76056	48776	41451	27779	86241	63562
1048	11961	67214	64606	51982	87588	67464
1049	28009	20616	02614	15759	65618	92266
1050	09194	83950	36645	00734	74652	87328

TABLE X.   TABLE OF RANDOM NUMBERS

Table X

297

(7)	(8)	(9)	(10)	(11)	(12)	(13)	(14)
52357	25532	46346	39806	70361	41836	33098	03790
58210	76835	72248	43023	85884	08785	41152	11830
63282	61454	27337	10560	76390	36976	62785	77529
48243	94318	85992	15716	48891	74601	90030	82376
65047	77873	02912	52675	62971	40993	79767	19178
77818	56893	06714	25147	69782	56399	55494	29547
58168	14205	87257	59443	29927	30667	99040	61060
60873	37423	62206	78386	10008	03027	69882	99663
18514	49158	93679	84854	27953	73627	45315	94563
16237	72296	86715	17588	24327	69905	43592	52283
87002	71667	21899	50605	44017	48019	01348	33230
26507	11208	85354	34288	33031	06081	45402	54274
91260	03079	17350	91991	46955	16951	42276	18627
30507	68475	49167	54279	89734	24807	23802	60439
11879	18115	91925	93506	10331	52809	06032	98707
22811	24863	62130	86396	12606	40352	44181	56860
06926	12672	23096	62738	48877	98889	77375	72317
01312	46689	56900	74653	65614	94333	44775	42054
97723	34469	08852	51465	38544	25247	00105	14825
39751	45760	76922	62810	47392	47765	69009	41520
49498	21871	80128	71588	34936	93724	47479	30366
38906	18779	96970	81430	77457	90793	53509	79535
11168	60260	97668	29687	78958	71480	75638	32812
77112	77126	46942	35873	21104	70323	07091	25408
96129	73594	67259	08958	45986	91442	28529	91358
61962	93280	54750	91572	22825	35238	34955	77692
32280	59823	97638	63913	81772	20605	13104	86762
17775	74169	04324	00195	97321	71104	70286	44675
41361	20732	39538	95746	91363	37416	08833	72639
60119	75385	83475	94949	94360	64076	24305	66042
60438	94263	21708	82714	33487	15713	24149	90641
02026	79893	28219	95848	29099	68424	97854	87967
38411	08823	83158	85734	18917	92697	20102	13956
18616	93244	67666	36344	60506	10937	69837	92511
61437	55258	36401	21686	20416	77429	99664	16803
61829	75281	73850	09094	10748	87175	35537	42489
73438	87774	65688	13701	38111	44531	10728	25248
50225	35742	46423	27373	95388	90238	81178	95956
32049	89647	81900	49167	81910	98790	66208	92721
20401	96558	22600	59000	44669	97303	25080	61986
84712	35431	27652	36130	67396	98295	58395	54493
48587	12514	20234	20769	86575	53071	84995	46531
64588	58016	26532	17957	59119	70048	14130	72848
04453	46783	97341	34526	02731	34808	65028	61054
75968	23807	70533	73805	64118	60682	17286	45369
66060	38915	55240	10473	52382	74523	13452	18918
98013	21109	57835	56161	76084	85338	65491	85719
54389	52910	37447	71173	85880	06581	73833	62987
13545	39578	36021	09262	71240	16627	28570	44003
53507	71924	47057	83163	16248	76396	17886	46311

TABLE X.   TABLE OF RANDOM NUMBERS

NATURAL SINES, COSINES, AND TANGENTS
OF ANGLES FROM 0° to 90°

0°–14.9°

Degs.	Function	0.0°	0.1°	0.2°	0.3°	0.4°	0.5°	0.6°	0.7°	0.8°	0.9°
0	sin	0.0000	0.0017	0.0035	0.0052	0.0070	0.0087	0.0105	0.0122	0.0140	0.0157
	cos	1.0000	1.0000	1.0000	1.0000	1.0000	1.0000	0.9999	0.9999	0.9999	0.9999
	tan	0.0000	0.0017	0.0035	0.0052	0.0070	0.0087	0.0105	0.0122	0.0140	0.0157
1	sin	0.0175	0.0192	0.0209	0.0227	0.0244	0.0262	0.0279	0.0297	0.0314	0.0332
	cos	0.9998	0.9998	0.9998	0.9997	0.9997	0.9997	0.9996	0.9996	0.9995	0.9995
	tan	0.0175	0.0192	0.0209	0.0227	0.0244	0.0262	0.0279	0.0297	0.0314	0.0332
2	sin	0.0349	0.0366	0.0384	0.0401	0.0419	0.0436	0.0454	0.0471	0.0488	0.0506
	cos	0.9994	0.9993	0.9993	0.9992	0.9991	0.9990	0.9990	0.9989	0.9988	0.9987
	tan	0.0349	0.0367	0.0384	0.0402	0.0419	0.0437	0.0454	0.0472	0.0489	0.0507
3	sin	0.0523	0.0541	0.0558	0.0576	0.0593	0.0610	0.0628	0.0645	0.0663	0.0680
	cos	0.9986	0.9985	0.9984	0.9983	0.9982	0.9981	0.9980	0.9979	0.9978	0.9977
	tan	0.0524	0.0542	0.0559	0.0577	0.0594	0.0612	0.0629	0.0647	0.0664	0.0682
4	sin	0.0698	0.0715	0.0732	0.0750	0.0767	0.0785	0.0802	0.0819	0.0837	0.0854
	cos	0.9976	0.9974	0.9973	0.9972	0.9971	0.9969	0.9968	0.9966	0.9965	0.9963
	tan	0.0699	0.0717	0.0734	0.0752	0.0769	0.0787	0.0805	0.0822	0.0840	0.0857
5	sin	0.0872	0.0889	0.0906	0.0924	0.0941	0.0958	0.0976	0.0993	0.1011	0.1028
	cos	0.9962	0.9960	0.9959	0.9957	0.9956	0.9954	0.9952	0.9951	0.9949	0.9947
	tan	0.0875	0.0892	0.0910	0.0928	0.0945	0.0963	0.0981	0.0998	0.1016	0.1033
6	sin	0.1045	0.1063	0.1080	0.1097	0.1115	0.1132	0.1149	0.1167	0.1184	0.1201
	cos	0.9945	0.9943	0.9942	0.9940	0.9938	0.9936	0.9934	0.9932	0.9930	0.9928
	tan	0.1051	0.1069	0.1086	0.1104	0.1122	0.1139	0.1157	0.1175	0.1192	0.1210
7	sin	0.1219	0.1236	0.1253	0.1271	0.1288	0.1305	0.1323	0.1340	0.1357	0.1374
	cos	0.9925	0.9923	0.9921	0.9919	0.9917	0.9914	0.9912	0.9910	0.9907	0.9905
	tan	0.1228	0.1246	0.1263	0.1281	0.1299	0.1317	0.1334	0.1352	0.1370	0.1388
8	sin	0.1392	0.1409	0.1426	0.1444	0.1461	0.1478	0.1495	0.1513	0.1530	0.1547
	cos	0.9903	0.9900	0.9898	0.9895	0.9893	0.9890	0.9888	0.9885	0.9882	0.9880
	tan	0.1405	0.1423	0.1441	0.1459	0.1477	0.1495	0.1512	0.1530	0.1548	0.1566
9	sin	0.1564	0.1582	0.1599	0.1616	0.1633	0.1650	0.1668	0.1685	0.1702	0.1719
	cos	0.9877	0.9874	0.9871	0.9869	0.9866	0.9863	0.9860	0.9857	0.9854	0.9851
	tan	0.1584	0.1602	0.1620	0.1638	0.1655	0.1673	0.1691	0.1709	0.1727	0.1745
10	sin	0.1736	0.1754	0.1771	0.1788	0.1805	0.1822	0.1840	0.1857	0.1874	0.1891
	cos	0.9848	0.9845	0.9842	0.9839	0.9836	0.9833	0.9829	0.9826	0.9823	0.9820
	tan	0.1763	0.1781	0.1799	0.1817	0.1835	0.1853	0.1871	0.1890	0.1908	0.1926
11	sin	0.1908	0.1925	0.1942	0.1959	0.1977	0.1994	0.2011	0.2028	0.2045	0.2062
	cos	0.9816	0.9813	0.9810	0.9806	0.9803	0.9799	0.9796	0.9792	0.9789	0.9785
	tan	0.1944	0.1962	0.1980	0.1998	0.2016	0.2035	0.2053	0.2071	0.2089	0.2107
12	sin	0.2079	0.2096	0.2113	0.2130	0.2147	0.2164	0.2181	0.2198	0.2215	0.2232
	cos	0.9781	0.9778	0.9774	0.9770	0.9767	0.9763	0.9759	0.9755	0.9751	0.9748
	tan	0.2126	0.2144	0.2162	0.2180	0.2199	0.2217	0.2235	0.2254	0.2272	0.2290
13	sin	0.2250	0.2267	0.2284	0.2300	0.2318	0.2334	0.2351	0.2368	0.2385	0.2402
	cos	0.9744	0.9740	0.9736	0.9732	0.9728	0.9724	0.9720	0.9715	0.9711	0.9707
	tan	0.2309	0.2327	0.2345	0.2364	0.2382	0.2401	0.2419	0.2438	0.2456	0.2475
14	sin	0.2419	0.2436	0.2453	0.2470	0.2487	0.2504	0.2521	0.2538	0.2554	0.2571
	cos	0.9703	0.9699	0.9694	0.9690	0.9686	0.9681	0.9677	0.9673	0.9668	0.9664
	tan	0.2493	0.2512	0.2530	0.2549	0.2568	0.2586	0.2605	0.2623	0.2642	0.2661
Degs.	Function	0'	6'	12'	18'	24'	30'	36'	42'	48'	54'

TABLE XI. NATURAL SINES, COSINES, AND TANGENTS OF ANGLES

TABLE XI. NATURAL SINES, COSINES AND TANGENTS OF ANGLES
From *Mathematics*, Vol. 1, Bureau of Naval Personnel, Navy Training Course, Navpers 10069–C, 1966, pp. 213–218.

Table XI 299

15°–29.9°

Degs.	Function	0.0°	0.1°	0.2°	0.3°	0.4°	0.5°	0.6°	0.7°	0.8°	0.9°
15	sin	0.2588	0.2605	0.2622	0.2639	0.2656	0.2672	0.2689	0.2706	0.2723	0.2740
	cos	0.9569	0.9655	0.9650	0.9646	0.9641	0.9636	0.9632	0.9627	0.9622	0.9617
	tan	0.2679	0.2698	0.2717	0.2736	0.2754	0.2773	0.2792	0.2811	0.2830	0.2849
16	sin	0.2756	0.2773	0.2790	0.2807	0.2823	0.2840	0.2857	0.2874	0.2890	0.2907
	cos	0.9613	0.9608	0.9603	0.9598	0.9593	9.9588	0.9583	0.9578	0.9573	0.9568
	tan	0.2867	0.2886	0.2905	0.2924	0.2943	0.2962	0.2981	0.3000	0.3019	0.3038
17	sin	0.2924	0.2940	0.2957	0.2974	0.2990	0.3007	0.3024	0.3040	0.3057	0.3074
	cos	0.9563	0.9558	0.9553	0.9548	0.9542	0.9537	0.9532	0.9527	0.9521	0.9516
	tan	0.3057	0.3076	0.3096	0.3115	0.3134	0.3153	0.3172	0.3191	0.3211	0.3230
18	sin	0.3090	0.3107	0.3123	0.3140	0.3156	0.3173	0.3190	0.3206	0.3223	0.3239
	cos	0.9511	0.9505	0.9500	0.9494	0.9489	0.9483	0.9478	0.9472	0.9466	0.9461
	tan	0.3249	0.3269	0.3288	0.3307	0.3327	0.3346	0.3365	0.3385	0.3404	0.3424
19	sin	0.3256	0.3272	0.3289	0.3305	0.3322	0.3338	0.3355	0.3371	0.3387	0.3404
	cos	0.9455	0.9449	0.9444	0.9438	0.9432	0.9426	0.9421	0.9415	0.9409	0.9403
	tan	0.3443	0.3463	0.3482	0.3502	0.3522	0.3541	0.3561	0.3581	0.3600	0.3620
20	sin	0.3420	0.3437	0.3453	0.3469	0.3486	0.3502	0.3518	0.3535	0.3551	0.3567
	cos	0.9397	0.9391	0.9385	0.9379	0.9373	0.9367	0.9361	0.9354	0.9348	0.9342
	tan	0.3640	0.3659	0.3679	0.3699	0.3719	0.3739	0.3759	0.3779	0.3799	0.3819
21	sin	0.3584	0.3600	0.3616	0.3633	0.3649	0.3665	0.3681	0.3697	0.3714	0.3730
	cos	0.9336	0.9330	0.9323	0.9317	0.9311	0.9304	0.9298	0.9291	0.9285	0.9278
	tan	0.3839	0.3859	0.3879	0.3899	0.3919	0.3939	0.3959	0.3979	0.4000	0.4020
22	sin	0.3746	0.3762	0.3778	0.3795	0.3811	0.3827	0.3843	0.3859	0.3875	0.3891
	cos	0.9272	0.9265	0.9259	0.9252	0.9245	0.9239	0.9232	0.9225	0.9219	0.9212
	tan	0.4040	0.4061	0.4081	0.4101	0.4122	0.4142	0.4163	0.4183	0.4204	0.4224
23	sin	0.3907	0.3923	0.3939	0.3955	0.3971	0.3987	0.4003	0.4019	0.4035	0.4051
	cos	0.9205	0.9198	0.9191	0.9184	0.9178	0.9171	0.9164	0.9157	0.9150	0.9143
	tan	0.4245	0.4265	0.4286	0.4307	0.4327	0.4348	0.4369	0.4390	0.4411	0.4431
24	sin	0.4067	0.4083	0.4099	0.4115	0.4131	0.4147	0.4163	0.4179	0.4195	0.4210
	cos	0.9135	0.9128	0.9121	0.9114	0.9107	0.9100	0.9092	0.9085	0.9078	0.9070
	tan	0.4452	0.4473	0.4494	0.4515	0.4536	0.4557	0.4578	0.4599	0.4621	0.4642
25	sin	0.4226	0.4242	0.4258	0.4274	0.4289	0.4305	0.4321	0.4337	0.4352	0.4368
	cos	0.9063	0.9056	0.9048	0.9041	0.9033	0.9026	0.9018	0.9011	0.9003	0.8996
	tan	0.4663	0.4684	0.4706	0.4727	0.4748	0.4770	0.4791	0.4813	0.4834	0.4856
26	sin	0.4384	0.4399	0.4415	0.4431	0.4446	0.4462	0.4478	0.4493	0.4509	0.4524
	cos	0.8988	0.8980	0.8973	0.8965	0.8957	0.8949	0.8942	0.8934	0.8926	0.8918
	tan	0.4877	0.4899	0.4921	0.4942	0.4964	0.4986	0.5008	0.5029	0.5051	0.5073
27	sin	0.4540	0.4555	0.4571	0.4586	0.4602	0.4617	0.4633	0.4648	0.4664	0.4679
	cos	0.8910	0.8902	0.8894	0.8886	0.8878	0.8870	0.8862	0.8854	0.8846	0.8838
	tan	0.5095	0.5117	0.5139	0.5161	0.5184	0.5206	0.5228	0.5250	0.5272	0.5295
28	sin	0.4695	0.4710	0.4726	0.4741	0.4756	0.4772	0.4787	0.4802	0.4818	0.4833
	cos	0.8829	0.8821	0.8813	0.8805	0.8796	0.8788	0.8780	0.8771	0.8763	0.8755
	tan	0.5317	0.5340	0.5362	0.5384	0.5407	0.5430	0.5452	0.5475	0.5498	0.5520
29	sin	0.4848	0.4863	0.4879	0.4894	0.4909	0.4924	0.4939	0.4955	0.4970	0.4985
	cos	0.8746	0.8738	0.8729	0.8721	0.8712	0.8704	0.8695	0.8686	0.8678	0.8669
	tan	0.5543	0.5566	0.5589	0.5612	0.5635	0.5658	0.5681	0.5704	0.5727	0.5750
Degs.	Function	0'	6'	12'	18'	24'	30'	36'	42'	48'	54'

TABLE XI. NATURAL SINES, COSINES, AND TANGENTS OF ANGLES

Table XI

30° 44.9°

Degs.	Function	0.0°	0.1°	0.2°	0.3°	0.4°	0.5°	0.6°	0.7°	0.8°	0.9°
30	sin	0.5000	0.5015	0.5030	0.5045	0.5060	0.5075	0.5090	0.5105	0.5120	0.5135
	cos	0.8660	0.8652	0.8643	0.8634	0.8625	0.8616	0.8607	0.8599	0.8590	0.8581
	tan	0.5774	0.5797	0.5820	0.5844	0.5867	0.5890	0.5914	0.5938	0.5961	0.5985
31	sin	0.5150	0.5165	0.5180	0.5195	0.5210	0.5225	0.5240	0.5255	0.5270	0.5284
	cos	0.8572	0.8563	0.8554	0.8545	0.8536	0.8526	0.8517	0.8508	0.8499	0.8490
	tan	0.6009	0.6032	0.6056	0.6080	0.6104	0.6128	0.6152	0.6176	0.6200	0.6224
32	sin	0.5299	0.5314	0.5329	0.5344	0.5358	0.5373	0.5388	0.5402	0.5417	0.5432
	cos	0.8480	0.8471	0.8462	0.8453	0.8443	0.8434	0.8425	0.8415	0.8406	0.8396
	tan	0.6249	0.6273	0.6297	0.6322	0.6346	0.6371	0.6395	0.6420	0.6445	0.6469
33	sin	0.5446	0.5461	0.5476	0.5490	0.5505	0.5519	0.5534	0.5548	0.5563	0.5577
	cos	0.8387	0.8377	0.8368	0.8358	0.8348	0.8339	0.8329	0.8320	0.8310	0.8300
	tan	0.6494	0.6519	0.6544	0.6569	0.6594	0.6619	0.6644	0.6669	0.6694	0.6720
34	sin	0.5592	0.5606	0.5621	0.5635	0.5650	0.5664	0.5678	0.5693	0.5707	0.5721
	cos	0.8290	0.8281	0.8271	0.8261	0.8251	0.8241	0.8231	0.8221	0.8211	0.8202
	tan	0.6745	0.6771	0.6796	0.6822	0.6847	0.6873	0.6899	0.6924	0.6950	0.6976
35	sin	0.5736	0.5750	0.5764	0.5779	0.5793	0.5807	0.5821	0.5835	0.5850	0.5864
	cos	0.8192	0.8181	0.8171	0.8161	0.8151	0.8141	0.8131	0.8121	0.8111	0.8100
	tan	0.7002	0.7028	0.7054	0.7080	0.7107	0.7133	0.7159	0.7186	0.7212	0.7239
36	sin	0.5878	0.5892	0.5906	0.5920	0.5934	0.5948	0.5962	0.5976	0.5990	0.6004
	cos	0.8090	0.8080	0.8070	0.8059	0.8049	0.9039	0.8028	0.8018	0.8007	0.7997
	tan	0.7265	0.7292	0.7319	0.7346	0.7373	0.7400	0.7427	0.7454	0.7481	0.7508
37	sin	0.6018	0.6032	0.6046	0.6060	0.6074	0.6088	0.6101	0.6115	0.6129	0.6143
	cos	0.7986	0.7976	0.7965	0.7955	0.7944	0.7934	0.7923	0.7912	0.7902	0.7891
	tan	0.7536	0.7563	0.7590	0.7618	0.7646	0.7673	0.7701	0.7729	0.7757	0.7785
38	sin	0.6157	0.6170	0.6184	0.6198	0.6211	0.6225	0.6239	0.6252	0.6266	0.6280
	cos	0.7880	0.7869	0.7859	0.7848	0.7837	0.7826	0.7815	0.7804	0.7793	0.7782
	tan	0.7813	0.7841	0.7869	0.7898	0.7926	0.7954	0.7983	0.8012	0.8040	0.8069
39	sin	0.6293	0.6307	0.6320	0.6334	0.6347	0.6361	0.6374	0.6388	0.6401	0.6414
	cos	0.7771	0.7760	0.7749	0.7738	0.7727	0.7716	0.7705	0.7694	0.7683	0.7672
	tan	0.8098	0.8127	0.8156	0.8185	0.8214	0.8243	0.8273	0.8302	0.8332	0.8361
40	sin	0.6428	0.6441	0.6455	0.6468	0.6481	0.6494	0.6508	0.6521	0.6534	0.6547
	cos	0.7660	0.7649	0.7638	0.7627	0.7615	0.7604	0.7593	0.7581	0.7570	0.7559
	tan	0.8391	0.8421	0.8451	0.8481	0.8511	0.8541	0.8571	0.8601	0.8632	0.8662
41	sin	0.6561	0.6574	0.6587	0.6600	0.6613	0.6626	0.6639	0.6652	0.6665	0.6678
	cos	0.7547	0.7536	0.7524	0.7513	0.7501	0.7490	0.7478	0.7466	0.7455	0.7443
	tan	0.8693	0.8724	0.8754	0.8785	0.8816	0.8847	0.8878	0.8910	0.8941	0.8972
42	sin	0.6691	0.6704	0.6717	0.6730	0.6743	0.6756	0.6769	0.6782	0.6794	0.6807
	cos	0.7431	0.7420	0.7408	0.7396	0.7385	0.7373	0.7361	0.7349	0.7337	0.7325
	tan	0.9004	0.9036	0.9067	0.9099	0.9131	0.9163	0.9195	0.9228	0.9260	0.9293
43	sin	0.6820	0.6833	0.6845	0.6858	0.6871	0.6884	0.6896	0.6909	0.6921	0.6934
	cos	0.7314	0.7302	0.7290	0.7278	0.7266	0.7254	0.7242	0.7230	0.7218	0.7206
	tan	0.9325	0.9358	0.9391	0.9424	0.9457	0.9490	0.9523	0.9556	0.9590	0.9623
44	sin	0.6947	0.6959	0.6972	0.6984	0.6997	0.7009	0.7022	0.7034	0.7046	0.7059
	cos	0.7193	0.7181	0.7169	0.7157	0.7145	0.7133	0.7120	0.7108	0.7096	0.7083
	tan	0.9657	0.9691	0.9725	0.9759	0.9793	0.9827	0.9861	0.9896	0.9930	0.9965
Degs.	Function	0'	6'	12'	18'	24'	30'	36'	42'	48'	54'

TABLE XI. NATURAL SINES, COSINES, AND TANGENTS OF ANGLES

Table XI 301

45°–59.9°

Degs.	Function	0.0°	0.1°	0.2°	0.3°	0.4°	0.5°	0.6°	0.7°	0.8°	0.9°
45	sin	0.7071	0.7083	0.7096	0.7108	0.7120	0.7133	0.7145	0.7157	0.7169	0.7181
	cos	0.7071	0.7059	0.7046	0.7034	0.7022	0.7009	0.6997	0.6984	0.6972	0.6959
	tan	1.0000	1.0035	1.0070	1.0105	1.0141	1.0176	1.0212	1.0247	1.0283	1.0319
46	sin	0.7193	0.7206	0.7218	0.7230	0.7242	0.7254	0.7266	0.7278	0.7290	0.7302
	cos	0.6947	0.6934	0.6921	0.6909	0.6896	0.6884	0.6871	0.6858	0.6845	0.6833
	tan	1.0355	1.0392	1.0428	1.0464	1.0501	1.0538	1.0575	1.0612	1.0649	1.0686
47	sin	0.7314	0.7325	0.7337	0.7349	0.7361	0.7373	0.7385	0.7396	0.7408	0.7420
	cos	0.6820	0.6807	0.6794	0.6782	0.6769	0.6756	0.6743	0.6730	0.6717	0.6704
	tan	1.0724	1.0761	1.0799	1.0837	1.0875	1.0913	1.0951	1.0990	1.1028	1.1067
48	sin	0.7431	0.7443	0.7455	0.7466	0.7478	0.7490	0.7501	0.7513	0.7524	0.7536
	cos	0.6691	0.6678	0.6665	0.6652	0.6639	0.6626	0.6613	0.6600	0.6587	0.6574
	tan	1.1106	1.1145	1.1184	1.1224	1.1263	1.1303	1.1343	1.1383	1.1423	1.1463
49	sin	0.7547	0.7559	0.7570	0.7581	0.7593	0.7604	0.7615	0.7627	0.7638	0.7649
	cos	0.6561	0.6547	0.6534	0.6521	0.6508	0.6494	0.6481	0.6468	0.6455	0.6441
	tan	1.1504	1.1544	1.1585	1.1626	1.1667	1.1708	1.1750	1.1792	1.1833	1.1875
50	sin	0.7660	0.7672	0.7683	0.7694	0.7705	0.7716	0.7727	0.7738	0.7749	0.7760
	cos	0.6428	0.6414	0.6401	0.6388	0.6374	0.6361	0.6347	0.6334	0.6320	0.6307
	tan	1.1918	1.1960	1.2002	1.2045	1.2088	1.2131	1.2174	1.2218	1.2261	1.2305
51	sin	0.7771	0.7782	0.7793	0.7804	0.7815	0.7826	0.7837	0.7848	0.7859	0.7869
	cos	0.6293	0.6280	0.6266	0.6252	0.6239	0.6225	0.6211	0.6198	0.6184	0.6170
	tan	1.2349	1.2393	1.2437	1.2482	1.2527	1.2572	1.2617	1.2662	1.2708	1.2753
52	sin	0.7880	0.7891	0.7902	0.7912	0.7923	0.7934	0.7944	0.7955	0.7965	0.7976
	cos	0.6157	0.6143	0.6129	0.6115	0.6101	0.6088	0.6074	0.6060	0.6046	0.6032
	tan	1.2799	1.2846	1.2892	1.2938	1.2985	1.3032	1.3079	1.3127	1.3175	1.3222
53	sin	0.7986	0.7997	0.8007	0.8018	0.8028	0.8039	0.8049	0.8059	0.8070	0.8080
	cos	0.6018	0.6004	0.5990	0.5976	0.5962	0.5948	0.5934	0.5920	0.5906	0.5892
	tan	1.3270	1.3319	1.3367	1.3416	1.3465	1.3514	1.3564	1.3613	1.3663	1.3713
54	sin	0.8090	0.8100	0.8111	0.8121	0.8131	0.8141	0.8151	0.8161	0.8171	0.8181
	cos	0.5878	0.5864	0.5850	0.5835	0.5821	0.5807	0.5793	0.5779	0.5764	0.5750
	tan	1.3764	1.3814	1.3865	1.3916	1.3968	1.4019	1.4071	1.4124	1.4176	1.4229
55	sin	0.8192	0.8202	0.8211	0.8221	0.8231	0.8241	0.8251	0.8261	0.8271	0.8281
	cos	0.5736	0.5721	0.5707	0.5693	0.5678	0.5664	0.5650	0.5635	0.5621	0.5606
	tan	1.4281	1.4335	1.4388	1.4442	1.4496	1.4550	1.4605	1.4659	1.4715	1.4770
56	sin	0.8290	0.8300	0.8310	0.8320	0.8329	0.8339	0.8348	0.8358	0.8368	0.8377
	cos	0.5592	0.5577	0.5563	0.5548	0.5534	0.5519	0.5505	0.5490	0.5476	0.5461
	tan	1.4826	1.4882	1.4938	1.4994	1.5051	1.5108	1.5166	1.5224	1.5282	1.5340
57	sin	0.8387	0.8396	0.8406	0.8415	0.8425	0.8434	0.8443	0.8453	0.8462	0.8471
	cos	0.5446	0.5432	0.5417	0.5402	0.5388	0.5373	0.5358	0.5344	0.5329	0.5314
	tan	1.5399	1.5458	1.5517	1.5577	1.5637	1.5697	1.5757	1.5818	1.5880	1.5941
58	sin	0.8480	0.8490	0.8499	0.8508	0.8517	0.8526	0.8536	0.8545	0.8554	0.8563
	cos	0.5299	0.5284	0.5270	0.5255	0.5240	0.5225	0.5210	0.5195	0.5180	0.5165
	tan	1.6003	1.6066	1.6128	1.6191	1.6255	1.6319	1.6383	1.6447	1.6512	1.6577
59	sin	0.8572	0.8581	0.8590	0.8599	0.8607	0.8616	0.8625	0.8634	0.8643	0.8652
	cos	0.5150	0.5135	0.5120	0.5105	0.5090	0.5075	0.5060	0.5045	0.5030	0.5015
	tan	1.6643	1.6709	1.6775	1.6842	1.6909	1.6977	1.7045	1.7113	1.7182	1.7251
Degs.	Function	0′	6′	12′	18′	24′	30′	36′	42′	48′	54′

TABLE XI. NATURAL SINES, COSINES, AND TANGENTS OF ANGLES

Table XI

60°–74.9°

Degs.	Function	0.0°	0.1°	0.2°	0.3°	0.4°	0.5°	0.6°	0.7°	0.8°	0.9°
60	sin	0.8660	0.8669	0.8678	0.8686	0.8695	0.8704	0.8712	0.8721	0.8729	0.8738
	cos	0.5000	0.4985	0.4970	0.4955	0.4939	0.4924	0.4909	0.4894	0.4879	0.4863
	tan	1.7321	1.7391	1.7461	1.7532	1.7603	1.7675	1.7747	1.7820	1.7893	1.7966
61	sin	0.8746	0.8755	0.8763	0.8771	0.8780	0.8788	0.8796	0.8805	0.8813	0.8821
	cos	0.4848	0.4833	0.4818	0.4802	0.4787	0.4772	0.4756	0.4741	0.4726	0.4710
	tan	1.8040	1.8115	1.8190	1.8265	1.8341	1.8418	1.8495	1.8572	1.8650	1.8728
62	sin	0.8829	0.8838	0.8846	0.8854	0.8862	0.8870	0.8878	0.8886	0.8894	0.8902
	cos	0.4695	0.4679	0.4664	0.4648	0.4633	0.4617	0.4602	0.4586	0.4571	0.4555
	tan	1.8807	1.8887	1.8967	1.9047	1.9128	1.9210	1.9292	1.9375	1.9458	1.9542
63	sin	0.8910	0.8918	0.8926	0.8934	0.8942	0.8949	0.8957	0.8965	0.8973	0.8980
	cos	0.4540	0.4524	0.4509	0.4493	0.4478	0.4462	0.4446	0.4431	0.4415	0.4399
	tan	1.9626	1.9711	1.9797	1.9883	1.9970	2.0057	2.0145	2.0233	2.0323	2.0413
64	sin	0.8988	0.8996	0.9003	0.9011	0.9018	0.9026	0.9033	0.9041	0.9048	0.9056
	cos	0.4384	0.4368	0.4352	0.4337	0.4321	0.4305	0.4289	0.4274	0.4258	0.4242
	tan	2.0503	2.0594	2.0686	2.0778	2.0872	2.0965	2.1060	2.1155	2.1251	2.1348
65	sin	0.9063	0.9070	0.9078	0.9085	0.9092	0.9100	0.9107	0.9114	0.9121	0.9128
	cos	0.4226	0.4210	0.4195	0.4179	0.4163	0.4147	0.4131	0.4115	0.4099	0.4083
	tan	2.1445	2.1543	2.1642	2.1742	2.1842	2.1943	2.2045	2.2148	2.2251	2.2355
66	sin	0.9135	0.9143	0.9150	0.9157	0.9164	0.9171	0.9178	0.9184	0.9191	0.9198
	cos	0.4067	0.4051	0.4035	0.4019	0.4003	0.3987	0.3971	0.3955	0.3939	0.3923
	tan	2.2460	2.2566	2.2673	2.2781	2.2889	2.2998	2.3109	2.3220	2.3332	2.3445
67	sin	0.9205	0.9212	0.9219	0.9225	0.9232	0.9239	0.9245	0.9252	0.9259	0.9265
	cos	0.3907	0.3891	0.3875	0.3859	0.3843	0.3827	0.3811	0.3795	0.3778	0.3762
	tan	2.3559	2.3673	2.3789	2.3906	2.4023	2.4142	2.4262	2.4383	2.4504	2.4627
68	sin	0.9272	0.9278	0.9285	0.9291	0.9298	0.9304	0.9311	0.9317	0.9323	0.9330
	cos	0.3746	0.3730	0.3714	0.3697	0.3681	0.3665	0.3649	0.3633	0.3616	0.3600
	tan	2.4751	2.4876	2.5002	2.5129	2.5257	2.5386	2.5517	2.5649	2.5782	2.5916
69	sin	0.9336	0.9342	0.9348	0.9354	0.9361	0.9367	0.9373	0.9379	0.9385	0.9391
	cos	0.3584	0.3567	0.3551	0.3535	0.3518	0.3502	0.3486	0.3469	0.3453	0.3437
	tan	2.6051	2.6187	2.6325	2.6464	2.6605	2.6746	2.6889	2.7034	2.7179	2.7326
70	sin	0.9397	0.9403	0.9409	0.9415	0.9421	0.9426	0.9432	0.9438	0.9444	0.9449
	cos	0.3420	0.3404	0.3387	0.3371	0.3355	0.3338	0.3322	0.3305	0.3289	0.3272
	tan	2.7475	2.7625	2.7776	2.7929	2.8083	2.8239	2.8397	2.8556	2.8716	2.8878
71	sin	0.9455	0.9461	0.9466	0.9472	0.9478	0.9483	0.9489	0.9494	0.9500	0.9505
	cos	0.3256	0.3239	0.3223	0.3206	0.3190	0.3173	0.3156	0.3140	0.3123	0.3107
	tan	2.9042	2.9208	2.9375	2.9544	2.9714	2.9887	3.0061	3.0237	3.0415	3.0595
72	sin	0.9511	0.9516	0.9521	0.9527	0.9532	0.9537	0.9542	0.9548	0.9553	0.9558
	cos	0.3090	0.3074	0.3057	0.3040	0.3024	0.3007	0.2990	0.2974	0.2957	0.2940
	tan	3.0777	3.0961	3.1146	3.1334	3.1524	3.1716	3.1910	3.2106	3.2305	3.2506
73	sin	0.9563	0.9568	0.9573	0.9578	0.9583	0.9588	0.9593	0.9598	0.9603	0.9608
	cos	0.2924	0.2907	0.2890	0.2874	0.2857	0.2840	0.2823	0.2807	0.2790	0.2773
	tan	3.2709	3.2914	3.3122	3.3332	3.3544	3.3759	3.3977	3.4197	3.4420	3.4646
74	sin	0.9613	0.9617	0.9622	0.9627	0.9632	0.9636	0.9641	0.9646	0.9650	0.9655
	cos	0.2756	0.2740	0.2723	0.2706	0.2689	0.2672	0.2656	0.2639	0.2622	0.2605
	tan	3.4874	3.5105	3.5339	3.5576	3.5816	3.6059	3.6305	3.6554	3.6806	3.7062
Degs.	Function	0′	6′	12′	18′	24′	30′	36′	42′	48′	54′

TABLE XI. NATURAL SINES, COSINES, AND TANGENTS OF ANGLES

Table XI

303

75°–89.9°

Degs.	Function	0.0°	0.1°	0.2°	0.3°	0.4°	0.5°	0.6°	0.7°	0.8°	0.9°
75	sin	0.9659	0.9664	0.9668	0.9673	0.9677	0.9681	0.9686	0.9690	0.9694	0.9699
	cos	0.2588	0.2571	0.2554	0.2538	0.2521	0.2504	0.2487	0.2470	0.2453	0.2436
	tan	3.7321	3.7583	3.7848	3.8118	3.8391	3.8667	3.8947	3.9232	3.9520	3.9812
76	sin	0.9703	0.9707	0.9711	0.9715	0.9720	0.9724	0.9728	0.9732	0.9736	0.9740
	cos	0.2419	0.2402	0.2385	0.2368	0.2351	0.2334	0.2317	0.2300	0.2284	0.2267
	tan	4.0108	4.0408	4.0713	4.1022	4.1335	4.1653	4.1976	4.2303	4.2635	4.2972
77	sin	0.9744	0.9748	0.9751	0.9755	0.9759	0.9763	0.9767	0.9770	0.9774	0.9778
	cos	0.2250	0.2232	0.2215	0.2198	0.2181	0.2164	0.2147	0.2130	0.2113	0.2096
	tan	4.3315	4.3662	4.4015	4.4374	4.4737	4.5107	4.5483	4.5864	4.6252	4.6646
78	sin	0.9781	0.9785	0.9789	0.9792	0.9796	0.9799	0.9803	0.9806	0.9810	0.9813
	cos	0.2079	0.2062	0.2045	0.2028	0.2011	0.1994	0.1977	0.1959	0.1942	0.1925
	tan	4.7046	4.7453	4.7867	4.8288	4.8716	4.9152	4.9594	5.0045	5.0504	5.0970
79	sin	0.9816	0.9820	0.9823	0.9826	0.9829	0.9833	0.9836	0.9839	0.9842	0.9845
	cos	0.1908	0.1891	0.1874	0.1857	0.1840	0.1822	0.1805	0.1788	0.1771	0.1754
	tan	5.1446	5.1929	5.2422	5.2924	5.3435	5.3955	5.4486	5.5026	5.5578	5.6140
80	sin	0.9848	0.9851	0.9854	0.9857	0.9860	0.9863	0.9866	0.9869	0.9871	0.9874
	cos	0.1736	0.1719	0.1702	0.1685	0.1668	0.1650	0.1633	0.1616	0.1599	0.1582
	tan	5.6713	5.7297	5.7894	5.8502	5.9124	5.9758	6.0405	6.1066	6.1742	6.2432
81	sin	0.9877	0.9880	0.9882	0.9885	0.9888	0.9890	0.9893	0.9895	0.9898	0.9900
	cos	0.1564	0.1547	0.1530	0.1513	0.1495	0.1478	0.1461	0.1444	0.1426	0.1409
	tan	6.3138	6.3859	6.4596	6.5350	6.6122	6.6912	6.7720	6.8548	6.9395	7.0264
82	sin	0.9903	0.9905	0.9907	0.9910	0.9912	0.9914	0.9917	0.9919	0.9921	0.9923
	cos	0.1392	0.1374	0.1357	0.1340	0.1323	0.1305	0.1288	0.1271	0.1253	0.1236
	tan	7.1154	7.2066	7.3002	7.3962	7.4947	7.5958	7.6996	7.8062	7.9158	8.0285
83	sin	0.9925	0.9928	0.9930	0.9932	0.9934	0.9936	0.9938	0.9940	0.9942	0.9943
	cos	0.1219	0.1201	0.1184	0.1167	0.1149	0.1132	0.1115	0.1097	0.1080	0.1063
	tan	8.1443	8.2636	8.3863	8.5126	8.6427	8.7769	8.9152	9.0579	9.2052	9.3572
84	sin	0.9945	0.9947	0.9949	0.9951	0.9952	0.9954	0.9956	0.9957	0.9959	0.9960
	cos	0.1045	0.1028	0.1011	0.0993	0.0976	0.0958	0.0941	0.0924	0.0906	0.0889
	tan	9.5144	9.6768	9.8448	10.02	10.20	10.39	10.58	10.78	10.99	11.20
85	sin	0.9962	0.9963	0.9965	0.9966	0.9968	0.9969	0.9971	0.9972	0.9973	0.9974
	cos	0.0872	0.0854	0.0837	0.0819	0.0802	0.0785	0.0767	0.0750	0.0732	0.0715
	tan	11.43	11.66	11.91	12.16	12.43	12.71	13.00	13.30	13.62	13.95
86	sin	0.9976	0.9977	0.9978	0.9979	0.9980	0.9981	0.9982	0.9983	0.9984	0.9985
	cos	0.0698	0.0680	0.0663	0.0645	0.0628	0.0610	0.0593	0.0576	0.0558	0.0541
	tan	14.30	14.67	15.06	15.46	15.89	16.35	16.83	17.34	17.89	18.46
87	sin	0.9986	0.9987	0.9988	0.9989	0.9990	0.9990	0.9991	0.9992	0.9993	0.9993
	cos	0.0523	0.0506	0.0488	0.0471	0.0454	0.0436	0.0419	0.0401	0.0384	0.0366
	tan	19.08	19.74	20.45	21.20	22.02	22.90	23.86	24.90	26.03	27.27
88	sin	0.9994	0.9995	0.9995	0.9996	0.9996	0.9997	0.9997	0.9997	0.9998	0.9998
	cos	0.0349	0.0332	0.0314	0.0297	0.0279	0.0262	0.0244	0.0227	0.0209	0.0192
	tan	28.64	30.14	31.82	33.69	35.80	38.19	40.92	44.07	47.74	52.08
89	sin	0.9998	0.9999	0.9999	0.9999	0.9999	1.000	1.000	1.000	1.000	1.000
	cos	0.0175	0.0157	0.0140	0.0122	0.0105	0.0087	0.0070	0.0052	0.0035	0.0017
	tan	57.29	63.66	71.62	81.85	95.49	114.6	143.2	191.0	286.5	573.0
Degs.	Function	0′	6′	12′	18′	24′	30′	36′	42′	48′	54′

TABLE XI. NATURAL SINES, COSINES, AND TANGENTS OF ANGLES

INDEX

A

Abscissa axis, 63–66
Acute angle, definition of, 146
Addition in nomographs, 67–68
Algebra, 127–134
    axioms of equality, 130–131
    laws and definitions, basic, 127
    polynomials, 128–129
        factoring, 129–130
    quadratic equation formula, 132–134
    simultaneous equations, 131–132
Alignment charts, 63–70
    (see also "Nomographs")
American Society of Mechanical Engineers, 81*n*
Angle, definition of, 139, 146
Annual compounding interest, table of interest
    factors for, 232–240
Annual cost method for economic replacement
    study, 44–45
Anti-logarithm, meaning of, 137
Arc of circle, 143
Arithmetic average, 77–78
Attributes, control charts by, 35–36
Averages, mathematical, 77–78
    arithmetic, 77–78
    median, 78
    mode, 78
Axioms of equality, 130–131

B

Break-even point, business volume, 43
BTA definition, 85–86
Budgetary cost control, 46–48
    conventional, 46–47
    flexible, 47–48
Building construction outline, 57–58
Building services, 115–124
    electricity, 117–121
        horsepower, 120–121
        Ohm's Law, 117–118

Building services (*cont.*)
        parallel connections, 118–119
        power, 119–120
        series connections, 118
    heat, water and air, 122–124
        fuel consumption, 123–124
        heating and cooling, 122–123
Bureau of Transport Economics and Statistics,
    95*n*, 256*n*
Business volume break-even point, 43

C

Carmichael, Robert D., 161*n*, 249*n*
Characteristic of logarithm, 136
Chemical Rubber Company, 202*n*
Chord of circle, 143
Circle, 143
    circumference, graphing, 65–66
Clerical jobs, merit rating in, 112–113
Common factors, 129
Common logarithms, 136
    table of, 213–217
Constants, important, table of, 161
Construction outline, 57–58
Contribution ratio, 43–44
Control charts by attributes, 35–36
Cosecant, 147
Cosine, 147
Cost factors, intangible, in choice of plant
    location, 51
Cotangent, 148
CPS, 41–42
Critical Path Scheduling, 41–42
"Critical Path Scheduling," 41*n*
Cross department relationships, 58
Cumulative Poisson distribution, 73
    tables of, 241–248
Cylinder, 144

307

D

Davis, Ralph C., 25n
Definitions, basic algebraic, 127
Delay, probability of, 75
Diameter of circle, 143
Division in nomographs, 68–69
Dodge-Romig Sampling Inspection Tables,
      36n, 37
Double sampling lot tolerance tables, 37

E

Electricity, 117–121
      horsepower, 120–121
      Ohm's Law, 117–118
      parallel connections, 118–119
      power, 119–120
      series connections, 118
Engineering Economy, 155n, 232n
Equal-payment-capital-recovery factor, 158
Equal-payment-compound-amount factor,
      156–157
Equal-payment-present-worth factor, 157–158
Equal-payment-sinking-fund factor, 157
Equality, axioms of, 130–131
Equations, simultaneous, 131–132
Exponential and hyperbolic functions,
      table of, 249–255
Exponential and logarithm relations, 136
Exponential Poisson distribution, 73

F

Fabrycky, W. J., 155n, 232n
Factorials, 212
Factoring polynomials, 129–130
Factors of site selection, 55
Factory, 41n, 51n
50–50 employee-employer saving share plan,
      105–106
Floor area requirements, estimating, 56–57
Flow process charts, 81–84
      definition and purpose, 81
      material type, 82
      symbols, 81
Fuel consumption, 123–124

G

Geometry, 139–145
      circle, 143
      cylinder, 144
      definition and scope, 139
      parallelogram, 141–142
      quadrilateral, 141
      sphere, 144–145
      trapezoid, 142–143
      triangle, 139–141
Graphs, 63–70
      circle circumference, 65–66
      definition, 63–66

H

Halsey plan as wage incentive, 105–106
*Handbook of Mathematical Functions With For-
      mulas, Graphs and Mathematical Tables,*
      162n, 213, 218
Heat, water and air, 122–124
      fuel consumption, 123–124
      heating and cooling, 122–123
Holding costs, average, 32
Holding and out of stock costs, weekly, 33–34
Hourly rated jobs, evaluation of, 109–111
Hypotenuse, definition of, 146

I

Industrial Management, 88n
Industrial Management Society, 88n
"Industrial Site Selection Today," 51n
"Influence of the Unit of Supervision and the
      Span of Executive Control on the Economy
      of Line Organization Structure," 25n
Inspection and quality control, 35–40
      attributes, control charts by, 35–36
      double sampling lot tolerance tables, 37
      outgoing quality, average, sampling tables for,
            37
      single sampling lot tolerance tables, 36–37
Interest factors, use of, 155–158
      equal-payment-capital-recovery factor, 158
      equal-payment-compound-amount factor,
            156–157
      equal-payment-present-worth factor, 157–158

Interest factors *(cont.)*
    equal-payment-sinking-fund factor, 157
    single-payment-compound-amount factor,
        155–156
    single-payment-present-worth factor, 156
    symbols and definitions, 155
Interest factors for annual compounding interest,
    table for, 232–240
Interpolation of tables, 137
Interstate Commerce Commission, 95n, 256n
Inventory Control, 31–34
    average, 31–32
    holding costs, average, 32
    out of stock costs, average, 32–33
    reorder point, 31
    weekly holding and out of stock costs, total,
        33–34

J

Job evaluation, 109–111
    hourly rated jobs, 109–111

K

Kelly, J. E. Jr., 41n
Keuffel & Esser Co., 153n

L

Laws of algebra, basic, 127
Layout, plant, 56–59
    building construction outline, 57–58
    cross department relationships, 58
    floor area requirements, estimating, 56–57
    space utilization, 59
Learning curves, 99–102
    definition, 99
Levels of line supervision, calculation of number
    of, 25–26
Line, definition of, 139
Location of plant, 51–55
    intangible cost factors in choice of, 51
    new plants, setting of, 55
    planning methods, rating of company's, 55
    reasons for constructing new plant, 55
    secondary factors of, 55

Location of plant *(cont.)*
    shifting selection factors, 55
    survey, 51–53
Logarithms, common, table of, 213–217
    natural, table of, 218–231
Logarithms, fundamentals of, 135–138
    components of, 136–137
    definition, 135
    exponential and logarithm relations, 136
    interpolation of tables, 137
    natural and common, 136
    uses of, 137–138
Low-high method of flexible budgeting, 47–48

M

Management controls, 23–48
    budgetary cost control, 46–48
        conventional, 46–47
        flexible, 47–48
    Critical Path Scheduling, 41–42
    inspection and quality control, 35–40
        attributes, control charts by, 35–36
        double sampling lot tolerance tables, 37
        outgoing quality, average, sampling tables
            for, 37
        single sampling lot tolerance tables, 36–37
    inventory control, 31–34
        average, 31–32
        holding costs, average, 32
        out of stock costs, average, 32–33
        reorder point, 31
        weekly holding and out of stock costs, 33–34
    manufacturing economics, 43–45
        annual cost method for economic replace-
            ment study, 44–45
        break-even point, business volume, 43
        contribution ratio, 43–44
    production control, 29–30
        and inventory costs, 29
        inventory costs and purchase quantity dis-
            counts, 30
    span control, 25–28
        levels of line supervision, calculation of
            number of, 25–26
        payroll savings, 27–28
        personnel, line supervision, calculation of
            number of, 26–27
Management jobs, merit rating for, 113–114
Manpower planning, 88–92

"Manpower Planning for Fluctuating Operations," 88n
Mantissa of logarithm, 136
Manufacturing economics, 43–45
    break-even point, business volume, 43
    contribution ratio, 43–44
    economic replacement study annual cost method, 44–45
Materials handling, 56–59
    cross department relationships, 58
Mathematical averages, 77–78
    arithmetic, 77–78
    median, 78
    mode, 78
Mathematical Tables and Formulas, 161n, 249n
Mazel, Joseph L., 51n
Median average, 78
Merit rating, 112–114
    clerical jobs, 112–113
    management jobs, 113–114
Mode average, 78
Modern Manufacturing, 41n, 51n
Multiplication in nomographs, 68–69

N

Napierian logarithms, 136
Natural logarithms, 136
    table of, 218–231
Natural sines, cosines, and tangents of angles, 298–303
New plant, reasons for constructing, 55
    setting of, 55
"Nomogram Helps Find Excess Air," 69n
Nomographs, 63–70
    addition and subtraction, 67–68
    definition, 66
    excess air in furnace, 69–70
    multiplication and division, 68–69

O

Obtuse angle, definition of, 146
Ohio State University, 25n
Ohm's Law, 117–118
Oil, Gas and Petrochem Equipment, 69n
100% premium plan as wage incentive, 104–105
Ordinate axis, 63–66

Origin, 63–66
Out of stock costs, average, 32–33
Outgoing quality, average, sampling tables for, 37

P

Parallelogram, 141–142
Payroll savings, 27–28
Perimeter of triangle, 139
Personnel, 107–114
    job evaluation, 109–111
        hourly rated jobs, 109–111
    merit rating, 112–114
        clerical jobs, 112–113
        management jobs, 113–114
Personnel, line supervision, calculation of number of, 26–27
Piece work plan as wage incentive, 103–104
Plane, definition of, 139
Planning methods, rating of company's, 55
Plant facilities, 49–59
    layout and materials handling, 56–59
        building construction outline, 57–58
        cross department relationships, 58
        floor area requirements, estimating, 56–57
        space utilization, 59
    location, 51–55
        intangible cost factors in choice of, 51
        planning methods, rating of company's, 55
        reasons for constructing new plant, 55
        secondary factors of, 55
        setting of new plants, 55
        shifting selection factors, 55
        survey, 51–53
Point, definition of, 139
Poisson distribution, 71–76
    (see also "Probability—Poisson distribution")
    cumulative, 241–248
Polynomials, 128–129
    factoring, 129–130
Port of New York Authority, 74n
Powers and roots, table of, 162–201
Prime factors, 129
Probability—Poisson distribution, 71–76
    Cumulative distribution, 73
    delay, probability of, 75
    Exponential distribution, 73

Probability—Poisson distribution *(cont.)*
  formula definitions, 71, 72, 73
  Poisson distribution, 71–76
  symbols and definitions, 72
  waiting line system, average delay in, 76
Production control, 29–30
  and inventory costs, 29
  inventory costs and purchase quantity
      discounts, 30
Pythagorean theorem, 146–147

Q

Quadratic equation formula, 132–134
Quadrilateral, 141
Quality control, 35–40
  attributes, control charts by, 35–36
  double sampling lot tolerance tables, 37
  outgoing quality, average, sampling tables
      for, 37
  single sampling lot tolerance tables, 36–37

R

Radius of circle, 143
Random numbers, table of, 256–297
  use of in work sampling, 95–98
Reasons for constructing new plant, 55
Rectangle, definition of, 141
Reorder point, 31
Rhombus, definition of, 141
Right angle, definition of, 146

S

Sayer, J. S., 41*n*
Secant, 148
Simultaneous equations, 131–132
Sine, 147
Sines, cosines, and tangents of angles, natural,
    table of, 297–303
Single-payment-compound-amount factor,
    155–156
Single-payment-present-worth factor, 156
Single sampling lot tolerance tables, 36–37
Sisson, Bill, 69*n*
Site of plant, 51–55

Site of plant *(cont.)*
  intangible cost factors in choice of, 51
  planning methods, rating of company's, 55
  reasons for constructing new plant, 55
  secondary factors of, 55
  setting of new plants, 55
  shifting selection factors, 55
  survey, 51–53
Slide rule, 152–154
  logarithm scale, 154
  multiplication and division, 152–153
  operation and scope, 152
  powers of number, 154
  reading scales, 152–153
  trigonometry—SRT, S and D scales, 154
Smith, Edwin R., 161*n*, 249*n*
Solid, definition of, 139
"Solution Methods for Waiting Line Problems,"
    74*n*
Space utilization, 59
Span of control, 25–28
  levels of line supervision, calculation of
      number of, 25–26
  payroll savings, 27–28
  personnel, line supervision, calculation of
      number of, 26–27
Sphere, 144–145
Square, definition of, 141
ST definition, 85
Standard Mathematical Tables, 202*n*
Statistical mathematics, 61–78
  graphs and nomographs, 63–70
    addition and subtraction in nomographs,
        67–68
    circle circumference, 65–66
    graphs, definition of, 63–66
    multiplication and division in nomographs,
        68–69
    nomograph, definition of, 66
    nomograph—excess air in furnace, 69–70
  mathematical averages, 77–78
    arithmetic, 77–78
    median, 78
    mode, 78
  probability—Poisson distribution, 71–76
    Cumulative distribution, 73
    delay, probability of, 75
    Exponential distribution, 73
    formula definitions, 71, 72, 73
    Poisson distribution, 71–76

Statistical mathematics *(cont.)*
    symbols and definitions, 72
    waiting line system, average delay in, 76
Subtraction in nomographs, 67–68
Surface, definition of, 139
Symbols on flow chart, 81

T

"Table of 105,000 Random Decimal Digits,"
    95*n*, 256*n*
Tangent, 147
Thuesen, H. G., 155*n*, 232*n*
Tolerance tables, 36–37
  double sampling lot, 37
  single sampling lot, 36–37
Trapezoid, 142–143
Triangle, 139–141
Trigonometry, 146–151
  definition and scope, 146
  Pythagorean theorem, 146–147
  ratios—right triangle, 147–149
  relationships—all triangles, 149–151

U

Use of book, proper, 19–22

W

Wage incentives, 103–106
  Halsey plan, 105–106
  100% premium plan, 104–105
  piece work plan, 103–104
Waiting line system, average delay in, 76
Walker, Morgan R., 41*n*
Work measurement, 79–106
  flow process charts, 81–84
    definition and purpose, 81
    material type, 82
    symbols, 81
    learning curves, 99–102
    definition, 99
  manpower planning, 88–92
  wage incentives, 103–106
    Halsey plan, 105–106
    100% premium plan, 104–105
    piece work plan, 103–104
  work sampling, 93–98
    definition, 93–95
    random numbers, use of, 95–98
  work standard, 85–87
    (see also "Work standard")
Work sampling, 93–98
  (see also "Work measurement")
Work standard, 85–87
  definition, 85–86
  examples, 86–87